Algebras of Sets and Combinatorics

Translations of
MATHEMATICAL MONOGRAPHS

Volume 214

Algebras of Sets and Combinatorics

L. Š. Grinblat

American Mathematical Society
Providence, Rhode Island

Л. Ш. Гринблат

АЛГЕБРЫ МНОЖЕСТВ И КОМБИНАТОРИКА

Translated from an original Russian manuscript by A. Stoyanovskiĭ.
Translation edited by A. B. Sossinsky.

2000 *Mathematics Subject Classification.* Primary 03E05; Secondary 28A05, 54D35.

Library of Congress Cataloging-in-Publication Data

Grinblat, L. S. (Leonid S.), 1944–
[Algebry mnozhestv i kombinatorika. English]
Algebras of sets and combinatorics / L. S. Grinblat.
p. cm. — (Translations of mathematical monographs, ISSN 0065-9282 ; v. 214)
Includes bibliographical references and index.
ISBN 0-8218-2765-0 (alk. paper)
1. Combinatorial set theory. I. Title. II. Series.
QA248.G753 2002
511.3′22–dc21 2002074584
CIP

∞ The paper used in this book is acid-free and falls within the guidelines
established to ensure permanence and durability.
Visit the AMS home page at http://www.ams.org/

10 9 8 7 6 5 4 3 2 1 07 06 05 04 03 02

Contents

CHAPTER 1

Introduction

1.1. This monograph contains the results of [**Gr**] as well as many new results I have obtained in recent years. The new results and the results from [**Gr**] are so intertwined that it did not seem reasonable to publish all the new results in a separate very large paper. In contrast to [**Gr**], in the present monograph the theory is complete. Thus, the statements to which we refer without proofs and which we use for proving the main results are widely known and are simple facts of modern mathematics. This also means that all the main results entirely belong to "naive" set theory (except for the results in the Appendix written by Saharon Shelah).

Let us start with some history. George Cantor, the creator of set theory, began studying sets in 1872. His starting point were papers on trigonometric series inspired by the works of Riemann. In 1878 Cantor proposed the continuum hypothesis ($\aleph_1 = 2^{\aleph_0}$) which was placed first in Hilbert's 1900 list of problems bequeathed by him to the 20th century (see [**Hi**]). Breaking the chronology of our story, let us mention that the solution of the first Hilbert problem turned out to be striking. First, in the paper [**Go**] published in 1939, Gödel proved the consistency of the continuum hypothesis with the axioms of set theory, and in the early 60's Paul Cohen developed a powerful method for constructing models, the forcing method, and constructed a model of set theory in which $\aleph_1 < 2^{\aleph_0}$ (see [**Co1**]).[1] It became clear that the persistent efforts of Cantor and other mathematicians to prove the continuum hypothesis were of no avail. The continuum hypothesis, like Euclid's fifth postulate, can be neither proved nor disproved. The last papers of Cantor date back to 1895–1897; they are mostly devoted to the theory of completely ordered sets and to the calculus of ordinal numbers.[2]

In 1902 Lebesgue published his thesis [**L1**], which became the starting point of modern measure theory. The notions of a Lebesgue measurable and nonmeasurable set became fundamental in mathematics. New set theory problems arose. In the book [**L2**], published in 1904, Lebesgue posed the following problem.

THE LEBESGUE PROBLEM. *Does there exist a nonnegative real function μ on the set of all subsets of the interval $X = (0, 1)$ that satisfies the following conditions*:

(a) *if M_1 and M_2 are isometric sets, then $\mu(M_1) = \mu(M_2)$*;

---

[1]Although the reader is not assumed to be familiar with the forcing method (except in the Appendix), it is related to our monograph. Some examples use results obtained by the forcing method. The important theorem of Gitik–Shelah stated below was originally proved by the forcing method. The infinite analog of the Lusin theorem obtained by Gitik and Shelah — a result close to our subjects and discussed in Chapter 6 — was also proved by the forcing method, and up to now its purely mathematical proof is not known. An interesting result cited in 5.28 and presented in the Appendix was obtained by Saharon Shelah by the forcing method as the answer to a question of the author. In 6.12 we discuss an interesting result of Shelah obtained using the forcing method.

[2]The above information about Cantor is taken from [**Bo1**] and [**Co2**].

(b) *if* $M_1, \ldots, M_k, \ldots$ *is an at most countable sequence of pairwise disjoint sets, then* $\mu(\bigcup M_k) = \sum \mu(M_k)$;
(c) $\mu(X) = 1$?

In 1905 Vitali [**Vi**] obtained the presently well-known negative solution of this problem.

1.2. The following problem was posed by Banach.

THE BANACH PROBLEM. *What is the solution of the Lebesgue problem with condition* (a) *replaced by the following*: $\mu(\{x\}) = 0$ *for each point* $x \in X$?

In 1929 Banach and Kuratowski published a negative solution of this problem under the assumption of the continuum hypothesis in [**B-Ku**].[3] Probably around that time, many specialists believed that, even without the continuum hypothesis assumption, the Banach problem has a negative solution. Thus, in a well-known book [**Bi**] by G. Birkhoff published in 1948, one of the problems is posed as follows: prove the nonexistence of a nontrivial measure with countable additivity[4] for all subsets of the continuum, such that any point has measure 0, without assuming the continuum hypothesis.

1.3. REMARK. Here it is appropriate to recall the history of the study of other problems whose starting point is the Lebesgue problem formulated in 1.1. (We use the historical survey from [**Bo2**].) In 1914 Hausdorff proved in [**Hau1**] that there does not exist a nontrivial finitely additive measure that is well defined for all subsets of the unit ball in the 3-dimensional Euclidean space and invariant with respect to isometries. It was natural to investigate whether the same is true for the line and the plane — the problem was solved brilliantly by Banach in 1923, who showed in [**B**] that, on the contrary, it is just in these two cases that such a measure does exist. In 1929 von Neumann showed in [**Ne1**] that the reason for the difference between the line and the plane on one hand, and the Euclidean space of dimension greater than 2 on the other hand, has to do with the commutativity of the group of rotations of the plane. Therefore the σ-additivity condition in the Lebesgue problem cannot be replaced by the finite additivity condition. Using the notion of ultrafilter, one easily shows that the σ-additivity condition in the Banach problem cannot be replaced by the finite additivity condition. Note that the idea of the "averaging" method from [**B**] was used by Haar, who proved in 1933 (see [**H**]) the existence of an invariant measure for locally compact groups. This discovery immediately allowed von Neumann to solve (see [**Ne2**]), for compact groups, the fifth Hilbert problem on the characterization of Lie groups by purely topological properties.

---

[3] The result of Banach–Kuratowski follows from Theorem 1.6. It also follows from Theorem 6.8 (see 6.9).

[4] We consider only real nonnegative finite measure (i.e., the measure of a measurable set is a real nonnegative finite number). A measure is called nontrivial if there exists a measurable set whose measure is greater than 0. As usual, a countable additivity property is also called σ-additivity, and a σ-additive measure is called, for short, a σ-measure. Sometimes we will consider not only σ-additive measures, but also finitely additive measures which are not, in general, σ-additive.

1.4. In 1971, in the period started by Cohen's great discovery, the paper [**So**] by R. Solovay appeared with the following remarkable result:

SOLOVAY'S RESULT. *In a certain model there exists a σ-additive extension of the Lebesgue measure to all subsets of* $(0, 1)$.

1.5. Let us return to the Banach–Kuratowski result mentioned in 1.2. After this, S. Ulam proved in [**U1**] the following important more general result:

ULAM'S MATRIX. *If a set X has cardinality $\aleph_1$, then one can construct a matrix of subsets of X,*

$$\begin{pmatrix} M_1^1 & M_2^1 & \dots & M_\alpha^1 & \dots \\ M_1^2 & M_2^2 & \dots & M_\alpha^2 & \dots \\ \dots & \dots & \dots & \dots & \dots \\ M_1^k & M_2^k & \dots & M_\alpha^k & \dots \\ \dots & \dots & \dots & \dots & \dots \end{pmatrix},$$

which has $\aleph_0$ rows and $\aleph_1$ columns, and

(a) $M_\alpha^k \cap M_\beta^k = \varnothing$ *if* $\alpha \neq \beta$;

(b) $\#(X \setminus \bigcup_k M_\alpha^k) \leq \aleph_0$.[5]

1.6. The following theorem, due to Ulam, is an obvious corollary of the existence of Ulam's matrix:

THEOREM. *Let μ be a nontrivial σ-measure on a set of cardinality $\aleph_1$, and let $\mu(\{x\}) = 0$ for each one-point set $\{x\}$. Then there exist $\aleph_1$ pairwise disjoint μ-nonmeasurable sets.*

1.7. DEFINITION. A finitely additive measure μ defined on a set X is said to be *two-valued* if $\mu(\{x\}) = 0$ for all $x \in X$, $\mu(X) = 1$, and for any μ-measurable set M either $\mu(M) = 0$ or $\mu(M) = 1$. If μ is also σ-additive, then it is said to be *σ-two-valued*.

1.8. The following Alouglu–Erdös theorem stated in [**Er**] follows rather easily from Theorem 1.6.

THE ALOUGLU–ERDÖS THEOREM. *Assume that a countable sequence of σ-two-valued measures on a set of cardinality $\aleph_1$ is given. Then there exists a set that is nonmeasurable with respect to each of these measures.*

It is clear that in the Alouglu–Erdös theorem the σ-two-valuedness condition is inessential. The theorem is valid under the assumption that each of the measures is nontrivial, σ-additive, and each one-point set has measure 0.

1.9. The Alouglu–Erdös theorem gives a partial answer to the following Ulam problem stated in [**Er**] (see also [**Er-Ha**]).

THE ULAM PROBLEM. *Find the minimal cardinal*[6] $\varkappa_*$ *such that, for any family (of cardinality less than $\varkappa_*$) of σ-two-valued measures defined on a set of cardinality $\aleph_1$, there exists a set that is nonmeasurable with respect to each of the measures in the family.*

---

[5]In 6.4 we give a construction of the sets M_α^k. The symbol $\#(M)$ denotes the cardinality of the set M. (In the Appendix the cardinality of a set M is denoted by $|M|$.)

[6]By a *cardinal* $\varkappa$ we mean an ordinal number such that if $\alpha < \varkappa$, then $\#(\alpha) < \#(\varkappa)$. In what follows, the cardinality of a set will be treated as a cardinal.

Ulam proved that $\varkappa_* \geq \aleph_0$. By the Alouglu–Erdös theorem, $\varkappa_* \geq \aleph_1$. In [**S2**] Shelah has shown that in a certain model $\varkappa_* = \aleph_1$. Note that in the Gödel model L, we have $\varkappa_* = \aleph_2$.

1.10. In [**Wo**] Woodin improved Shelah's result. He showed that in a certain model there exists a σ-two-valued measure μ on a set X of cardinality $\aleph_1$ and a family $\{M_\alpha\}_{\alpha<\omega_1}$ of μ-nonmeasurable sets such that

$$\{\alpha \mid \mu(M_\alpha \setminus M) = 0\} \neq \varnothing$$

for each $M \subset X$ with $\mu(M) \neq 0$.

For each M_α let us define a σ-two-valued measure μ_α on X as follows: $\mu_\alpha(M) = 0$ if and only if $\mu(M \cap M_\alpha) = 0$. It is evident that each subset of X is measurable with respect to some measure μ_α. Therefore in the model considered by Woodin we have $\varkappa_* = \aleph_1$.

1.11. We will return to the history of the subject later. Now let us pass to the main theme of our monograph.

DEFINITION. By an *algebra $\mathcal{A}$ on a set X* we mean a nonempty system of subsets of X possessing the following property: if $M_1, M_2 \in \mathcal{A}$, then $M_1 \cup M_2$, $M_1 \setminus M_2 \in \mathcal{A}$.

Unless specified otherwise, all algebras and measures are considered on some abstract set $X \neq \varnothing$. Unless the contrary follows from the context, by a set we always mean a subset of X.

1.12. The following "general question" arises in connection with the Alouglu–Erdös theorem and the Ulam problem. Assume we are given a family of measures $\{\mu_\lambda\}_{\lambda\in\Lambda}$ (since the question is "general", we do not specify whether they are σ-additive and so forth), and for each $\lambda \in \Lambda$ there exists a μ_λ-nonmeasurable set. Does there exist a set nonmeasurable with respect to each of these measures? Shelah's and Woodin's results are nontrivial examples giving a negative answer to our question. But in these examples we deal with families such that $\#(\Lambda) = \aleph_1$. And what will be the answer if one considers families with $\#(\Lambda) \leq \aleph_0$? Let us ask our question in a more general and, I believe, more natural form. Let us consider algebras instead of measures. Thus, consider an at most countable family of algebras, each of which is not equal to $\mathfrak{P}(X)$ (as usual, $\mathfrak{P}(X)$ denotes the set of all subsets of X). Does there exist a set which does not belong to any of these algebras? If we are given two algebras $\mathcal{A}_1$ and $\mathcal{A}_2$, and each of them does not coincide with $\mathfrak{P}(X)$, then it is easy to show that

$$\mathcal{A}_1 \cup \mathcal{A}_2 \neq \mathfrak{P}(X).$$

1.13. However, if we take not two but three algebras each of which is not equal to $\mathfrak{P}(X)$, then it can happen that a set not belonging to any one of these algebras does not exist.

EXAMPLE. $\#(X) = 3$ and $X = \{x_1, x_2, x_3\}$. Define the algebras $\mathcal{A}_1', \mathcal{A}_2', \mathcal{A}_3'$ by listing their elements:

$$\mathcal{A}_1' = \{\varnothing, \{x_1\}, \{x_2, x_3\}, X\},$$
$$\mathcal{A}_2' = \{\varnothing, \{x_2\}, \{x_1, x_3\}, X\},$$
$$\mathcal{A}_3' = \{\varnothing, \{x_3\}, \{x_1, x_2\}, X\}.$$

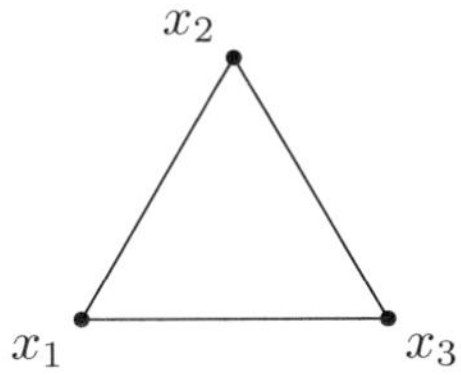

It is obvious that $\mathcal{A}'_k \neq \mathfrak{P}(X)$ and $\mathcal{A}'_1 \cup \mathcal{A}'_2 \cup \mathcal{A}'_3 = \mathfrak{P}(X)$.

1.14. The considerations with two algebras in 1.12 and with three algebras in 1.13 were given in [**Gr**]; they were probably known before [**Gr**]. Let us now give a natural generalization of Example 1.13. Take a natural number $m > 1$ and put $\#(X) = 2m - 1$. Let $M \subset X$ and $\#(M) = m$. Define the algebra

$$\mathcal{A}_M = \mathfrak{P}(X \setminus M) \cup \{S \subset X \mid S \supset M\}.$$

Obviously, $\bigcup \mathcal{A}_M = \mathfrak{P}(X)$. If $m = 2$, we obtain three algebras which are identical to the algebras $\mathcal{A}'_1, \mathcal{A}'_2, \mathcal{A}'_3$; for each of them there are two disjoint sets that do not belong to it. In the general case, if $x \in M$, $\{x\} \notin \mathcal{A}_M$, and hence there exist m pairwise disjoint sets that do not belong to $\mathcal{A}_M$. Clearly, there are $\binom{2m-1}{m}$ algebras $\mathcal{A}_M$. Due to the Stirling formula,

$$n = \binom{2m-1}{m} = \frac{1}{\sqrt{2\pi}} \frac{\sqrt{2m-1}}{\sqrt{m} \cdot \sqrt{m-1}} \cdot \frac{(2m-1)^{2m-1}}{m^m \cdot (m-1)^{m-1}} \cdot e^{\left(\frac{\theta_1(m)}{12(2m-1)} - \frac{\theta_2(m)}{12m} - \frac{\theta_3(m)}{12(m-1)}\right)},$$

where $0 < \theta_i(m) < 1$. Hence

$$\begin{aligned} \log_4 n &\approx -\frac{1}{2}\log_4 \pi + \frac{1}{2}\log_4\left(m - \frac{1}{2}\right) - \frac{1}{2}\log_4 m - \frac{1}{2}\log_4(m-1) \\ &\quad + \log_4\left(\frac{4^m}{2} \cdot \left(1 - \frac{1}{2m}\right)^m \cdot \left(1 + \frac{1}{2(m-1)}\right)^{m-1}\right) \\ &\approx m - \frac{1}{2}\log_4 m - \frac{1}{2}(1 + \log_4 \pi). \end{aligned}$$

Define a function $\varphi(n)$ on the set of natural numbers ≥ 3: $\varphi(n) = m$ if

$$\binom{2m-1}{m} \leq n < \binom{2m+1}{m+1}.$$

For any natural $n \geq 3$ one can construct algebras $\mathcal{A}_1, \ldots, \mathcal{A}_n$ such that $\bigcup_{k=1}^n \mathcal{A}_k = \mathfrak{P}(X)$, and for any of these algebras there exist $\varphi(n)$ pairwise disjoint sets that do not belong to it. We have

$$\varphi(n) - \log_4 n = \frac{1}{2}\log_4 \varphi(n) + c(n)$$

for n large enough and $|c(n)| < 1$.

Let us describe another generalization of Example 1.13, which yields a stronger result. Take a natural number m and put $\#(X) = 3m$. Let us divide X into m triples: $\{x_1^1, x_2^1, x_3^1\}, \ldots, \{x_1^m, x_2^m, x_3^m\}$. For each set

$$L = \{x_{i_1}^1, x_{i_2}^2, \ldots, x_{i_m}^m\}$$

define an algebra $\mathcal{A}_L$: a set $P \subset X$ does not belong to $\mathcal{A}_L$ if and only if there exists a natural number $k \leq m$ such that

$$\#(P \cap (\{x_1^k, x_2^k, x_3^k\} \setminus \{x_{i_k}^k\})) = 1.$$

It is clear that there exist $2m$ pairwise disjoint sets not belonging to $\mathcal{A}_L$. Clearly, there are 3^m algebras $\mathcal{A}_L$. Obviously,

$$\bigcup_L \mathcal{A}_L = \mathfrak{P}(X).$$

If $m = 1$, we obtain three algebras which are just $\mathcal{A}'_1, \mathcal{A}'_2, \mathcal{A}'_3$. Define a function $\varphi^*(n)$ on the set of natural numbers ≥ 3: $\varphi^*(n) = 2m$ if

$$3^m \leq n < 3^{m+1}.$$

For any natural $n \geq 3$ one can construct algebras $\mathcal{A}_1, \dots, \mathcal{A}_n$ such that $\bigcup_{k=1}^n \mathcal{A}_k = \mathfrak{P}(X)$, and for each of these algebras there exist $\varphi^*(n)$ pairwise disjoint sets not belonging to it. We have

$$0 \leq 2 \cdot \log_3 n - \varphi^*(n) < 2.$$

1.15. After Example 1.13, the "general question" that is the subject of many pages of the present monograph can be naturally stated as follows. Let $\mathcal{A}_1, \dots, \mathcal{A}_k, \dots$ be an at most countable sequence of algebras. What conditions on these algebras ensure that $\bigcup \mathcal{A}_k \neq \mathfrak{P}(X)$?

1.16. Since in modern set theory $\mathbb{N}$ denotes the set of nonnegative integers, let us denote by $\mathbb{N}^+$ the set of natural numbers. One more notation: if $n_1, n_2 \in \mathbb{N}^+$ and $n_1 \leq n_2$, then

$$[n_1, n_2] = \{k \in \mathbb{N}^+ \mid n_1 \leq k \leq n_2\}.$$

1.17. In [**Gr**] it is shown that for each $n \in \mathbb{N}^+$ there exists a function $\psi\colon [1, n] \to \mathbb{N}^+$ such that if a sequence of algebras $\mathcal{A}_1, \dots, \mathcal{A}_n$ is given and for each $k \in [1, n]$ there exist $\psi(k)$ pairwise disjoint sets not belonging to $\mathcal{A}_k$, then $\bigcup_{k=1}^n \mathcal{A}_k \neq \mathfrak{P}(X)$. *For n fixed, denote by Ψ^n the totality of all such functions ψ.*[7] It is natural to pose the "general problem" of describing Ψ^n for each n. Seemingly, this problem is very difficult. In the present book its solution is given for $n \leq 7$.

Denote by Ψ the totality of all functions $\psi\colon \mathbb{N}^+ \to \mathbb{N}^+$ possessing the following property: for each $n \in \mathbb{N}^+$ there exists a function $\psi_n \in \Psi^n$ such that $\psi_n(k) \leq \psi(k)$ for all $k \in [1, n]$. We prove the following theorem.

THEOREM. *Let $\psi \in \Psi$, and let $\mathcal{A}_1, \dots, \mathcal{A}_k, \dots$ be a countable sequence of σ-algebras*[8] *such that for each k there exist $\psi(k)$ pairwise disjoint sets not belonging to $\mathcal{A}_k$. Then $\bigcup \mathcal{A}_k \neq \mathfrak{P}(X)$.*

Note that a considerable part of this monograph is devoted to delicate theorems on algebras from a wider class than σ-algebras, the class of almost σ-algebras, defined below.

---

[7]It is natural to pose the following question. Assume that $\varphi\colon [1, n] \to \mathbb{N}^+$ has the following property: if $\mathcal{A}_1, \dots, \mathcal{A}_n$ is a sequence of algebras and for each $k \in [1, n]$ there exist exactly $\varphi(k)$ pairwise disjoint sets not belonging to $\mathcal{A}_k$, then $\bigcup_{k=1}^n \mathcal{A}_k \neq \mathfrak{P}(X)$. Is it true that $\varphi \in \Psi^n$? At the end of Chapter 11 it is shown that in general the answer is negative.

[8]An algebra $\mathcal{A}$ is called a σ-algebra if for any countable sequence $M_1, \dots, M_k, \dots \in \mathcal{A}$ we have $\bigcup_{k=1}^\infty M_k \in \mathcal{A}$.

1.18. In the proof of Theorem 1.17, the following Gitik–Shelah theorem, whose generalization is Theorem 1.17, is used.

THE GITIK–SHELAH THEOREM. *Assume that a countable sequence of σ-two-valued measures on a set of cardinality $\mathfrak{c}$*[9] *is given. Then there exists a set that is nonmeasurable with respect to each of these measures.*[10]

As already mentioned in footnote 1.1, the first proof of the Gitik–Shelah theorem in [**Gi-S1**] is metamathematical and uses the forcing method. Purely mathematical proofs have been proposed by Fremlin in [**Frem1**] and [**Frem2**] and by Kamburelis in [**Ka**]. We give a proof following [**Ka**].

1.19. We use a rather clear equivalent of the Gitik–Shelah theorem.

THEOREM. *The following two statements are equivalent:*

(a) *The Gitik–Shelah theorem holds.*
(b) *Let $\mu_1, \dots, \mu_k, \dots$ be a countable sequence of σ-two-valued measures and $\#(X) = \mathfrak{c}$. Then there exists a countable sequence of pairwise disjoint sets $M_1, \dots, M_k, \dots$ such that M_k is a μ_k-nonmeasurable set.*[11]

PROOF. (a) $\Longrightarrow$ (b). By the Gitik–Shelah theorem, there exists a set M_1 nonmeasurable with respect to each measure μ_k. Define σ-two-valued measures $\mu_1^1, \dots, \mu_k^1, \dots$ on the set $M' = X \setminus M_1$. By definition, $\mu_k^1(M) = 0$ if and only if $\mu_k(M) = 0$. By the Gitik–Shelah theorem, there exists a set $M_2 \subset M'$ nonmeasurable with respect to each of the measures $\mu_1^1, \dots, \mu_k^1, \dots$. Obviously, M_2 is nonmeasurable with respect to each of the measures $\mu_1, \dots, \mu_k, \dots$. Naturally continuing our construction, we obtain a countable sequence of pairwise disjoint sets, each of which is nonmeasurable with respect to each of the measures μ_k.

(b) $\Longrightarrow$ (a). For each k let us take two μ_k-nonmeasurable sets M_k', M_k'' ($M_k' \cup M_k'' \subset M_k$, $M_k' \cap M_k'' = \varnothing$). Obviously, the set $\bigcup_k M_k'$ is μ_n-nonmeasurable for each $n \in \mathbb{N}^+$. $\square$

1.20. We deal with different variations of the "general question" formulated above. Let us give two examples of these "variations".

Consider an at most countable sequence of algebras $\mathcal{A}_1, \dots, \mathcal{A}_k, \dots$ and assume that there exists a matrix

$$\begin{pmatrix} U_1^1 & \dots & U_{m_1}^1 \\ \dots & \dots & \dots \\ U_1^k & \dots & U_{m_k}^k \\ \dots & \dots & \dots \end{pmatrix}$$

of pairwise disjoint sets[12] such that $U_i^k \notin \mathcal{A}_k$. Under what conditions does $\bigcup \mathcal{A}_k \neq \mathfrak{P}(X)$?[13]

---

[9]As usual, by $\mathfrak{c}$ we denote the continuum cardinal.

[10]The Gitik–Shelah theorem is an obvious corollary of Theorem 1.17 since for any σ-two-valued measure on a set of cardinality $\mathfrak{c}$ there exist $\aleph_0$ pairwise disjoint sets nonmeasurable with respect to this measure. It is evident that the Gitik–Shelah theorem is a generalization of the Alouglu–Erdös theorem. This generalization is deep and nontrivial.

[11]As far as I know, the idea of the proof of this theorem will not surprise specialists in set theory. I learned about this theorem from Moti Gitik.

[12]This means that $U_i^k \cap U_j^l = \varnothing$ unless $k = l$ and $i = j$.

[13]This question has been put to me by Fremlin.

Consider an at most countable sequence of algebras $\mathcal{A}_1, \dots, \mathcal{A}_k, \dots$. What conditions should be imposed on these algebras so as to deduce that there exist pairwise disjoint sets $U_1, \dots, U_k, \dots$ such that if $U_k \subset Q \subset \bigcup_i U_i$, then $Q \notin \mathcal{A}_k$?

1.21. Let us return to the history. I started to study the questions under discussion in January 1987. Then, in conversations with Moti Gitik a problem arose, whose positive solution is the Gitik–Shelah theorem published in 1989.[14] But the starting point of the history of the subject of our research is the statement of the Ulam problem (Problem 1.9) and the publication of Erdös' paper [**Er**]. The publication of [**Gr**] was preceded by the above-mentioned results from [**S2**] and [**Wo**][15] and the papers [**Fr-Gut**], [**J-Pr**], [**S1**], [**T**], and [**Ta**] cited below. There are many other papers by specialists in set theory. The referee of [**Gr**] has kindly drawn my attention to Grzegorek's paper [**Grz**]. Let us quote the abstract of this paper:

"Our Theorem 5 implies the following corollary: Let F be a family of σ-fields[16] on the real line $\mathbb{R}$ such that for every $\mathcal{A} \in F$ all one-element subsets of $\mathbb{R}$ belong to $\mathcal{A}$ and $\mathcal{A} \neq \mathfrak{P}(\mathbb{R})$. Then any of the conditions (i) $\#(F) < \omega$, (ii) $\#(F) < 2^\omega$ and $2^\omega = \omega_1$, (iii) $\#(F) < 2^{2^\omega}$ and Gödel's axiom of constructibility implies $\bigcup F \neq \mathfrak{P}(\mathbb{R})$. Theorem 5 is a generalization of some results of Ulam, Alouglu–Erdös, Jensen, Prikry, and Taylor connected with Ulam's problem about sets of measures."

Obviously, the result on finite sequences of algebras from 1.17 and Theorem 1.17 are generalizations of Grzegorek's result under conditions (i) and (ii).

1.22. Although in the prehistory of our research we cannot cite any papers on combinatorics,[17] a considerable part of our monograph is devoted to pure combinatorics. Thus, in the next chapter, a function $\mathfrak{g}(n)$ on $\mathbb{N}^+$ is defined, and in Chapter 9 it is proved that $\mathfrak{g}(10) \leq 5$. Let us translate this statement into the simplest combinatorial language. Assume that ten subsets

$$A_1, \dots, A_{10}$$

of a set A are given and

$$5 \leq \#(A_k) \leq 6$$

for all $k \in [1, 10]$. If $\#(A_k) = 5$, then A_k is divided into two k-classes: one k-class consists of two elements, and another consists of three elements. If $\#(A_k) = 6$, then A_k is divided into three k-classes, each containing two elements. In order to prove that $\mathfrak{g}(10) \leq 5$, it is necessary and sufficient to prove the existence of a set $Q \subset A$ such that for each $k \in [1, 10]$ there exists a k-class Q_k and

(a) $Q_k \cap Q \neq \varnothing$,
(b) $Q_k \setminus Q \neq \varnothing$.

[14] It turned out that this problem had been posed earlier by Taylor in [**Ta**] and Krawczyk and Pelc in [**Kr-P**].

[15] Although Woodin's monograph appeared in 1999 his result from 1.10 had been obtained much earlier and was known to specialists.

[16] By a field one means an algebra containing X (see [**Sik**]). We will not use the notion of field.

[17] Theorems like the Szemerédi–Furstenberg theorem and theorems on systems of distinct representatives cited below are not related to the main results of our monograph.

1.23. The next chapter gives a clear overview of the results of the monograph. One should note that in 3.32, rather than algebras of sets we consider Banach algebras of continuous bounded functions defined on a completely regular topological space. In Chapter 19 we deal with objects more general than algebras of sets — semistructures and structures of sets. After the main part of the book was already written and no major changes in the text were possible, I obtained a generalization of the conceptually important Theorem 2.1. This generalization is presented in the final Chapter 20.

1.24. A few words about references: A reference to, say, 6.8 means that we refer to Section 6.8, whereas a reference to footnote 6.8 means that we refer to footnote 8 in Chapter 6.

1.25. ACKNOWLEDGMENTS. It is a pleasure to thank those whose comments and advice helped during the work: Moti Gitik (Israel), Martin Goldstern (Austria), Vyacheslav Malykhin (Russia), David Fremlin (England), Saharon Shelah (Israel). Particular gratitude is due to Moti Gitik. I am grateful to Saharon Shelah who agreed to put his article as an Appendix to the book. Also, I am grateful to the AMS editor Sergei Gelfand and to the translator of the book.

CHAPTER 2

Main Results[1]

2.1. THEOREM. *If $\psi \in \Psi^n$, then the function*[2]

$$\psi'(k) = \begin{cases} \psi(k) & \text{if } k \in [1,n], \\ \left[\frac{4}{3}n\right] + 1 & \text{if } k = n+1 \end{cases}$$

belongs to Ψ^{n+1}.

2.2. The first part of the following theorem is an obvious corollary of Theorem 2.1.

THEOREM. (1) *Let $\mathcal{A}_1, \dots, \mathcal{A}_n$ be a finite sequence of algebras, and for each $k \in [1,n]$ let there exist more than $\frac{4}{3}(k-1)$ pairwise disjoint sets not belonging to $\mathcal{A}_k$. Then*

$$\bigcup_{k=1}^{n} \mathcal{A}_k \neq \mathfrak{P}(X).$$

(2) *The estimate $\frac{4}{3}(k-1)$ is the best possible in the following sense. For any natural $n > 1$ one can construct a sequence of algebras $\mathcal{A}_1, \dots, \mathcal{A}_n, \mathcal{A}_{n+1}$ such that for each $k \in [1,n]$ there exist more than $\frac{4}{3}(k-1)$ pairwise disjoint sets not belonging to $\mathcal{A}_k$, there exist $\left[\frac{4}{3}n\right]$ pairwise disjoint sets not belonging to $\mathcal{A}_{n+1}$, and*

$$\bigcup_{k=1}^{n+1} \mathcal{A}_k = \mathfrak{P}(X).$$

2.3. REMARK. Clearly, if $n \geq 2$ in Theorem 2.2(1), then for $\mathcal{A}_2$ it is sufficient to assume that $\mathcal{A}_2 \neq \mathfrak{P}(X)$.

2.4. THEOREM. (1) *One can construct a countable sequence of two-valued measures $\mu_1, \dots, \mu_k, \dots$ such that for any k there exists a family $\{M_\alpha^k\}$ consisting of $\mathfrak{c}$ μ_k-nonmeasurable sets with the following property:*

$$\mu_k(M_\alpha^k \cap M_\beta^k) = 0 \quad \text{if } \alpha \neq \beta,$$

and each subset of X is μ_k-measurable for some k. (Clearly, for each k there exist $\aleph_0$ pairwise disjoint μ_k-nonmeasurable sets.)

(2) *Let $\mathcal{A}_1, \dots, \mathcal{A}_k, \dots$ be a countable sequence of algebras, and assume that for each k there exist more than $\aleph_0$ pairwise disjoint sets not belonging to $\mathcal{A}_k$. Then $\bigcup \mathcal{A}_k \neq \mathfrak{P}(X)$.*

[1]This chapter contains the statements of theorems on finite sequences of algebras and their infinite analogs for σ-algebras and almost σ-algebras. The theorems on finite sequences of algebras are, formally, particular cases of their infinite analogs (an explanation is given in the next chapter). But since the theorems on finite sequences of algebras are proved separately, it is natural to state them separately also.

[2]By $[\rho]$ we denote the maximal integer $\leq \rho$.

(3) *Let $\mathcal{A}_1, \dots, \mathcal{A}_k, \dots$ be a countable sequence of algebras, and assume that for each k there exists a family $\{M_\alpha^k\}$ consisting of more than $\mathfrak{c}$ sets with the following property*:

$$\mathfrak{P}\big((M_\alpha^k \setminus M_\beta^k) \cup (M_\beta^k \setminus M_\alpha^k)\big) \setminus \mathcal{A}_k \neq \varnothing$$

if $\alpha \neq \beta$. Then

$$\bigcup \mathcal{A}_k \neq \mathfrak{P}(X).^3$$

2.5. Due to Theorem 2.4(1), one cannot generalize Theorem 2.2(1) to a countable number of algebras without imposing additional restrictions.

2.6. Definition. An algebra $\mathcal{A}$ is said to be an *almost σ-algebra* if for any countable sequence of sets $M_1, \dots, M_k, \dots$ such that $\mathfrak{P}(M_k) \subset \mathcal{A}$ for each k, we have $\bigcup M_k \in \mathcal{A}$.

2.7. Theorem. *Let $\psi \in \Psi$ be such that $\psi(n_1) - \psi(n_2) \to \infty$ as $n_1 - n_2 \to \infty$. Let $\mathcal{A}_1, \dots, \mathcal{A}_k, \dots$ be a countable sequence of almost σ-algebras such that for each k there exist $\psi(k)$ pairwise disjoint sets not belonging to $\mathcal{A}_k$. Then there exists $\mathbb{N}^* \subset \mathbb{N}^+$ such that*

$$\#(\mathbb{N}^+ \setminus \mathbb{N}^*) < \aleph_0 \quad \text{and} \quad \bigcup_{k \in \mathbb{N}^*} \mathcal{A}_k \neq \mathfrak{P}(X).^4$$

2.8. Due to Theorem 1.17, one can generalize Theorem 2.2(1) to a countable number of σ-algebras. Moreover, a stronger result is true.

Theorem. *Let $\mathcal{A}_1, \dots, \mathcal{A}_k, \dots$ be a countable sequence of almost σ-algebras, and for each k let there exist more than $\frac{4}{3}(k-1)$ pairwise disjoint sets not belonging to $\mathcal{A}_k$. Then*

$$\bigcup \mathcal{A}_k \neq \mathfrak{P}(X).$$

2.9. Let us give a refinement of Theorem 2.8.

Theorem. *Let $\mathcal{A}_1, \dots, \mathcal{A}_k, \dots$ be a countable sequence of almost σ-algebras, and assume that for each k there exists a finite sequence of sets*

$$U_1^k, \dots, U_{m_k}^k \notin \mathcal{A}_k$$

and $m_k > \frac{4}{3}(k-1)$. Then for each U_i^k one can choose $\widehat{U}_i^k \subset U_i^k$ such that:

(i) $\widehat{U}_i^k \notin \mathcal{A}_k$;
(ii) *either* $\widehat{U}_i^k = \widehat{U}_j^l$ *or* $\widehat{U}_i^k \cap \widehat{U}_j^l = \varnothing$;

---

[3] One can construct a σ-two-valued measure μ' such that the algebra of all μ'-measurable sets possesses the property of the algebras from the second part of the Theorem but does not possess the property of the algebras from the third part of the Theorem. One can construct a σ-two-valued measure μ'' such that the algebra of all μ''-measurable sets possesses the property of algebras from the third part of the Theorem but does not possess the property of algebras from the second part of the Theorem.

[4] Note the following fact. Let $\psi \in \Psi$, $\psi(n_1) - \psi(n_2) \to \infty$ as $n_1 - n_2 \to \infty$, and let $\psi' \colon \mathbb{N}^+ \to \mathbb{N}^+$ be a function such that $\psi'(k) \geq \psi(k)$ for all $k \in \mathbb{N}^+$ and $\psi'(k) - \psi(k) \to \infty$. Let $\mathcal{A}_1, \dots, \mathcal{A}_k, \dots$ be a countable sequence of almost σ-algebras such that for each k there exist $\psi'(k)$ pairwise disjoint sets not belonging to $\mathcal{A}_k$. Then $\bigcup \mathcal{A}_k \neq \mathfrak{P}(X)$.

(iii) *there exists a set* Q, *which is a union of sets of the type of* $\widehat{U}^l_j$, *such that* $Q \notin \bigcup \mathcal{A}_k$.[5]

2.10. DEFINITION. For each $n \in \mathbb{N}^+$ denote by $\mathfrak{g}(n)$ the minimal natural number such that the function $\psi(k) = \mathfrak{g}(n)$ for all $k \in [1, n]$ belongs to Ψ^n.

2.11. STATEMENT. *For each* $n \in \mathbb{N}^+$ *we have*:

$$1 + 4 \cdot [\log_5 n] \leq \mathfrak{g}(n) \leq \max\{1, 2 \cdot [\log_2 n] + 2 \cdot \operatorname{sign}(\log_2 n - [\log_2 n]) - 1\}.$$[6]

2.12. Due to Example 1.13, we have $\mathfrak{g}(3) \geq 3$. Examples constructed below show that $\mathfrak{g}(12) \geq 6$, $\mathfrak{g}(13) \geq 7$. Below it is shown that

$$\mathfrak{g}(m \cdot n) \geq \mathfrak{g}(m) + \mathfrak{g}(n) - 1$$

for any $m, n \in \mathbb{N}^+$. Therefore, one can improve the lower estimate for $\mathfrak{g}(n)$ from Statement 2.11. One can also improve the upper estimate for $\mathfrak{g}(n)$ from Statement 2.11. We have obtained a difficult result:

$$\mathfrak{g}(2^m + 2) \leq 2m - 1 \quad \text{for all natural } m \geq 3.$$

We obtain the following estimate for $\mathfrak{g}(n)$ on $[1, 258]$:

$$\begin{aligned}
&\mathfrak{g}(n) = 1 \quad \text{if } n \in [1, 2];\\
&\mathfrak{g}(n) = 3 \quad \text{if } n \in [3, 4];\\
&\mathfrak{g}(n) = 5 \quad \text{if } n \in [5, 10];\\
&5 \leq \mathfrak{g}(11) \leq 7;\\
&6 \leq \mathfrak{g}(12) \leq 7;\\
&\mathfrak{g}(n) = 7 \quad \text{if } n \in [13, 18];\\
&7 \leq \mathfrak{g}(n) \leq 9 \quad \text{if } n \in [19, 24];\\
&\mathfrak{g}(n) = 9 \quad \text{if } n \in [25, 34];\\
&9 \leq \mathfrak{g}(n) \leq 11 \quad \text{if } n \in [35, 59];\\
&10 \leq \mathfrak{g}(n) \leq 11 \quad \text{if } n \in [60, 64];\\
&\mathfrak{g}(n) = 11 \quad \text{if } n \in [65, 66];\\
&11 \leq \mathfrak{g}(n) \leq 13 \quad \text{if } n \in [67, 124];\\
&\mathfrak{g}(n) = 13 \quad \text{if } n \in [125, 130];\\
&13 \leq \mathfrak{g}(n) \leq 15 \quad \text{if } n \in [131, 258].
\end{aligned}$$

Thus, the difference between the upper and the lower estimate for $\mathfrak{g}(n)$ on $[1, 258]$ does not exceed 2. Besides, we know 34 exact values of $\mathfrak{g}(n)$, and for six values of

---

[5]Let $\psi\colon \mathbb{N}^+ \to \mathbb{N}^+$ be a function such that for an arbitrary sequence $\mathcal{A}_1, \ldots, \mathcal{A}_k, \ldots$ of almost σ-algebras, we have $\bigcup \mathcal{A}_k \neq \mathfrak{P}(X)$ provided for each k there exist $\psi(k)$ pairwise disjoint sets not belonging to $\mathcal{A}_k$. Then a statement similar to Theorem 2.9 is true, with the estimate $m_k > \frac{4}{3}(k-1)$ replaced by the estimate $m_k = \psi(k)$. Theorem 8.11 (cited in 2.34) is stronger than Theorem 2.9; and the estimate $m_k > \frac{4}{3}(k-1)$ is essential in that theorem.

[6]Recall that sign is the Kronecker function:

$$\operatorname{sign}(x) = \begin{cases} 1 & \text{if } x > 0,\\ -1 & \text{if } x < 0,\\ 0 & \text{if } x = 0. \end{cases}$$

n the difference between the upper and the lower estimate for $\mathfrak{g}(n)$ is 1. One more result: if $l \geq 3$ is an integer, then there exists $m_l \in \mathbb{N}^+$ such that if $m > m_l$ is an integer, then

$$\mathfrak{g}(2^m + l) \leq 2m - 1.$$

2.13. Let us return to the "general problem" from 1.17. A nondecreasing function $\psi \in \Psi^n$ is said to be *basic* if for any $\psi' \in \Psi^n$ the inequality $\psi'(k) \leq \psi(k)$ for all $k \in [1, n]$ implies $\psi(k) \equiv \psi'(k)$. For n fixed, the set of basic functions is finite. Denote this set by Ψ^n_*. Another statement of the "general problem" in 1.17 is to describe Ψ^n_* for each n. If $\psi \in \Psi^n_*$, then

$$\mathfrak{g}(n) \leq \psi(n) \leq \left[\frac{4}{3}(n-1)\right] + 1.$$

Due to Theorem 2.2 and Remark 2.3, the function

$$\psi_*(k) = \begin{cases} \left[\frac{4}{3}(k-1)\right] + 1 & \text{if } k \neq 2, \\ 1 & \text{if } k = 2 \end{cases}$$

defined on $[1, n]$ belongs to Ψ^n_*. Let us give the table of all basic functions for all $n \leq 7$. The functions of the type $\psi_*(k)$ are marked by a star.

(1–1)	1	*						
(2–1)	1	1	*					
(3–1)	1	1	3	*				
(4–1)	1	3	3	3				
(4–2)	1	1	3	5	*			
(5–1)	1	3	3	3	5			
(5–2)	1	1	5	5	5			
(5–3)	1	1	3	5	6	*		
(6–1)	3	3	3	3	5	5		
(6–2)	1	3	3	5	5	5		
(6–3)	1	1	5	5	6	6		
(6–4)	1	3	3	3	5	7		
(6–5)	1	1	5	5	5	7		
(6–6)	1	1	3	5	6	7	*	
(7–1)	3	3	3	5	5	5	5	
(7–2)	1	3	5	5	5	5	5	
(7–3)	3	3	3	3	5	5	7	
(7–4)	1	3	3	5	5	5	7	
(7–5)	1	1	5	6	6	6	7	
(7–6)	1	3	3	3	7	7	7	
(7–7)	1	1	5	5	7	7	7	
(7–8)	1	1	3	6	7	7	7	
(7–9)	1	1	5	5	6	6	8	
(7–10)	1	3	3	3	5	7	8	
(7–11)	1	1	5	5	5	7	8	
(7–12)	1	1	3	5	7	7	8	
(7–13)	1	1	3	5	6	7	9	*

2.14. Let us fix $l \in \mathbb{N}^+$, and let k be a variable with values in $[1, l]$. By definition, $\xi(l; k)$ *is any function defined on* $[1, l]$*, taking values in* $\mathbb{N}^+$*, and satisfying*

$$\sum_{k=1}^{l} 2^{-\left[\frac{\xi(l;k)+1}{2}\right]} \le 1.$$

STATEMENT. $\xi(l; k) \in \Psi^l$.

2.15. Using Statement 2.14 the following interesting theorem is proved.

THEOREM. *Let* $\mathcal{A}_1, \dots, \mathcal{A}_k, \dots$ *be a countable sequence of almost* σ*-algebras such that for each* k *there exist more than* $4 \cdot \log_2 k$ *pairwise disjoint sets not belonging to* $\mathcal{A}_k$*. Then*

$$\bigcup \mathcal{A}_k \neq \mathfrak{P}(X).$$ [7]

2.16. REMARK. Previously, we had the estimate $\frac{4}{3}(k-1)$, which was the best possible in a certain sense, whereas the above theorem gives an estimate of completely different order $4 \cdot \log_2 k$. The reason is that

$$[4 \cdot \log_2 k] > \left[\frac{4}{3}(k-1)\right]$$

for all $k \in [2, 10]$. This is an interesting "instability" effect.

2.17. Assume that $n \in \mathbb{N}^+$, and for each $k \in [1, n]$ a nonempty class $\widehat{\Psi}^k \subset \Psi^k$ is chosen. Let $m \in \mathbb{N}$, $p \in \mathbb{N}^+$, and $n \ge p - m > 0$. Define two functions (of the variable k) on $[m+1, p]$ taking values in $\mathbb{N}^+$ as follows:

$$\psi_1(m, p; k) = 2m + \psi(k - m), \quad \text{where } \psi \in \widehat{\Psi}^{p-m},$$
$$\psi_2(m, p; k) = \left[\frac{m}{2}\right] + p.$$

The following statement is presented in a form convenient for future use.

STATEMENT. *Let* $m \le n$ *be natural numbers. The function*

$$\psi(k) = \begin{cases} \psi_1(0, m; k) & \textit{if } k \in [1, m], \\ \psi_i(m, n+1; k) & \textit{if } k \in [m+1, n+1] \end{cases}$$

belongs to Ψ^{n+1}.

2.18. Assume that, as in the previous section, for each $k \in [1, n]$ a nonempty class $\widehat{\Psi}^k$ is chosen. Let us describe an inductive way of constructing a certain class $\widehat{\Psi}^{n+1} \subset \Psi^{n+1}$. The class $\widehat{\Psi}^{n+1}$ is formed by:

1) the functions of the type

$$\psi(k) = \begin{cases} \psi_1(0, n; k) & \text{if } k \in [1, n], \\ \left[\frac{4}{3}n\right] + 1 & \text{if } k = n + 1; \end{cases}$$

2) the function $\psi(k) \equiv \mathfrak{g}(n+1)$;
3) the functions of the type $\xi(n+1; k)$;

---

[7]Note that the function $\psi(k) = [4 \cdot \log_2 k] + 1$ does not possess the limit property $\psi(n_1) - \psi(n_2) \to \infty$ if $n_1 - n_2 \to \infty$.

4) the functions of the type

$$\psi(k) = \begin{cases} \psi_1(0, m; k) & \text{if } k \in [1, m], \\ \psi_i(m, n+1; k) & \text{if } k \in [m+1, n+1]. \end{cases}$$

2.19. Assume that for each $n \in \mathbb{N}^+$ a nonempty class $\widehat{\Psi}^n \subset \Psi^n$ is chosen. Assume also that:

1) all functions in $\widehat{\Psi}^n$ are nondecreasing;
2) if $\psi, \psi' \in \widehat{\Psi}^n$ and $\psi'(k) \leq \psi(k)$ for all $k \in [1, n]$, then $\psi(k) \equiv \psi'(k)$;
3) if $\psi \in \widehat{\Psi}^n$ and $n' < n$ is a natural number, then there exists a function $\psi' \in \widehat{\Psi}^{n'}$ such that $\psi'(k) \leq \psi(k)$ for all $k \in [1, n']$;
4) if φ is a nondecreasing function defined on $[1, n+1]$ constructed inductively as described in 2.18, then there exists a function $\psi \in \widehat{\Psi}^{n+1}$ such that $\psi(k) \leq \varphi(k)$ for all $k \in [1, n+1]$;
5) there exists $n_0 \in \mathbb{N}^+$ such that all functions in $\widehat{\Psi}^n$ are constructed in the inductive way if $n > n_0$.

Define the maximal class $\widehat{\Psi}$ of nondecreasing functions on $\mathbb{N}^+$ with values in $\mathbb{N}^+$: $\psi \in \widehat{\Psi}$ if and only if for each $n \in \mathbb{N}^+$ there exists a function $\psi_n \in \widehat{\Psi}^n$ such that $\psi_n(k) \leq \psi(k)$ for all $k \in [1, n]$. Define the class $\widehat{\Psi}_* \subset \widehat{\Psi}$ as follows: $\psi \in \widehat{\Psi}_*$ if and only if there does not exist a function $\psi' \in \widehat{\Psi}$ such that $\psi'(k) \leq \psi(k)$ for all k and $\psi'(k_0) < \psi(k_0)$ for at least one k_0.

One more notion: *$\xi^*(k)$ is any function defined on $\mathbb{N}^+$, taking values in $\mathbb{N}^+$, and satisfying*

$$\sum_{k=1}^{\infty} 2^{-\left[\frac{\xi^*(k)+1}{2}\right]} = 1.$$

STATEMENT. *Let $\psi \in \widehat{\Psi}_*$. Then one of the following two statements is true:*

(1) *there exists a countable sequence of integers*

$$0 = \pi_0 < \cdots < \pi_i < \cdots$$

such that the following hold:
(1–a) $\psi(\pi_{i+1}) = \left[\frac{4}{3}\pi_i\right] + 1$ *if* $\pi_{i+1} - \pi_i = 1$ *and* $\pi_i \neq 1$;
(1–b) $\psi(\pi_i) = 1$ *if* $\pi_i = 2$;
(1–c) $\psi(k) = \psi_{j(i)}(\pi_i, \pi_{i+1}; k)$ *for all* $k \in [\pi_i + 1, \pi_{i+1}]$ *if* $\pi_{i+1} - \pi_i > 1$; $j(0) = 1$;[8]

(2) *there exists an $m \in \mathbb{N}$ such that:*
(2–a) *if* $m > 0$, *then* $\psi(k) = \psi_1(0, m; k)$ *for all* $k \in [1, m]$;
(2–b) $\psi(k) = 2m + \xi^*(k - m)$ *for all* $k > m$.

2.20. Using Statement 2.19, one proves the following theorem, which generalizes Theorem 2.8 as well as Theorem 2.15.

THEOREM. *Let $\psi \in \widehat{\Psi}$, and let $\mathcal{A}_1, \ldots, \mathcal{A}_k, \ldots$ be a countable sequence of almost σ-algebras such that for any k there exist $\psi(k)$ pairwise disjoint sets not belonging to $\mathcal{A}_k$. Then*

$$\bigcup \mathcal{A}_k \neq \mathfrak{P}(X).$$

[8] One can assume that π_1 is greater than an arbitrary given natural number.

2.21. This and the following three sections contain a series of new results.

THEOREM. (1) *Let* $\mathcal{A}_1, \dots, \mathcal{A}_k, \dots$ *be a countable sequence of algebras, and let there exist an infinite matrix of pairwise disjoint sets*[9]

$$\begin{pmatrix} U_1^1 & & \\ U_1^2 & & \\ \dots & \dots & \dots \\ U_1^k & \dots & U_{m_k}^k \\ \dots & \dots & \dots \end{pmatrix}$$

such that:

(i) $U_i^k \notin \mathcal{A}_k$;
(ii) $m_1 = m_2 = 1$;
(iii) $m_k > 1$ *if* $k > 2$;
(iv) $m_k \to \infty$.

Then $\bigcup \mathcal{A}_k \neq \mathfrak{P}(X)$.

(2) *If the above conditions are changed so that we allow* $m_3 = 1$, *then it may happen that* $\bigcup \mathcal{A}_k = \mathfrak{P}(X)$.

(3) *If a matrix satisfies all the conditions of the first part except the last one (i.e.,* $\underline{\lim}\, m_k < \infty$*), then it may happen that* $\bigcup \mathcal{A}_k = \mathfrak{P}(X)$.

2.22. Theorem 2.21(2) does not require a long proof: for the algebras $\mathcal{A}_1', \mathcal{A}_2', \mathcal{A}_3'$ considered in Example 1.13, there exist pairwise disjoint sets U_1, U_2, U_3 such that $U_k \notin \mathcal{A}_k'$, and $\mathcal{A}_1' \cup \mathcal{A}_2' \cup \mathcal{A}_3' = \mathfrak{P}(X)$.

2.23. The following theorem is an obvious corollary of Theorem 2.21(1).

THEOREM. *Let* $\mathcal{A}_1, \dots, \mathcal{A}_n$ *be a finite sequence of algebras. Assume that there exists a matrix of pairwise disjoint sets*

$$\begin{pmatrix} U_1^1 & \\ U_1^2 & \\ U_1^3 & U_2^3 \\ \dots & \dots \\ U_1^n & U_2^n \end{pmatrix}$$

(each row starting with the third contains two elements) such that $U_i^k \notin \mathcal{A}_k$. *Then*

$$\bigcup_{k=1}^{n} \mathcal{A}_k \neq \mathfrak{P}(X).$$

2.24. After the proof of Theorem 2.21(1) it is rather easy to prove the following theorem.

THEOREM. *Let* $\mathcal{A}_1, \dots, \mathcal{A}_k, \dots$ *be a countable sequence of* σ*-algebras, and let there exist an infinite matrix of pairwise disjoint sets*

$$\begin{pmatrix} U_1^1 & \\ U_1^2 & \\ U_1^3 & U_2^3 \\ \dots & \dots \\ U_1^k & U_2^k \\ \dots & \dots \end{pmatrix}$$

---

[9]By definition, an infinite matrix consists of $\aleph_0$ rows, each row with a finite number of elements.

(each row starting with the third contains two elements) such that $U_i^k \notin \mathcal{A}_k$. Then $\bigcup \mathcal{A}_k \neq \mathfrak{P}(X)$.

2.25. DEFINITION. Assume that $\mathbb{N}^* \subset \mathbb{N}^+$, and let n be an arbitrary natural number. The number

$$\varlimsup_{n\to\infty} \frac{\#(\{k \le n \mid k \in \mathbb{N}^*\})}{n},$$

denoted $p(\mathbb{N}^*)$, is called the *density* of $\mathbb{N}^*$. The number

$$\lim_{n\to\infty} \frac{\#(\{k \le n \mid k \in \mathbb{N}^*\})}{n},$$

denoted $p_a(\mathbb{N}^*)$, is called the *absolute density* of $\mathbb{N}^*$.

2.26. One can ask whether Theorem 2.24 will remain true if in its statement one replaces σ-algebras with almost-σ-algebras. It is not even clear how one can prove the following result. If in Theorem 2.24 one replaces σ-algebras with almost-σ-algebras, does there exist $\mathbb{N}^* \subset \mathbb{N}^+$ such that

$$p(\mathbb{N}^*) = 1 \quad \text{and} \quad \bigcup_{k\in\mathbb{N}^*} \mathcal{A}_k \neq \mathfrak{P}(X)?$$

2.27. This section opens a new series of results.

THEOREM. (1) *Let $\mathcal{A}_1, \dots, \mathcal{A}_n$ be a finite sequence of algebras, and assume that for each $k \in [1, n]$ there exist more than $\frac{5}{2}(k-1)$ pairwise disjoint sets not belonging to $\mathcal{A}_k$. Then there exist pairwise disjoint sets $U_1, \dots, U_n$ such that*

$$\text{if } U_k \subset Q \subset \bigcup_{i=1}^{n} U_i, \text{ then } Q \notin \mathcal{A}_k.$$

(2) *The estimate $\frac{5}{2}(k-1)$ is the best possible in the following sense. For any $n \in \mathbb{N}^+$ one can construct algebras $\mathcal{A}_1, \dots, \mathcal{A}_n, \mathcal{A}_{n+1}$ such that:*

(i) *$\mathcal{A}_2 \neq \mathfrak{P}(X)$;*

(ii) *for each $k \in [1, n] \setminus \{2\}$ there exist more than $\frac{5}{2}(k-1)$ pairwise disjoint sets not belonging to $\mathcal{A}_k$;*

(iii) *if $n > 1$, then there exist three pairwise disjoint sets not belonging to $\mathcal{A}_1 \cap \mathcal{A}_2$;*

(iv) *there exist $\left[\frac{5}{2}n\right]$ pairwise disjoint sets not belonging to $\mathcal{A}_{n+1}$;*

(v) *for the algebras $\mathcal{A}_1, \dots, \mathcal{A}_n, \mathcal{A}_{n+1}$ the corresponding sets $U_1, \dots, U_n, U_{n+1}$ do not exist.*

2.28. REMARK. From the proof of the first part of the theorem one clearly gets the following result, which can be compared with Theorem 2.27(2). Let $\mathcal{A}_1, \dots, \mathcal{A}_n$ be a finite sequence of algebras such that $\mathcal{A}_2 \neq \mathfrak{P}(X)$ if $n > 1$. Put $\mathcal{A}'_k = \mathcal{A}_k$ if $k \in [1, n] \setminus \{2\}$, and $\mathcal{A}'_2 = \mathcal{A}_1 \cap \mathcal{A}_2$ if $n > 1$. Assume that for each $k \in [1, n]$ there exist more than $\frac{5}{2}(k-1)$ pairwise disjoint sets not belonging to $\mathcal{A}'_k$. Then there exist pairwise disjoint sets $U_1, \dots, U_n$ such that if $U_k \subset Q \subset \bigcup_{i=1}^n U_i$, then $Q \notin \mathcal{A}_k$.

2.29. THEOREM. *Let $\mathcal{A}_1, \dots, \mathcal{A}_k, \dots$ be a countable sequence of σ-algebras such that for each k there exist more than $\frac{5}{2}(k-1)$ pairwise disjoint sets not belonging to $\mathcal{A}_k$. Then there exists a countable sequence of pairwise disjoint sets $U_1, \dots, U_k, \dots$ such that*

$$\text{if } U_k \subset Q \subset \bigcup U_i, \text{ then } Q \notin \mathcal{A}_k.$$

2.30. Let us return to Theorem 2.27(1). By its assumption, for each $k \in [1, n]$ there exist pairwise disjoint sets

$$U_1^k, \dots, U_{m_k}^k \notin \mathcal{A}_k$$

and $m_k > \frac{5}{2}(k-1)$. One can then show that there exist pairwise disjoint sets $U_1, \dots, U_n$ such that:

(i) $U_k \subset U_{j_k}^{i_k}$;

(ii) if $U_k \subset Q \subset \bigcup_{i=1}^n U_i$, then $Q \notin \mathcal{A}_k$.

Two questions arise. The first one is: what is the best possible (in some sense) estimate for m_k under the additional condition $i_k \leq k$ for all $k \leq n$? The second one is: what is the best possible (in some sense) estimate for m_k under the stronger additional condition $i_k = k$ for all $k \leq n$? Both questions have the same answer:

$$m_k > 3(k-1).$$

The corresponding result for a countable sequence of σ-algebras is also true.

2.31. THEOREM. (1) *Let $\mathcal{A}_1, \dots, \mathcal{A}_n$ be a finite sequence of algebras, and assume that for each $k \in [1, n]$ there exist more than $4(k-1)$ pairwise disjoint sets not belonging to $\mathcal{A}_k$. Then there exist pairwise disjoint sets $U_1, \dots, U_n, V_1, \dots, V_n$ such that if $U_k \subset Q$ and $V_k \cap Q = \varnothing$, then $Q \notin \mathcal{A}_k$.*

(2) *The estimate $4(k-1)$ is the best possible in the following sense. For any $n \in \mathbb{N}^+$ one can construct algebras $\mathcal{A}_1, \dots, \mathcal{A}_n, \mathcal{A}_{n+1}$ such that for each $k \in [1, n]$ there exist more than $4(k-1)$ pairwise disjoint sets not belonging to $\mathcal{A}_k$, there exist $4n$ pairwise disjoint sets not belonging to $\mathcal{A}_{n+1}$, and for the algebras $\mathcal{A}_1, \dots, \mathcal{A}_n, \mathcal{A}_{n+1}$ the corresponding sets $U_1, \dots, U_n, U_{n+1}, V_1, \dots, V_n, V_{n+1}$ do not exist.*

2.32. THEOREM. *Let $\mathcal{A}_1, \dots, \mathcal{A}_k, \dots$ be a countable sequence of σ-algebras, and assume that for each k there exist more than $4(k-1)$ pairwise disjoint sets not belonging to $\mathcal{A}_k$. Then there exist pairwise disjoint sets $U_1, \dots, U_k, \dots, V_1, \dots, V_k, \dots$ such that for each $k \in \mathbb{N}^+$ the following holds: if $U_k \subset Q$ and $V_k \cap Q = \varnothing$, then $Q \notin \mathcal{A}_k$.*

2.33. REMARK. Since Theorem 2.8 admits a *refinement* (Theorem 2.9), an *a priori* similar refinement is true for Theorem 2.2(1). Natural refinements are true also for Theorems 2.27(1) and 2.29, the results at the end of 2.30, and Theorems 2.31(1) and 2.32. In refining the last two theorems one can take into account Remark 14.9.

2.34. The proof of Theorem 2.2(1) obviously implies that under the assumptions of Theorem 2.2(1) there exist sets $U_1, \dots, U_n$ with the following properties:

(i) either $U_i = U_j$ or $U_i \cap U_j = \varnothing$;

(ii) if $U_k \subset Q \subset \bigcup_{i=1}^n U_i$, then $Q \notin \mathcal{A}_k$.

It turns out that a stronger result holds: the sets $U_1, \dots, U_n$ can be constructed in such a way that

$$\#\big(\{i \in [1, n] \mid U_i = U_k\}\big) \leq 4$$

for each $k \in [1, n]$.[10] The following theorem is also true.

---

[10] Possibly this statement can be strengthened by replacing the estimate 4 by 3. But one cannot replace 4 by 2.

THEOREM. *Let $\mathcal{A}_1, \dots, \mathcal{A}_k, \dots$ be a countable sequence of σ-algebras, and assume that for each k there exists a finite sequence of sets*

$$U_1^k, \dots, U_{m_k}^k \notin \mathcal{A}_k$$

and $m_k > \frac{4}{3}(k-1)$. Then for each U_i^k one can take $\widehat{U}_i^k \subset U_i^k$ such that the following is true:

(i) $\widehat{U}_i^k \notin \mathcal{A}_k$;
(ii) *either* $\widehat{U}_i^k = \widehat{U}_j^l$ *or* $\widehat{U}_i^k \cap \widehat{U}_j^l = \varnothing$;
(iii) *for each $k \in \mathbb{N}^+$ there exists $\widehat{U}_{\psi_k}^{\xi_k}$ such that*

$$\text{if } \widehat{U}_{\psi_k}^{\xi_k} \subset Q \subset \bigcup_{j \in \mathbb{N}^+} \widehat{U}_{\psi_j}^{\xi_j}, \text{ then } Q \notin \mathcal{A}_k;$$

(iv) $\#(\{i \in \mathbb{N}^+ \mid \widehat{U}_{\psi_i}^{\xi_i} = \widehat{U}_{\psi_k}^{\xi_k}\}) \leq 4$ *for each $k \in \mathbb{N}^+$.*

Thus, the estimate $\frac{5}{2}(k-1)$ instead of $\frac{4}{3}(k-1)$ in Theorems 2.27 and 2.29 is necessary only for the requirement that the sets U_k be pairwise disjoint. Hence, it is natural to pass to Theorems 2.27 and 2.29 after considering the theorems stating only existence of a set not belonging to an at most countable number of algebras. It is also natural to pass to Theorems 2.31 and 2.32 after Theorems 2.27 and 2.29. The reason is that the statement of Theorem 2.27(1) is equivalent to the following statement: there exist pairwise disjoint sets $U_1, \dots, U_n, V$ such that if $U_k \subset Q$ and $V \cap Q = \varnothing$, then $Q \notin \mathcal{A}_k$. (It is clear that the corresponding statement about a countable sequence of σ-algebras is equivalent to the statement of Theorem 2.29.)

In Theorem 2.9 and its generalization in footnote 2.5, one can construct the sets $\widehat{Q}_i^k$ and Q in such a way that for each $k \in \mathbb{N}^+$ there exists $\widehat{U}_{\psi_k}^{\xi_k} \subset Q$ such that $\widehat{U}_{\psi_k}^{\xi_k} \notin \mathcal{A}_k$. (It is possible that $\widehat{U}_{\psi_k}^{\xi_k} = \widehat{U}_{\psi_l}^{\xi_l}$ although $k \neq l$; the existence of the sets $\widehat{U}_{\psi_k}^{\xi_k}$ is not obvious since we deal with almost σ-algebras rather than with σ-algebras.) If one considerably increases the bound for m_k in the assumptions of Theorem 2.9 (e.g. $m_k = k^2$), then one can construct $\widehat{U}_i^k$ and for each $k \in \mathbb{N}^+$ take $\widehat{U}_{\psi_k}^{\xi_k}$ so that the following holds: if $\widehat{U}_{\psi_k}^{\xi_k} \subset \widehat{Q} \subset \bigcup_{j \in \mathbb{N}^+} \widehat{U}_{\psi_j}^{\xi_j}$, then $\widehat{Q} \notin \mathcal{A}_k$. However, under the assumptions of Theorem 2.9, one can construct $\widehat{U}_i^k$ and for each $k \in \mathbb{N}^+$ choose $\widehat{U}_{\psi_k}^{\xi_k}$ so that the following holds: for any real number $\rho < 1$ and for any finite $\mathbb{N}_* \subset \mathbb{N}^+$ there is a subset $\mathbb{N}^* \subset \mathbb{N}^+$ such that $\mathbb{N}_* \subset \mathbb{N}^*$, $p_a(\mathbb{N}^*) > \rho$, and if

$$\widehat{U}_{\psi_k}^{\xi_k} \subset \widehat{Q} \subset \bigcup_{j \in \mathbb{N}^*} \widehat{U}_{\psi_j}^{\xi_j},$$

then $\widehat{Q} \notin \mathcal{A}_k$; moreover, there exists a set Q which is a union of sets of the type of $\widehat{U}_{\psi_l}^{\xi_l}$ such that $Q \notin \bigcup \mathcal{A}_k$. Also, if one replaces the estimate $\frac{4}{3}(k-1)$ by $\frac{3}{2}(k-1)$, then one can require in addition that $\xi_k \in \mathbb{N}^*$ for all $k \in N^*$. I believe that this and the previous results are very subtle. If we have the estimate $m_k > \frac{3}{2}(k-1)$ but deal with σ-algebras instead of almost σ-algebras, then Theorem 2.34 holds, with the estimate 4 replaced by the estimate 3. And if one replaces the estimate $\frac{4}{3}(k-1)$ by $2(k-1)$ in the assumptions of Theorem 2.34, then one can require that $\xi_k = k$ and

$$\#(\{i \in \mathbb{N}^+ \mid \widehat{U}_{\psi_i}^{i} = \widehat{U}_{\psi_k}^{k}\}) \leq 2$$

for each $k \in \mathbb{N}^+$.

2.35. Similarly to the definition of the function $\mathfrak{g}(n)$, it is natural to give the following definition.

DEFINITION. For each $n \in \mathbb{N}^+$ denote by $\mathfrak{h}(n)$ the minimal number such that if $\mathcal{A}_1, \dots, \mathcal{A}_n$ is a sequence of algebras and for each $k \in [1, n]$ there exist $\mathfrak{h}(n)$ pairwise disjoint sets not belonging to $\mathcal{A}_k$, then there exist pairwise disjoint sets $U_1, \dots, U_n$ such that

$$\text{if } U_k \subset Q \subset \bigcup_{i=1}^{n} U_i, \text{ then } Q \notin \mathcal{A}_k.$$

2.36. THEOREM. $\mathfrak{h}(n) = 2n - 1$.

2.37. Now it is natural to give the following definition.

DEFINITION. For each $n \in \mathbb{N}^+$ denote by $\mathfrak{v}(n)$ the minimal number such that if $\mathcal{A}_1, \dots, \mathcal{A}_n$ is a sequence of algebras and for each $k \in [1, n]$ there exist $\mathfrak{v}(n)$ pairwise disjoint sets not belonging to $\mathcal{A}_k$, then there exist pairwise disjoint sets $U_1, \dots, U_n, V_1, \dots, V_n$ such that if $U_k \subset Q$ and $V_k \cap Q = \varnothing$, then $Q \notin \mathcal{A}_k$.

2.38. The following theorem summarizes the analysis of the function $\mathfrak{v}(n)$.

THEOREM.

(1) $\mathfrak{v}(n) = 4n - 3$ *for* $n \leq 3$.
(2) $\mathfrak{v}(n) \leq 4n - 5$ *for* $n > 3$.
(3) $\mathfrak{v}(n) \leq 4n - \left[\frac{n+3}{2}\right]$.
(4) $3n - 2 \leq \mathfrak{v}(n)$.[11]

2.39. Let us pass to the statements of the theorems similar to Theorems 2.23 and 2.24.

THEOREM. (1) *Let $\mathcal{A}_1, \dots, \mathcal{A}_n$ be a finite sequence of algebras, and let there exist a matrix of pairwise disjoint sets*

$$\begin{pmatrix} U_1^1 & & \\ \dots & \dots & \dots \\ U_1^k & \dots & U_{m_k}^k \\ \dots & \dots & \dots \\ U_1^n & \dots & U_{m_n}^n \end{pmatrix}$$

such that $m_k = k$, $U_i^k \notin \mathcal{A}_k$. Then there exist pairwise disjoint sets $U_1, \dots, U_n$ such that

$$\text{if } U_k \subset Q \subset \bigcup_{i=1}^{n} U_i, \text{ then } Q \notin \mathcal{A}_k.$$

(2) *The estimate $m_k = k$ is the best possible in the following sense. For any natural $n > 1$ one can construct algebras $\mathcal{A}_1, \dots, \mathcal{A}_n$ and a matrix of pairwise disjoint sets*

$$\begin{pmatrix} U_1^1 & & \\ \dots & \dots & \dots \\ U_1^k & \dots & U_{m_k}^k \\ \dots & \dots & \dots \\ U_1^n & \dots & U_{m_n}^n \end{pmatrix}$$

such that $m_k = k$ if $k < n$, $m_n = n-1$ and $U_i^k \notin \mathcal{A}_k$, and such that for the algebras $\mathcal{A}_1, \dots, \mathcal{A}_n$ the corresponding sets $U_1, \dots, U_n$ do not exist.

---

[11] By Theorem 2.38 $10 \leq \mathfrak{v}(4) \leq 11$. Apparently, $\mathfrak{v}(4) = 10$. Then $\mathfrak{v}(n) \leq 4n - 6$ for $n > 3$.

2.40. Let us refine the statement of Theorem 2.39(1). One can choose the sets $U_1, \dots, U_n$ so that $U_k \subset U_{j_k}^{i_k}$, and $U_{j_k}^{i_k} \neq U_{j_l}^{i_l}$ if $k \neq l$. Moreover, one can achieve that

$$\#(\{k \in [1,n] \mid i_k > k\}) \leq \frac{n}{2}.$$

The estimate $n/2$ is the best possible.

2.41. In Theorem 2.39(1) put $m_1 = 1$ and $m_k = k+1$ if $k > 1$. Then one can choose $U_1, \dots, U_n$ so that:

(i) $U_k \subset U_{j_k}^{i_k}$;
(ii) $U_{j_k}^{i_k} \neq U_{j_l}^{i_l}$ if $k \neq l$;
(iii) $i_k \leq k$.

This estimate for m_k is in a sense the best possible.

2.42. In Theorem 2.39(1) put

$$m_k = \begin{cases} 1 & \text{if } k = 1, \\ 3 & \text{if } k = 2, \\ k+2 & \text{if } k > 2. \end{cases}$$

Then one can choose $U_1, \dots, U_n$ so that $U_k \subset U_{j_k}^{k}$. This estimate for m_k can also be called the best possible (in a certain sense).

2.43. The results in 2.39–2.42 are corollaries of the general theorems. Thus, in Theorem 2.39(1) and its refinement in 2.40 it is inessential that $m_k = k$. It is essential only that $m_1, \dots, m_n$ be a nondecreasing sequence and

$$\sum_{i=1}^{k} m_i \geq \frac{k(k+1)}{2}$$

for each $k \in [1,n]$. In the statement of 2.41 it is essential that $m_1, \dots, m_n$ be a nondecreasing sequence and

$$\sum_{i=1}^{k} m_i \geq \frac{k^2+3k-2}{2}$$

for each $k \in [1,n]$. If in the conditions of Theorem 2.39(1) $m_1, \dots, m_n$ is a nondecreasing sequence and

$$\sum_{i=1}^{k} m_i \geq \frac{k^2+5k-4}{2}$$

for each $k \in [1,n]$, then the corresponding sets $U_1, \dots, U_n$ exist, and $U_k \subset U_{j_k}^{k}$. With this known, one can easily prove the statement in 2.42.

2.44. The results of 2.39, 2.41, 2.42 and the general theorems implying these results admit a generalization to countable sequences of σ-algebras.

The result in 2.40 is stated in a form which emphasizes that the estimate given in it is the best possible. A more delicate result is true: one can choose corresponding sets $U_1, \dots, U_n$ so that if $i_k > k$, then there exists $r_k \geq i_k$ such that $U_{r_k} \subset U_{j_{r_k}}^{r_k}$; moreover, $k \to r_k$ is a one-to-one correspondence. The following corresponding general theorem and its generalization for a countable sequence of σ-algebras are also true.

THEOREM. *Let $\mathcal{A}_1, \dots, \mathcal{A}_k, \dots$ be a countable sequence of σ-algebras, and let there exist an infinite matrix of pairwise disjoint sets*

$$\begin{pmatrix} U_1^1 & \dots & U_{m_1}^1 \\ \dots & \dots & \dots \\ U_1^k & \dots & U_{m_k}^k \\ \dots & \dots & \dots \end{pmatrix}$$

such that $m_1, \dots, m_k, \dots$ is a nondecreasing sequence, $\sum_{i=1}^k m_i \geq \frac{k(k+1)}{2}$ for each $k \in \mathbb{N}^+$, and $U_i^k \notin \mathcal{A}_k$. Then there exists a countable sequence of sets $U_1, \dots, U_k, \dots$ such that if $U_k \subset Q \subset \bigcup U_i$, then $Q \notin \mathcal{A}_k$, and, in addition:

(i) $U_k \subset U_{j_k}^{i_k}$;

(ii) $U_{j_k}^{i_k} \neq U_{j_l}^{i_l}$ *if* $k \neq l$;

(iii) *if $i_k > k$, then there exists $r_k \geq i_k$ such that $U_{r_k} \subset U_{j_{r_k}}^{r_k}$;*

(iv) *$k \to r_k$ is a one-to-one correspondence.*

2.45. THEOREM. (1) *Let $\mathcal{A}_1, \dots, \mathcal{A}_n$ be a finite sequence of algebras, and let there exist a matrix of pairwise disjoint sets*

$$\begin{pmatrix} U_1^1 & & \\ \dots & \dots & \dots \\ U_1^k & \dots & U_{m_k}^k \\ \dots & \dots & \dots \\ U_1^n & \dots & U_{m_n}^n \end{pmatrix}$$

such that $m_1 = 1$, $m_k = 2k$ if $k \in [2, n]$, and $U_i^k \notin \mathcal{A}_k$. Then there exist pairwise disjoint sets $U_1, \dots, U_n, V_1, \dots, V_n$ such that

$$\text{if } U_k \subset Q \text{ and } V_k \cap Q = \varnothing, \text{ then } Q \notin \mathcal{A}_k.$$

(2) *The estimate for m_k is the best possible in the following sense. For any natural $n > 1$ one can construct algebras $\mathcal{A}_1, \dots, \mathcal{A}_n$ and a matrix of pairwise disjoint sets*

$$\begin{pmatrix} U_1^1 & & \\ \dots & \dots & \dots \\ U_1^k & \dots & U_{m_k}^k \\ \dots & \dots & \dots \\ U_1^n & \dots & U_{m_n}^n \end{pmatrix}$$

such that $m_1 = 1$, $m_k = 2k$ if $2 \leq k \leq n-1$, $m_n = 2n-1$, and $U_i^k \notin \mathcal{A}_k$; and for the algebras $\mathcal{A}_1, \dots, \mathcal{A}_n$ the corresponding sets $U_1, \dots, U_n, V_1, \dots, V_n$ do not exist.

2.46. THEOREM. *Let $\mathcal{A}_1, \dots, \mathcal{A}_k, \dots$ be a countable sequence of σ-algebras, and assume that there exists an infinite matrix of pairwise disjoint sets*

$$\begin{pmatrix} U_1^1 & & \\ \dots & \dots & \dots \\ U_1^k & \dots & U_{m_k}^k \\ \dots & \dots & \dots \end{pmatrix}$$

such that $m_1 = 1$, $m_k = 2k$ if $k > 1$, and $U_i^k \notin \mathcal{A}_k$. Then there exist pairwise disjoint sets $U_1, \dots, U_k, \dots, V_1, \dots, V_k, \dots$ such that for each $k \in \mathbb{N}^+$ the following is true:

$$\text{if } U_k \subset Q \text{ and } V_k \cap Q = \varnothing, \text{ then } Q \notin \mathcal{A}_k.$$

2.47. Consider objects more general than algebras.

DEFINITION. A set $\mathcal{A} \subset \mathfrak{P}(X)$ containing $\varnothing$ and possessing the property

$$\text{if } M_1, M_2 \in \mathcal{A}, \text{ then } M_1 \cup M_2,\ M_1 \cap M_2 \in \mathcal{A},$$

is called a *structure*. If, in addition, we have $\bigcup M_k \in \mathcal{A}$ for any countable sequence of sets $M_1, \dots, M_k, \dots$ such that $\mathfrak{P}(M_k) \subset \mathcal{A}$ for each k, then $\mathcal{A}$ is called an *almost* σ*-structure*.[12]

2.48. THEOREM. *Let $\mathcal{A}_1, \dots, \mathcal{A}_k, \dots$ be a countable sequence of almost σ-structures, each being different from $\mathfrak{P}(X)$ and containing all one-point sets. Assume that $\#(X) = \mathfrak{c}$. Then there exists a set $V = \bigcup_{k=1}^{\infty} V_k$ such that:*

(i) *the sets $V_1, \dots, V_k, \dots$ are pairwise disjoint and $V_k \notin \mathcal{A}_k$;*
(ii) $\bigcup_{l \geq k} V_l \notin \mathcal{A}_k$;
(iii) *if $V \in \mathcal{A}_k$, then $\bigcup_{l<k} V_l \in \mathcal{A}_k$.*[13]

2.49. REMARK. The Gitik–Shelah theorem is used in the proof of all the above results concerning countable sequences of σ-algebras and almost σ-algebras except Theorem 2.24; it is also used in the proof of Theorem 2.48. It is clear that each of the Theorems 2.8, 2.9, 2.15, 2.20, 2.29, 2.32, 2.34, 2.48 is a direct generalization of the Gitik–Shelah theorem (see footnote 1.10).

---

[12]If for any countable sequence $M_1, \dots, M_k, \dots \in \mathcal{A}$ we have $\bigcup M_k$, $\bigcap M_k \in \mathcal{A}$, then $\mathcal{A}$ is called a σ*-structure*.

[13]It is evident that if $\mathcal{A}_k$ are almost σ-algebras, then $V \notin \mathcal{A}_k$ for all k.

CHAPTER 3

The Main Idea

3.1. Consider an algebra $\mathcal{A}$ and a set $U \notin \mathcal{A}$. There are two possible cases:

Case (1). If $U \subset D$, then $D \notin \mathcal{A}$.

Case (2). There exists a set $D \supset U$ such that $D \in \mathcal{A}$.

Consider the first case. Let $U_1, \dots, U_m$ be a finite sequence of sets such that $U = \bigcup_{i=1}^m U_i$. Then there exists U_{i_0} such that if $U_{i_0} \subset M$, then $M \notin \mathcal{A}$.

Consider the second case. Let $U_1, \dots, U_m$ and $U'_1, \dots, U'_n$ be two finite sequences of sets such that

$$U = \bigcup_{i=1}^m U_i, \quad D \setminus U = \bigcup_{j=1}^n U'_j.$$

Then there exist sets U_{i_0}, U'_{j_0} such that if $U_{i_0} \subset M$ and $M \cap U'_{j_0} = \varnothing$, then $M \notin \mathcal{A}$.[1] Obviously, if one assumes that

$$X \setminus U = \bigcup_{j=1}^n U'_j$$

instead of

$$D \setminus U = \bigcup_{j-1}^n U'_j,$$

then one can also find the corresponding sets U_{i_0}, U'_{j_0}.

3.2. From the considerations in 3.1 one easily deduces the following statement, which is weaker than Theorem 2.2(1).

THEOREM. *Let $\mathcal{A}_1, \dots, \mathcal{A}_n$ be a finite sequence of algebras, and for each $k \in [1, n]$ let there exist more than $2(k-1)$ pairwise disjoint sets not belonging to $\mathcal{A}_k$. Then*

$$\bigcup_{k=1}^n \mathcal{A}_k \neq \mathfrak{P}(X).$$

PROOF. For $n = 1$ the theorem is true. Assume that it is true for $n = k$, and let us prove it for $n = k + 1$, i.e., find $Q' \notin \mathcal{A}_i$ for all $i \in [1, k+1]$. So, there exists $Q \notin \mathcal{A}_i$ for all $i \in [1, k]$, and there exist pairwise disjoint sets

$$Q_1, \dots, Q_{2k+1} \notin \mathcal{A}_{k+1}.$$

Take a finite sequence of pairwise disjoint sets $U_1, \dots, U_m$ such that each of the sets $X, Q, Q_1, \dots, Q_{2k+1}$ is the union of some U_i. From considerations in 3.1 it follows that there exist sets

$$U_{i_1}, \dots, U_{i_k} \subset Q$$

[1] If $U'_{j_0} \subset M'$ and $M' \cap U_{i_0} = \varnothing$, then, obviously, $M' \notin \mathcal{A}$.

such that for each $p \in [1, k]$ at least one of the following two conditions holds:

(i) if $U_{i_p} \subset D$, then $D \notin \mathcal{A}_p$;

(ii) there exists $U_{i'_p} \subset X \setminus Q$ such that $U_{i_p} \subset M$ and $M \cap U_{i'_p} = \varnothing$ imply $M \notin \mathcal{A}_p$.[2]

Since there are $2k+1$ sets Q_j and at most $2k$ sets U_{i_p} and $U_{i'_p}$, there exists Q_{j_0} such that

$$Q_{j_0} \cap \bigcup_{p=1}^{k} U_{i_p} = \varnothing$$

and $Q_{j_0} \cap U_{i'_p} = \varnothing$ for all sets $U_{i'_p}$. If $\mathcal{A}_{k+1} \not\ni \bigcup_{p=1}^{k} U_{i_p}$, then for the set Q' we take $Q' = \bigcup_{p=1}^{k} U_{i_p}$. If $\mathcal{A}_{k+1} \ni \bigcup_{p=1}^{k} U_{i_p}$, then we take

$$Q' = Q_{j_0} \cup \bigcup_{p=1}^{k} U_{i_p}.$$ [3] □

3.3. Remark. In the proof of the theorems on finite sequences of algebras one can assume that $\#(X) < \aleph_0$. Indeed, let $\mathcal{A}_1, \dots, \mathcal{A}_n$ be a finite sequence of algebras. In our considerations, we operate only with sets that constitute a matrix

$$\mathcal{U} = \begin{pmatrix} U_1^1 & \dots & U_{m_1}^1 \\ \dots & \dots & \dots \\ U_1^n & \dots & U_{m_n}^n \end{pmatrix},$$

where $U_i^k \notin \mathcal{A}_k$, $m_k < \infty$, and the sets from each row are pairwise disjoint. As in the proof of Theorem 3.2, let $U_1, \dots, U_m$ be a finite sequence of pairwise disjoint sets such that each of the sets U_i^k and X is the union of some U_i. Consider a set with m elements

$$X' = \{x_1, \dots, x_m\}$$

and the one-to-one mapping $f\colon x_i \to U_i$. For each $k \in [1, n]$ define an algebra $\mathcal{A}'_k$ on X' as follows: $\mathcal{A}'_k \ni M$ if and only if

$$\bigcup \{f(x_i) \mid x_i \in M\} \in \mathcal{A}_k.$$

Instead of the algebras $\mathcal{A}_1, \dots, \mathcal{A}_n$, one can study the algebras $\mathcal{A}'_1, \dots, \mathcal{A}'_n$, considering instead of the matrix $\mathcal{U}$ the matrix

$$\begin{pmatrix} \widetilde{U}_1^1 & \dots & \widetilde{U}_{m_1}^1 \\ \dots & \dots & \dots \\ \widetilde{U}_1^n & \dots & \widetilde{U}_{m_n}^n \end{pmatrix},$$

where $\widetilde{U}_i^k = \{f^{-1}(U_j) \mid U_j \subset U_i^k\}$.

This implies that the algebras considered in the theorems on finite sequences of algebras can be assumed to be σ-algebras.

---

[2] In general, the sets U_{i_p} for different p can coincide; also, in general, the sets $U_{i'_p}$ for different p can coincide.

[3] Let us note that Theorem 3.2 is a direct consequence of Statement 2.14 (see 9.13).

3.4. Let us recall that by a *filter* $\mathfrak{f}$ on X we mean a family of subsets of X satisfying the following conditions:

(1) if $M \in \mathfrak{f}$ and $M' \supset M$, then $M' \in \mathfrak{f}$;

(2) if $M_1, M_2 \in \mathfrak{f}$, then $M_1 \cap M_2 \in \mathfrak{f}$;

(3) $\varnothing \notin \mathfrak{f}$.

A filter $\mathfrak{f}$ is called a *maximal filter* (an *ultrafilter*) if for any filter $\mathfrak{f}'$ such that $\mathfrak{f}' \supset \mathfrak{f}$ we have $\mathfrak{f}' = \mathfrak{f}$. Filters were first introduced by F. Riesz in 1908 (see [**R**]). One should also mention the fundamental papers [**C1**], [**C2**] by H. Cartan. Unless specified otherwise, by an ultrafilter we always mean an ultrafilter on X.

3.5. For the proof of the theorems on countable sequences of algebras, one should consider ultrafilters.

Let us return to the two cases considered in 3.1. In case (1), there exists an ultrafilter $a \ni U$ such that if $M \in a$, then $M \notin \mathcal{A}$. In case (2), there exist ultrafilters a, b such that $U \in a$, $D \setminus U \in b$, and if M belongs to one of them but not to the other, then $M \notin \mathcal{A}$.

3.6. The arguments in 3.5 lead to the following definition.

DEFINITION. Consider an algebra $\mathcal{A}$. An ultrafilter a is said to be $\mathcal{A}$-*special* if $a \cap \mathcal{A} = \varnothing$. Two ultrafilters a, b are said to be $\mathcal{A}$-*similar* if $a \neq b$, there exists $D \in \mathcal{A}, a, b$, and, whenever M belongs to one of these ultrafilters but not to the other, we have $M \notin \mathcal{A}$.

3.7. Let us say a few words about $\mathcal{A}$-special and $\mathcal{A}$-similar ultrafilters.

The ultrafilters a, b are $\mathcal{A}$-similar if and only if $a \neq b$ and

$$a \cap b \supset (a \cup b) \cap \mathcal{A} \neq \varnothing.$$

If a, b are $\mathcal{A}$-similar ultrafilters, then we say that *a has an $\mathcal{A}$-similar ultrafilter b*, or *a is $\mathcal{A}$-similar to b*.

The $\mathcal{A}$-similarity relation is symmetric.

The $\mathcal{A}$-similarity relation is also transitive, i.e., if a is $\mathcal{A}$-similar to b, b is $\mathcal{A}$-similar to c, and $a \neq c$, then a is $\mathcal{A}$-similar to c.

An $\mathcal{A}$-special ultrafilter exists if and only if $X \notin \mathcal{A}$.

3.8. Let $q_1, \dots, q_n$ be a finite sequence of pairwise distinct ultrafilters, let M be a set of ultrafilters, $\#(M) < n$, and for each $k \in [1, n]$ there exist $\mathcal{A}$-similar ultrafilters q_k, q'_k with $q'_k \in M$. Then there exist ultrafilters q_{k_1}, q_{k_2}, q' such that $k_1 \neq k_2$, $q' \in M$, and (q_{k_1}, q'), (q_{k_2}, q') are two pairs of $\mathcal{A}$-similar ultrafilters. Since the relation of $\mathcal{A}$-similarity is transitive, q_{k_1}, q_{k_2} are $\mathcal{A}$-similar ultrafilters. These arguments will be used below.

3.9. The set X is considered as a topological space with discrete topology.[4] As usual, by βX we denote the Stone–Čech compactification of X, which is the set of all ultrafilters on X with the natural topology.[5] Since ultrafilters are points of βX, they are denoted by small letters as points usually are. If $M \subset \beta X$ (in particular if $M \subset X$), then by $\overline{M}$ we denote the closure of M in βX.

---

[4]Only in 3.32 is X considered as an arbitrary completely regular space.

[5]The construction of the Stone–Čech compactification of an arbitrary completely regular space is described in 3.33.

3.10. Main Statement. *Let $\{\mathcal{A}_\lambda\}$ be a family of algebras.*[6] *Then $\bigcup \mathcal{A}_\lambda \neq \mathfrak{P}(X)$ if and only if there exist sets $S, T \subset \beta X$ such that $\overline{S} \cap \overline{T} = \varnothing$ and for each λ at least one of the following two conditions holds*:

(1) *there exists an $\mathcal{A}_\lambda$-special ultrafilter $z_\lambda \in S$*;
(2) *there exist $\mathcal{A}_\lambda$-similar ultrafilters s_λ, t_λ such that $s_\lambda \in S$ and $t_\lambda \in T$.*

Proof. If $\bigcup \mathcal{A}_\lambda \neq \mathfrak{P}(X)$, i.e., if there exists $Q \notin \mathcal{A}_\lambda$ for all λ, then there exist sets $S, T \subset \beta X$ such that $S \subset \overline{Q}$, $T \subset \overline{X \setminus Q}$, and for each λ at least one of the two conditions of the Main Statement holds. It is clear that $\overline{S} \cap \overline{T} = \varnothing$. On the other hand, if the corresponding S and T exist, then there exists $Q \subset X$ such that $S \subset \overline{Q}$ and $T \cap \overline{Q} = \varnothing$. Hence $Q \notin \mathcal{A}_\lambda$ for all λ. □

If the family $\{\mathcal{A}_\lambda\}$ is at most countable, then the problem of the existence of a set $Q \notin \mathcal{A}_\lambda$ for all λ reduces to the problem of the existence of at most countable sets S and T satisfying all the conditions of the Main Statement.

3.11. This and the next seven sections give an idea of how the Main Statement 3.10 is used to obtain results on sets not belonging to algebras.

Statement. *Let $\mathcal{A}_1, \dots, \mathcal{A}_k, \dots$ be a countable sequence of algebras, and let $\mathcal{A}_k \neq \mathfrak{P}(X)$ for all k. Then there exists a countable subsequence $\mathcal{A}_{n_1}, \dots, \mathcal{A}_{n_k}, \dots$ $(n_1 < n_2 < \dots < n_k < \cdots)$ such that $\bigcup_k \mathcal{A}_{n_k} \neq \mathfrak{P}(X)$.*

Proof. Since $\mathcal{A}_k \neq \mathfrak{P}(X)$, there exist either an $\mathcal{A}_k$-special ultrafilter s_k or $\mathcal{A}_k$-similar ultrafilters s_k, t_k. Elementary topological arguments imply the existence of a countable subsequence of algebras

$$\mathcal{A}_{n_1}, \dots, \mathcal{A}_{n_k}, \dots \quad (n_1 < n_2 < \cdots < n_k < \cdots)$$

and open subsets U, V of βX such that

1) $\overline{U} \cap \overline{V} = \varnothing$;
2) $s_{n_k} \in U$;
3) if the ultrafilter t_{n_k} is considered, then $t_{n_k} \in V$.

By the Main Statement 3.10, $\bigcup_k \mathcal{A}_{n_k} \neq \mathfrak{P}(X)$. □

3.12. Let us give the statement of a remarkable theorem due to Szemerédi–Furstenberg (see [**Sz**], [**Fu**]).

Theorem. *Let $0 < \varepsilon < 1$. For any $n \in \mathbb{N}^+$ there exists $\lambda(\varepsilon, n) \in \mathbb{N}^+$ such that if*

$$\mathbb{N}^* \subset [a, b] \subset \mathbb{N}^+, \quad \#([a, b]) \geq \lambda(\varepsilon, n), \quad \#(\mathbb{N}^*) \geq \varepsilon \cdot \#([a, b]),$$

then there exists an arithmetic progression of n elements contained in $\mathbb{N}^*$.

In what follows, we assume that $\lambda(\varepsilon, n)$ is the minimal number satisfying the conditions of the previous theorem.

[6] It is mentioned in the Introduction that the subject matter of our study is at most countable families of algebras. However, in the Main Statement this restriction is irrelevant.

3.13. THEOREM. *For any $n \in \mathbb{N}^+$ there exists $l(n) \in \mathbb{N}^+$ such that if $\mathfrak{A} = \{\mathcal{A}_1, \dots, \mathcal{A}_{l(n)}\}$ is a sequence of algebras, $\mathcal{U} = \{U_1, \dots, U_{l(n)}\}$ is a sequence of pairwise disjoint sets, and $U_k \notin \mathcal{A}_k$ for all $k \le l(n)$, then there exists a set $P(\mathfrak{A}, \mathcal{U}) \subset [1, l(n)]$ with the following properties:*

(i) *the numbers in $P(\mathfrak{A}, \mathcal{U})$ form an arithmetic progression;*
(ii) $\#(P(\mathfrak{A}, \mathcal{U})) = n$;
(iii) *if $k \in P(\mathfrak{A}, \mathcal{U})$, then $\mathcal{A}_k \not\ni \bigcup\{U_i \mid i \in P(\mathfrak{A}, \mathcal{U})\}$.*

PROOF. Put $l(n) = \lambda\left(\frac{1}{3}, n\right)$[7] and consider the corresponding sequences $\mathfrak{A}$ and $\mathcal{U}$. To each set U_k let us associate either an $\mathcal{A}_k$-special ultrafilter $s_k \ni U_k$ or $\mathcal{A}_k$-similar ultrafilters s_k, t_k ($s_k \ni U_k$, $t_k \not\ni U_k$). If $t_1 \ni U_{k_1}$, then we call the set U_{k_1} *marked*, and the algebra $\mathcal{A}_{k_1}$ *covered*. If U_2 is a marked set, then consider U_3. Assume that U_2 is not a marked set. If $t_2 \ni U_1$, then we call the set U_2 marked but we will not call the algebra $\mathcal{A}_2$ covered. If $t_2 \ni U_{k_2}$ and $k_2 \neq 1$, then we will call the set U_{k_2} marked and call the algebra $\mathcal{A}_{k_2}$ covered. (It may happen that $k_1 = k_2$ and the set U_{k_2} has already been called marked, and the algebra $\mathcal{A}_{k_2}$ has been called covered.) Continuing our constructions in this way, we will come to a nonmarked set U_m. If $t_m \ni U_{k_m}$, $k_m < m$, and U_{k_m} is not a marked set, then we will call the set U_m marked, but we will not call the algebra $\mathcal{A}_m$ covered. And if $t_m \ni U_{k_m}$ and $k_m > m$, then we will call the set U_{k_m} marked and the algebra $\mathcal{A}_{k_m}$ covered. Having considered all the sets U_k, we put

$$N_1 = \{k \le l(n) \mid \text{the set } U_k \text{ is not marked}\},$$
$$N_2 = \{k \le l(n) \mid \text{the set } U_k \text{ is marked but the algebra } \mathcal{A}_k \text{ is not covered}\}.$$

By construction,

$$\frac{\max(\#(N_1), \#(N_2))}{l(n)} \ge \frac{1}{3}.$$

This inequality, our construction, the Main Statement 3.10, and the Szemerédi Furstenberg theorem imply our theorem. □

3.14. REMARK. The van der Waerden theorem [**W**] whose generalization is the Szemerédi–Furstenberg theorem and whose proof is much simpler than the proof of the Szemerédi–Furstenberg theorem, says the following: if $n, r \in \mathbb{N}^+$ are fixed, $[a, b] \subset \mathbb{N}^+$, and $b - a$ is a sufficiently large number $\mathfrak{r}$, then for each partition of $[a, b]$ into r classes there exists a class containing an arithmetic progression of n elements. (The estimate for $\mathfrak{r}$ was studied in [**S3**].) By the van der Waerden theorem (for $r = 3$), the arguments in the proof of Theorem 3.13, and the Main Statement 3.10, for any $n \in \mathbb{N}^+$ there exists $\gamma(n) \in \mathbb{N}^+$ such that if $\mathcal{A}_1, \dots, \mathcal{A}_{\gamma(n)}$ is a sequence of algebras, let $U_1, \dots, U_{\gamma(n)}$ is a sequence of pairwise disjoint sets, and let $U_k \notin \mathcal{A}_k$ for all $k \le \gamma(n)$, then there exists $P \subset [1, \gamma(n)]$ with the following properties:

(i) the numbers in P form an arithmetic progression;
(ii) $\#(P) = n$;
(iii) $\#(\{k \le \gamma(n) \mid \mathcal{A}_k \not\ni \bigcup_{i \in P} U_i\}) \ge n$.

---

[7] Here it is appropriate to give the main result of the remarkable paper of Gowers [**Gow**]. But first recall the notation of the power of a number from that paper: $a \uparrow b = a^b$; also $a \uparrow b \uparrow c = a \uparrow (b \uparrow c)$. And here is the result: $\lambda(\varepsilon, n) \le 2 \uparrow 2 \uparrow \varepsilon^{-1} \uparrow 2 \uparrow 2 \uparrow (n + 9)$ for $\varepsilon \le \frac{1}{2}$. Hence $l(n) \le 2 \uparrow 2 \uparrow 3 \uparrow 2 \uparrow 2 \uparrow (n + 9)$.

3.15. REMARK. Of course the main point of Theorem 3.13 is the following. Let $\mathcal{A}_1, \ldots, \mathcal{A}_n$ be a finite sequence of algebras, let $U_1, \ldots, U_n$ be a sequence of pairwise disjoint sets, and let $U_k \notin \mathcal{A}_k$ for all $k \leq n$. Then there exists a subset $\mathbb{N}^* \subset [1, n]$ such that the following is true:

(a) if $k \in \mathbb{N}^*$ and $U_k \subset Q \subset \bigcup_{j \in \mathbb{N}^*} U_j$, then $Q \notin \mathcal{A}_k$;

(b) $\frac{\#(\mathbb{N}^*)}{n} \geq \frac{1}{3}$.

Let us turn to the construction in 1.13. There we have the algebras $\mathcal{A}'_1, \mathcal{A}'_2, \mathcal{A}'_3$, pairwise disjoint sets $U_1 = \{x_2\}$, $U_2 = \{x_3\}$, $U_3 = \{x_1\}$, and $U_k \notin \mathcal{A}'_k$ for all $k \leq 3$. A nonempty set $\mathbb{N}^* \subset \{1, 2, 3\}$ possessing property (a) is either $\{1\}$, or $\{2\}$, or $\{3\}$, and $\#(\mathbb{N}^*)/3 = 1/3$. By analogy, for each $n = 3m$ ($m \in \mathbb{N}^+$) one can construct algebras $\mathcal{A}_1, \ldots, \mathcal{A}_n$ and pairwise disjoint sets $U_1, \ldots, U_n$ ($U_k \notin \mathcal{A}_k$ for all $k \leq n$) such that there exists an $\mathbb{N}^*$ possessing property (a), and $\#(\mathbb{N}^*)/n = 1/3$. On the other hand, there does not exist an $\mathbb{N}^*$ possessing property (a) with $\#(\mathbb{N}^*)/n > 1/3$.

3.16. Consider a finite sequence of algebras $\mathcal{A}_1, \ldots, \mathcal{A}_m$, and assume that there exist pairwise disjoint sets $U_1, \ldots, U_m$ such that $U_k \notin \mathcal{A}_k$ for all $k \leq m$. As above, for each U_k consider either an $\mathcal{A}_k$-special ultrafilter s_k, or $\mathcal{A}_k$-similar ultrafilters s_k, t_k. If s_k is an $\mathcal{A}_k$-special ultrafilter, then let us mark U_k by the number 1. After the following construction, each of the sets U_k will be marked by one of the numbers 0, 1, or 2. A sequence $U_{i_1}, \ldots, U_{i_l}$ is called a *chain* if the following conditions hold:

(i) all elements of the chain are nonmarked and pairwise distinct;

(ii) $t_{i_j} \ni U_{i_{j+1}}$ for $j < l$;

(iii) if $t_{i_l} \ni U_p$, then either U_p is an element $U_{i_{j_0}}$ of the chain, or U_p is a marked set.

Having constructed a chain (see below), let us mark its elements.

Case 1. *Either* $t_{i_l} \not\ni \bigcup_{k=1}^m U_k$, *or* U_p *is marked by the number* 2, *or* $l - j_0$ *is an odd number*. Let us mark U_{i_j} by the number 1 if j is an odd number; let us mark U_{i_j} by the number 0 if j is an even number.

Case 2. $l - j_0$ *is an even number*. Mark U_{i_j} by the number 1 if j is odd and $j < l$; mark U_{i_j} by the number 0 if j is even and $j < l$. Mark U_{i_l} by the number 2.

Case 3. U_p *is marked by the number* δ *equal to* 0 *or* 1. Mark U_{i_j} by the number δ if $l - j$ is odd; mark U_{i_j} by the number $1 - \delta$ if $l - j$ is even.

Now let us start constructing chains, if necessary. If there are nonmarked sets among the sets U_k, then take one of them, say U_{k^*}, and construct a chain in which $U_{i_1} = U_{k^*}$. If among the sets U_k there are still nonmarked sets, take one of them, e.g. U_{k_*}, and construct a chain with $U_{i_1} = U_{k_*}$, and so on. After all the sets U_k are marked, set

$$L = \{k \in [1, m] \mid U_k \text{ is not marked by } 2\},$$
$$L_0 = \{k \in [1, m] \mid U_k \text{ is marked by } 0\},$$
$$L_1 = \{k \in [1, m] \mid U_k \text{ is marked by } 1\}.$$

Since $l \geq 3$ in Case 2, we have

$$\frac{\#(L)}{m} \geq \frac{2}{3}.$$

Since $L_0 \cup L_1 = L$, we have

$$\frac{\max(\#(L_0), \#(L_1))}{m} \geq \frac{1}{3}.$$

Hence if we set $m = l(n)$, then by construction, the Main Statement 3.10, and the Szemerédi–Furstenberg theorem, we obtain again a proof of Theorem 3.13. But the arguments of this section also yield another interesting result, whose main point is that the estimate for $l(n)$ from the proof of Theorem 3.13 can be replaced by $l'(n) = \lambda(2/3, n)$, although with a not so simple structure of a set not belonging to all the algebras whose indices form an arithmetic progression of n elements.[8]

STATEMENT. *Let $\mathcal{A}_1, \dots, \mathcal{A}_{l'(n)}$ be a sequence of algebras, let $U_1, \dots, U_{l'(n)}$ be a sequence of pairwise disjoint sets, and let $U_k \notin \mathcal{A}_k$ for all $k \leq l'(n)$. Then there exist sets $P_* \subset P^* \subset [1, l'(n)]$ with the following properties*:

(i) *the numbers from P^* form an arithmetic progression*;
(ii) $\#(P^*) = n$;
(iii) *if $k \in P^*$, then $\mathcal{A}_k \not\ni \bigcup_{i \in P_*} U_i$.*

PROOF. Put $m = l'(n)$. By the Szemerédi–Furstenberg theorem, there exists an arithmetic progression $P^* \subset L$ consisting of n elements. If $k \in P^*$, U_k is marked by 0, and

$$t_k \not\ni \bigcup_{i \in P^*} U_i,$$

then let us mark U_k by the new number 3. If $k \in P^*$, $t_k \ni U_{k'}$, and $U_{k'}$ is marked by 3, then let us mark U_k by the new number 4. If $k \in P^*$, $t_k \ni U_{k'}$, and $U_{k'}$ is marked by 4, then let us mark U_k by the new number 5, and so on. Put

$$P_* = \{k \mid k \in P^* \text{ and } U_k \text{ is marked by an odd number}\}. \qquad \square$$

3.17. REMARK. Of course, the main point of Statement 3.16 is the following. Let $\mathcal{A}_1, \dots, \mathcal{A}_n$ be a finite sequence of algebras, let $U_1, \dots, U_n$ be a sequence of pairwise disjoint sets, and let $U_k \notin \mathcal{A}_k$ for all $k \leq n$. Then there exists $\mathbb{N}^* \subset [1, n]$ such that the following is true:

(a) if $\varnothing \neq \mathbb{N}^{**} \subset \mathbb{N}^*$, then there exists $\mathbb{N}_0^{**} \subset \mathbb{N}^{**}$ such that

$$\mathcal{A}_k \not\ni \bigcup_{i \in \mathbb{N}_0^{**}} U_i \quad \text{for all } k \in \mathbb{N}^{**};$$

(b) $\frac{\#(\mathbb{N}^*)}{n} \geq \frac{2}{3}$.

By the construction in 1.13, for each $n = 3m$ ($m \in \mathbb{N}^+$) one can construct algebras $\mathcal{A}_1, \dots, \mathcal{A}_n$ and pairwise disjoint sets $U_1, \dots, U_n$, $U_k \notin \mathcal{A}_k$ for all $k \leq n$, such that there exists $\mathbb{N}^*$ satisfying condition (a), and $\#(\mathbb{N}^*)/n = 2/3$. But there does not exist $\mathbb{N}^* \subset [1, n]$ such that $\#(\mathbb{N}^*)/n > 2/3$ and $\bigcup_{k \in \mathbb{N}^*} \mathcal{A}_k \neq \mathfrak{P}(X)$.

3.18. STATEMENT. *Let $\mathcal{A}_1, \dots, \mathcal{A}_k, \dots$ be a countable sequence of algebras, and let $U_1, \dots, U_k, \dots$ be a countable sequence of pairwise disjoint sets. Assume that $U_k \notin \mathcal{A}_k$ for all k. Then for any real number $\rho < 2/3$ there exist $\mathbb{N}_* \subset \mathbb{N}^* \subset \mathbb{N}^+$ such that*

(i) $p(\mathbb{N}^*) \geq \frac{2}{3}$,
(ii) $p_a(\mathbb{N}^*) > \rho$;
(iii) *if $k \in \mathbb{N}^*$, then $\mathcal{A}_k \not\ni \bigcup_{i \in \mathbb{N}_*} U_i$.*[9]

---

[8]Clearly, $l'(n) \leq l(n)$.

[9]In conditions (i), (ii) the notation from Definition 2.25 is used.

PROOF. As above, consider the ultrafilters s_k and t_k. Set

$$\mathbb{N}' = \Big\{k \mid t_k \ni \bigcup_{i\in\mathbb{N}^+} U_i \text{ but } t_k \not\ni U_i\Big\}.$$

If $p(\mathbb{N}') = 0$ and $k \in \mathbb{N}'$, we mark U_k by 2. Assume that $p(\mathbb{N}') > 0$. For each $k \in \mathbb{N}'$ take $\mathbb{N}(k) \subset \mathbb{N}^+$ such that

$$t_k \ni \bigcup_{i\in\mathbb{N}(k)} U_i.$$

Set

$$\mathbb{N}'' = \bigcup_{k\in\mathbb{N}'} \mathbb{N}(k).$$

One can assume that $p(\mathbb{N}'')$ is an arbitrarily small number. If $k \in \mathbb{N}''$, we mark U_k by 2. If $k \in \mathbb{N}' \setminus \mathbb{N}''$, then we mark U_k by 1. If $p(\mathbb{N}') = 0$, then put $\mathbb{N}'' = \varnothing$. If s_k is an $\mathcal{A}_k$-special ultrafilter and $k \notin \mathbb{N}''$, then we mark U_k by 1. In what follows, each of the sets U_k will be marked by one of the numbers $0, 1$, or 2. As in 3.16, we will consider chains. It is clear that we can meet infinite chains

$$U_{i_1}, \dots, U_{i_j}, \dots$$

(the sets U_{i_j} are nonmarked, pairwise distinct, and $t_{i_j} \ni U_{i_{j+1}}$). In what follows, having constructed an infinite chain, let us mark its elements as follows: if j is odd, then we mark U_{i_j} by 1, and if j is even, then we mark U_{i_j} by 0. The definition of a finite chain here is the same as the definition of a finite chain in 3.16. The only difference is in how the elements of a chain are marked.

Case 1^*. *Either* $t_{i_l} \not\ni \bigcup_{k\in\mathbb{N}^+} U_k$, *or* U_p *is marked by a number not equal to* 1, *or* $l - j_0$ *is odd.* Let us mark U_{i_j} by 1 if $l - j$ is even and by 0 if $l - j$ is odd.

Case 2^*. $l - j_0$ *is even.* Let us choose the maximal number among the numbers $i_{j_0}, \dots, i_l$ and denote it by i_{j^*}. Let us mark the set $U_{i_{j^*}}$ by 2. If $j \in [1, j^* - 1]$[10] and $j^* - j$ is odd, then let us mark U_{i_j} by 1; if $j \in [1, j^* - 1]$ and $j^* - j$ is even, then let us mark U_{i_j} by 0. If $U_{i_{j_0}}$ is not marked by 1, $j \in [j^* + 1, l]$, and $l - j$ is even, then let us mark U_{i_j} by 1; if $U_{i_{j_0}}$ is not marked by 1, $j \in [j^* + 1, l]$, and $l - j$ is odd, then let us mark U_{i_j} by 0. If $U_{i_{j_0}}$ is marked by 1, $j \in [j^* + 1, l]$, and $l - j$ is odd, then let us mark U_{i_j} by 1; if $U_{i_{j_0}}$ is marked by 1, $j \in [j^* + 1, l]$, and $l - j$ is even, then let us mark U_{i_j} by 0.[11]

Case 3^*. U_p *is marked by* 1. Let us mark U_{i_j} by 1 if $l - j$ is odd, and by 0 if $l - j$ is even.

Now, if necessary, let us start constructing chains. If among the sets U_k there are nonmarked sets, we put

$$k^* = \min\{k \mid U_k \text{ is a nonmarked set}\},$$

and construct a chain in which $U_{i_1} = U_{k^*}$. If among the sets U_k there are still nonmarked ones, we put

$$k_* = \min\{k \mid U_k \text{ is an nonmarked set}\},$$

[10] From the constructions below, it will become clear that $j^* > 1$.

[11] Thus, among the sets $U_{i_{j_0}}, \dots, U_{i_l}$ there is one set marked by 2, and there are at least two sets marked by $0, 1$. Because of this, the estimate $2/3$ is valid.

and construct a chain with $U_{i_1} = U_{k_*}$, and so on. After all the sets U_k are marked, we put

$$\mathbb{N}_* = \{k \mid U_k \text{ is marked by } 1\},$$
$$\mathbb{N}^* = \{k \mid U_k \text{ is not marked by } 2\}.$$

If $p(\mathbb{N}') = 0$, then $p_a(\mathbb{N}^*) \geq 2/3$. If $p(\mathbb{N}') > 0$ and $p(\mathbb{N}') \geq 3 \cdot p(\mathbb{N}'')$, then

$$p(\mathbb{N}^*) \geq \frac{2}{3} \quad \text{and} \quad p_a(\mathbb{N}^*) \geq \frac{2}{3} - p(\mathbb{N}'').$$

Since $p(\mathbb{N}'')$ can be arbitrarily small, the statement is proved. □

3.19. Let us give one of the main definitions.

DEFINITION. Consider an algebra $\mathcal{A}$ and put

$$\mathbb{K}_1 = \{a \in \beta X \mid a \text{ is an } \mathcal{A}\text{-special ultrafilter}\},$$
$$\mathbb{K}_2 = \{a \in \beta X \mid a \text{ has an } \mathcal{A}\text{-similar ultrafilter}\}.$$

The set $\mathbb{K}_1 \cup \mathbb{K}_2$ is called the *kernel of the algebra* $\mathcal{A}$ and denoted $\ker \mathcal{A}$.

One can easily show that

$$\ker \mathcal{A} = \{a \in \beta X \mid \text{there exists } Q \in a \text{ such that } (a \cap Q) \cap \mathcal{A} = \varnothing\}.$$

3.20. Let us say a few words about the kernel of an algebra.

$\mathcal{A} = \mathfrak{P}(X)$ if and only if $\ker \mathcal{A} = \varnothing$.

$\mathfrak{P}(M) \subset \mathcal{A}$ if and only if $\overline{M} \cap \ker \mathcal{A} = \varnothing$.

For an algebra $\mathcal{A}$ there exists a cardinal $A^{\mathcal{A}} > 0$ such that $\ker \mathcal{A}$ is the union of pairwise disjoint closed sets $M_\alpha^{\mathcal{A}}$, $\alpha < A^{\mathcal{A}}$, with $M_0^{\mathcal{A}} = \mathbb{K}_1$; if $\alpha > 0$, then $\#(M_\alpha^{\mathcal{A}}) > 1$; if $a \in M_\alpha^{\mathcal{A}}$ and $\alpha > 0$, then $M_\alpha^{\mathcal{A}}$ contains, in addition to a, all the ultrafilters $\mathcal{A}$-similar to a and only them. The family

$$\{M_\alpha^{\mathcal{A}}\}_{\alpha < A^{\mathcal{A}}}$$

is called the *structure of the kernel of the algebra* $\mathcal{A}$, or, for short, the *structure* $\ker \mathcal{A}$. The set $M_\alpha^{\mathcal{A}}$ is said to be $\mathcal{A}$-*similar* if $\alpha > 0$.

$X \in \mathcal{A}$ if and only if $M_0^{\mathcal{A}} = \varnothing$.

An algebra is determined by the structure of its kernel. This means the following. For the equality of algebras $\mathcal{A} = \mathcal{B}$ a necessary and sufficient condition is that $M_0^{\mathcal{A}} = M_0^{\mathcal{B}}$ and there exists a one-to-one correspondence between the sets $\{\alpha \mid 0 < \alpha < A^{\mathcal{A}}\}$ and $\{\beta \mid 0 < \beta < A^{\mathcal{B}}\}$ such that if α corresponds to β, then $M_\alpha^{\mathcal{A}} = M_\beta^{\mathcal{B}}$.

Below, the following simple fact will be used. Consider a finite sequence of pairwise disjoint finite sets $M_0, M_1, \dots, M_n \subset \beta X$ such that if $k \in [1, n]$, then $\#(M_k) > 1$. Then there exists an algebra $\mathcal{A}$ such that

$$\ker \mathcal{A} = \bigcup_{k=0}^{n} M_k, \quad A^{\mathcal{A}} = n + 1, \quad M_k^{\mathcal{A}} = M_k.$$

Here is one more fact, which, although evident, will be used later. Let $1 < \varkappa \leq \aleph_0$. For the existence of $\varkappa$ pairwise disjoint sets not belonging to $\mathcal{A}$ it is necessary and sufficient to have $\#(\ker \mathcal{A}) \geq \varkappa$.

Here is some more information about the cardinality of a kernel. If μ is a measure, then by $\mathcal{A}_\mu$ we denote the algebra of all μ-measurable sets. If $0 < \varkappa < \aleph_0$, then, obviously, there exist two-valued measures μ such that $\#(\ker \mathcal{A}_\mu) = \varkappa$. Let us formulate the following result, whose proof will be discussed below.

There exist two-valued measures μ (and even σ-two-valued measures with a certain separability property in the case $\#(X) < \mathfrak{c}$) such that $\#(\ker \mathcal{A}_\mu) = 2^{\mathfrak{c}}$; there does not exist an algebra $\mathcal{A}$ such that

$$\aleph_0 \leq \#(\ker \mathcal{A}) < 2^{\mathfrak{c}}.$$

3.21. Consider an algebra $\mathcal{A}$ on a finite set X (it is clear that in this case $\beta X = X$). To avoid using two different languages, the points of X will be called ultrafilters. Then everything becomes simpler; thus,

$$\ker \mathcal{A} = \big\{x \in X \mid \{x\} \notin \mathcal{A}\big\}.$$

If $x \in \ker \mathcal{A}$ and there does not exist a set $M \in \mathcal{A}$ such that $x \in M$, then x is an $\mathcal{A}$-*special ultrafilter*. Let M be a minimal nonempty subset of $\ker \mathcal{A}$ belonging to $\mathcal{A}$. This means that if $M' \subset M$, $M' \neq \varnothing$, and $M' \in \mathcal{A}$, then $M' = M$. Such a set M is called $\mathcal{A}$-*similar*. Clearly, $\#(M) \geq 2$. Two distinct points of M are called $\mathcal{A}$-*similar ultrafilters*.

3.22. Remark. Assume that

$$\#(\ker \mathcal{A}) < \aleph_0$$

and $M_0^{\mathcal{A}}, M_1^{\mathcal{A}}, \dots, M_n^{\mathcal{A}}$ is the structure of $\ker \mathcal{A}$. If

$$\#(M_k^{\mathcal{A}}) > 3 \quad \text{and} \quad k > 0,$$

then let us divide $M_k^{\mathcal{A}}$ into pairwise disjoint sets, each containing at least two and at most three elements. Thus, $\ker \mathcal{A} \setminus M_0^{\mathcal{A}}$ is divided into pairwise disjoint sets $M_1', \dots, M_p'$, and

$$2 \leq \#(M_k') \leq 3 \qquad \text{for all } k \in [1, p].$$

Let us construct a new algebra $\mathcal{A}'$ such that $M_0^{\mathcal{A}}$ is the set of all $\mathcal{A}'$-special ultrafilters and $M_1', \dots, M_p'$ are all $\mathcal{A}'$-similar sets. Clearly,

$$\#(\ker \mathcal{A}) = \#(\ker \mathcal{A}').$$

It is important that if $M \notin \mathcal{A}'$, then $M \notin \mathcal{A}$; i.e., $\mathcal{A} \subset \mathcal{A}'$. We will see that for some problems in which for each algebra $\mathcal{A}$ we have

$$\#(\ker \mathcal{A}) < \aleph_0,$$

it is essential to assume that each $\mathcal{A}$-similar set contains at most three elements. However, in the final Chapter 20, when generalizing Theorem 2.1, it will be important that $\#(M_k^{\mathcal{A}})$ can be greater than 3 for all $k \in [1, n]$.

3.23. Remark. Consider an algebra $\mathcal{A}$ and the structure of its kernel $\{M_\alpha^{\mathcal{A}}\}_{\alpha < A^{\mathcal{A}}}$. If $\#(M_0^{\mathcal{A}}) \leq 1$, then put $\mathcal{A}' = \mathcal{A}$. If $\#(M_0^{\mathcal{A}}) > 1$, then define an algebra $\mathcal{A}'$ as follows: $M_0^{\mathcal{A}'} = \varnothing$ and $\big\{M_\alpha^{\mathcal{A}} \mid 0 \leq \alpha < A^{\mathcal{A}}\big\}$ is the family of all $\mathcal{A}'$-similar sets. The algebra $\mathcal{A}'$ possesses the following properties:

(a) there exists at most one $\mathcal{A}'$-special ultrafilter;
(b) $\ker \mathcal{A} = \ker \mathcal{A}'$;
(c) $\mathcal{A} \subset \mathcal{A}'$;

(d) if $\{U_\iota\}_{\iota\in I}$ is a family of pairwise disjoint sets not belonging to $\mathcal{A}$, then there exists a family of sets $\{V_\iota\}_{\iota\in I}$ not belonging to $\mathcal{A}'$ such that $V_\iota \subset U_\iota$ for all $\iota \in I$ and

$$\#\big(\{\iota \in I \mid U_\iota \neq V_\iota\}\big) \leq 1.$$

Properties (a), (b), and (c) are obvious. As for property (d), there can exist only one set $U_{\iota_0} \in \mathcal{A}'$. In this case $\overline{U_{\iota_0}} \supset M_0^{\mathcal{A}}$, $\#(M_0^{\mathcal{A}}) > 1$, and one can choose $V_{\iota_0} \subset U_{\iota_0}$ so that $V_{\iota_0} \notin \mathcal{A}'$. By the above, in the proof of the following theorem it will already be assumed that we consider an algebra $\mathcal{B}_k$ having at most one $\mathcal{B}_k$-special ultrafilter. However, in some situations it is convenient to consider algebras $\mathcal{A}$ for which there is more than one $\mathcal{A}$-special ultrafilter. Let us now consider an algebra $\mathcal{A}$ such that a_0 is the unique $\mathcal{A}$-special ultrafilter. Construct an algebra $\mathcal{B}$ as follows:

$$\ker \mathcal{B} = \ker \mathcal{A} \setminus \{a_0\}$$

and any two $\mathcal{A}$-similar ultrafilters are $\mathcal{B}$-similar. It is clear that $X \in \mathcal{B}$. In certain cases where $\#(\ker \mathcal{A}) > \aleph_0$, we can replace $\mathcal{A}$ with $\mathcal{B}$. The algebra $\mathcal{A}'$ (when $\#(M_0^{\mathcal{A}}) > 1$) and the algebra $\mathcal{B}$ (when $\#(M_0^{\mathcal{A}}) = 1$) are constructed in a very simple way: one takes all sets belonging to $\mathcal{A}$ and the complements of all these sets in X.

3.24. In 1.17 we defined the class of functions Ψ^n. The following theorem is used below.

THEOREM. *Let $\psi\colon [1,n] \to \mathbb{N}^+$ be a function satisfying the following property: if $\mathcal{A}_1, \dots, \mathcal{A}_n$ is a sequence of algebras and for each $k \in [1,n]$ we have $X \in \mathcal{A}_k$ and there exist $\psi(k)$ pairwise disjoint sets not belonging to $\mathcal{A}_k$, then*

$$\bigcup_{k=1}^{n} \mathcal{A}_k \neq \mathfrak{P}(X).$$

Then $\psi \in \Psi^n$.

PROOF. Let $\mathcal{B}_1, \dots, \mathcal{B}_n$ be a sequence of algebras such that for each $k \in [1,n]$ there exist $\psi(k)$ pairwise disjoint sets not belonging to $\mathcal{B}_k$. We must prove that

$$\bigcup_{k=1}^{n} \mathcal{B}_k \neq \mathfrak{P}(X).$$

By Remark 3.3, we can assume that $\#(X) < \aleph_0$. We will assume that there exists a point

$$q \in X \setminus \bigcup_{k=1}^{n} \ker \mathcal{B}_k.$$

If a $\mathcal{B}_k$-special ultrafilter does not exist, then we put $\mathcal{B}'_k = \mathcal{B}_k$. On the other hand, if there exists a $\mathcal{B}_k$-special ultrafilter r_k (by Remark 3.23, we can assume that there are no other $\mathcal{B}_k$-special ultrafilters), then we construct an algebra $\mathcal{B}'_k \ni X$ such that each $\mathcal{B}_k$-similar set is a $\mathcal{B}'_k$-similar set, and there is one more $\mathcal{B}'_k$-similar set $\{q, r_k\}$. By the Main Statement 3.10, for the algebras $\mathcal{B}'_1, \dots, \mathcal{B}'_n$ there exist corresponding sets of ultrafilters S and T. Define S^* and T^* as follows:

$$\text{if } q \in S, \text{ put } S^* = T, \quad T^* = S \setminus \{q\};$$
$$\text{if } q \notin S, \text{ put } S^* = S, \quad T^* = T \setminus \{q\}.$$

Clearly, $S^* \cap T^* = \varnothing$ (since the sets are finite, this means that $\overline{S^*} \cap \overline{T^*} = \varnothing$). For each $k \in [1, n]$ at least one of the following two conditions holds:

(1) there exists a $\mathcal{B}_k$-special ultrafilter $z_k \in S^*$;

(2) there exist $\mathcal{B}_k$-similar ultrafilters s_k, t_k such that $s_k \in S^*$, $t_k \in T^*$.

By the Main Statement 3.10, the theorem is proved. □

3.25. REMARK. Suppose that $\psi\colon [1, n] \to \mathbb{N}^+$ is given, and one must check whether $\psi \in \Psi^n$ or $\psi \notin \Psi^n$. To do that, one considers an arbitrary sequence of algebras $\mathcal{A}_1, \ldots, \mathcal{A}_n$ such that $\#(\ker \mathcal{A}_k) \geq \psi(k)$. If $\psi(k)$ is even, then one can assume that $\#(\ker \mathcal{A}_k) = \psi(k)$; and if $\psi(k)$ is odd, then one can assume that $\#(\ker \mathcal{A}_k) \leq \psi(k) + 1$. Similar assumptions hold for the algebras from Theorems 2.27, 2.31, and so forth.

3.26. Let us recall a well-known definition.

DEFINITION. Let $\{\Theta_\lambda\}_{\lambda \in \Lambda}$ be a family of sets. A set $\{\tau_\lambda\}_{\lambda \in \Lambda}$ is called a *system of distinct representatives* of this family if $\tau_\lambda \in \Theta_\lambda$ for all $\lambda \in \Lambda$ and $\tau_{\lambda_1} \neq \tau_{\lambda_2}$ for $\lambda_1 \neq \lambda_2$.

3.27. The following well-known theorem is proved by Ph. Hall in [**Hall**] (see also [**Halm-V**], [**We**]).

THEOREM. *Let $\Theta_1, \ldots, \Theta_n$ be a finite family of sets. This family has a system of distinct representatives if and only if the following condition is satisfied: for any $M \subset [1, n]$ we have*

$$\#\left(\bigcup_{\lambda \in M} \Theta_\lambda\right) \geq \#(M).$$

3.28. DEFINITION. Let $\mathcal{A}_1, \ldots, \mathcal{A}_k, \ldots$ be an at most countable sequence of algebras. Pairwise disjoint sets $U_1, \ldots, U_k, \ldots$ are said to form a *system of distinct representatives of this sequence* if $U_k \notin \mathcal{A}_k$ for each k.

3.29. The following theorem is not among the main results of this monograph. However, this theorem and the related results given below are of certain interest. After the notion of kernel of an algebra has been introduced, the proof of the present theorem is obvious in view of Theorem 3.27.

THEOREM. *Let $\mathcal{A}_1, \ldots, \mathcal{A}_n$ be a finite sequence of algebras. This sequence has a system of distinct representatives if and only if for any nonempty set $M \subset [1, n]$ there exist pairwise disjoint sets*

$$V_M^1, \ldots, V_M^{\#(M)} \in \bigcup_{j \in M} (\mathfrak{P}(X) \setminus \mathcal{A}_j).$$

3.30. The following result is a corollary of Theorems 3.13 and 3.29.

THEOREM. *For each $n \in \mathbb{N}^+$ there is an $l(n) \in \mathbb{N}^+$ such that if $\mathcal{A}_1, \ldots, \mathcal{A}_{l(n)}$ is a sequence of algebras and for each nonempty set $M \subset [1, l(n)]$ there exist pairwise disjoint sets*

$$V_M^1, \ldots, V_M^{\#(M)} \in \bigcup_{j \in M} (\mathfrak{P}(X) \setminus \mathcal{A}_j),$$

then one can find a family of pairwise disjoint sets

$$\mathcal{U} = \{U_1, \ldots, U_{l(n)}\} \quad (U_k \notin \mathcal{A}_k \textit{ for all } k \in [1, l(n)])$$

and a set $P(\mathcal{U}) \subset [1, l(n)]$, *with the following properties*:

(i) *the numbers from* $P(\mathcal{U})$ *form an arithmetic progression*;
(ii) $\#(P(\mathcal{U})) = n$;
(iii) *if* $k \in P(\mathcal{U})$, *then* $\mathcal{A}_k \not\ni \bigcup_{i \in P(\mathcal{U})} U_i$.

3.31. The following statement is a corollary of Statement 3.16 and Theorem 3.29. (The number $l'(n)$ in this statement was defined in 3.16.)

STATEMENT. *Let* $\mathcal{A}_1, \ldots, \mathcal{A}_{l'(n)}$ *be a sequence of algebras, and for each nonempty set* $M \subset [1, l'(n)]$ *let there exist pairwise disjoint sets*

$$V_M^1, \ldots, V_M^{\#(M)} \in \bigcup_{j \in M} (\mathfrak{P}(X) \setminus \mathcal{A}_j).$$

Then there exist a family of pairwise disjoint sets

$$\mathcal{U} = \{U_1, \ldots, U_{l'(n)}\} \quad (U_k \notin \mathcal{A}_k \text{ for all } k \in [1, l'(n)])$$

and sets $P_*(\mathcal{U}) \subset P^*(\mathcal{U}) \subset [1, l'(n)]$, *with the following properties*:

(i) *the numbers from* $P^*(\mathcal{U})$ *form an arithmetic progression*;
(ii) $\#(P^*(\mathcal{U})) = n$;
(iii) *if* $k \in P^*(\mathcal{U})$, *then* $\mathcal{A}_k \not\ni \bigcup_{i \in P_*(\mathcal{U})} U_i$.

3.32. In this section the main idea of our theory is applied to Banach algebras of continuous bounded functions. Let $\mathcal{A}$ be an algebra, and assume that $X \in \mathcal{A}$. For each $M \in \mathcal{A}$ define a function on X:

$$f_M(x) = \begin{cases} 1 & \text{if } x \in M, \\ 0 & \text{if } x \notin M. \end{cases}$$

Denote by $\mathcal{A}(X)$ the Banach algebra of real bounded functions on X generated by all the functions f_M. As usual, for $f \in \mathcal{A}(X)$ define

$$\|f\| = \sup_{x \in X} |f(x)|.$$

Since X is taken with the discrete topology, each function on X is continuous. Considering Banach algebras of the type of $\mathcal{A}(X)$ instead of the algebras $\mathcal{A}$ is hardly sensible, though possible. But considering such objects is quite sensible when X has a nondiscrete topology.

Thus, let X be an arbitrary completely regular space with the topology $\mathcal{O}$ (i.e., $\mathcal{O}$ is the family of all open subsets of X), and let $\mathcal{A}(X)$ be a Banach algebra consisting of real continuous bounded functions on X. Denote by $C(X)$ the Banach algebra of all real continuous bounded functions defined on X. As usual, by βX we denote the Stone–Čech compactification of X as a completely regular space with the topology $\mathcal{O}$. Define the *kernel* of $\mathcal{A}(X)$ as follows:

$$\ker \mathcal{A}(X) = \big\{x \in \beta X \mid \text{there exists a point } y \in \beta X \setminus \{x\} \text{ such that } f(x) = f(y) \text{ for all } f \in \mathcal{A}(X)\big\}.$$

Two distinct points $x, y \in \ker \mathcal{A}(X)$ are called $\mathcal{A}(X)$-*similar* if $f(x) = f(y)$ for all $f \in \mathcal{A}(X)$. A closed set $M \subset X$ is called $\mathcal{A}(X)$-*free* if there exists a closed set $M' \subset X \setminus M$ such that

$$\text{if } f \in C(X),\ f(M) = 0, \text{ and } f(M') = 1, \text{ then } f \notin \mathcal{A}(X).$$

THEOREM. *Let $\mathcal{A}_1(X), \ldots, \mathcal{A}_n(X)$ be a finite sequence of Banach algebras of real continuous bounded functions, and let $\psi \in \Psi^n$. Assume that for each $k \in [1, n]$ there exist $\psi(k)$ pairwise disjoint $\mathcal{A}_k(X)$-free sets.*[12] *Then there exists a set that is $\mathcal{A}_k(X)$-free for all $k \in [1, n]$.*

PROOF. It is clear that

$$\#(\ker \mathcal{A}_k(X)) \geq \psi(k).$$

The rest of the proof is easy. □

3.33. REMARK. Since in our monograph we "speak the language of ultrafilters" and have defined the Stone–Čech compactification of a discrete space as the space of ultrafilters, let us describe a construction of βX as a space of ultrafilters in the general case also. (Compactification by means of a space of ultrafilters was first constructed by Wallman in [**Wa**]; there the *Wallman* compactification of a topological space is constructed for any topological space possessing only the property that each point is the intersection of its neighborhoods.) The construction given below has been proposed by L. Gillman and M. Jerison in [**G-Je**] and is similar to the construction of the Wallman compactification. A set $M \subset X$ is called *functionally closed* if there exists a function $f \in C(X)$ such that

$$M = \{x \in X \mid f(x) = 0\}.$$

The complement in X to a functionally closed set is called a *functionally open* set. The family of all ultrafilters on the family of all functionally closed sets is βX. Let U be a functionally open set. By definition,

$$U^* = \{x \in \beta X \mid \text{there exists } M \subset U \text{ such that } M \in x\}$$

is an open set in βX. The sets of the type of U^* form a base in βX.

[12] The condition of existence of $\psi(k)$ pairwise disjoint $\mathcal{A}_k(X)$-free sets can be replaced by an equivalent condition: there exist functions $f_1^k, \ldots, f_{\psi(k)}^k \in C(X) \setminus \mathcal{A}_k(X)$ such that $f_i^k \cdot f_j^k \equiv 0$ if $i \neq j$.

CHAPTER 4

Finite Sequences of Algebras (1). Proof of Theorems 2.1 and 2.2

4.1. Consider a finite sequence of algebras $\mathcal{A}_1, \ldots, \mathcal{A}_n$. By the Main Statement 3.10,

$$\bigcup_{k=1}^{n} \mathcal{A}_k \neq \mathfrak{P}(X)$$

if and only if there exist two sets $S_n, T_n \subset \beta X$ such that

- $S_n = \{s_1, \ldots, s_i, \ldots, s_n\}$, where in general it is possible that $s_i = s_j$ $(i \neq j)$;
- each ultrafilter s_i is either $\mathcal{A}_i$-special or there exists a pair of $\mathcal{A}_i$-similar ultrafilters s_i, t_i;
- T_n is the set of all ultrafilters t_i, and in general it is possible that $t_i = t_j$ $(i \neq j)$;
- if s_i is an $\mathcal{A}_i$-special ultrafilter, then, obviously, the ultrafilter t_i is not considered.

It is required that $S_n \cap T_n = \varnothing$. (The Main Statement 3.10 requires that $\overline{S}_n \cap \overline{T}_n = \varnothing$. But since S_n and T_n are finite sets, we have $\overline{S}_n = S_n$, $\overline{T}_n = T_n$, hence $S_n \cap T_n = \varnothing$ implies $\overline{S}_n \cap \overline{T}_n = \varnothing$.)

4.2. REMARK. Among all ultrafilters from $\ker \mathcal{A}_i$ we consider either the $\mathcal{A}_i$-special ultrafilter s_i or the $\mathcal{A}_i$-similar ultrafilters s_i, t_i.

4.3. DEFINITION. A *bush* K of (S_n, T_n) is a minimal nonempty set $K \subset S_n \cup T_n$ such that:

1) if $s_i \in K$ and t_i is considered, then $t_i \in K$;
2) if $t_i \in K$, then $s_i \in K$.

4.4. There are four kinds of bushes. Take $s \in S_n$ and assume that there does not exist a pair s_i, t_i such that $s = s_i$. Then the one-point set $\{s\}$ is called a *bush of the first kind of* (S_n, T_n). Obviously, there exists an s_i such that $s = s_i$ and s_i is an $\mathcal{A}_i$-special ultrafilter. It is possible that $s = s_j$ $(i \neq j)$ and s_j is an $\mathcal{A}_j$-special ultrafilter.

4.5. Take $t \in T_n$ and assume that there do not exist pairs s_i, t_i and s_j, t_j such that $t = t_i \neq t_j$, $s_i = s_j$. Then the set

$$\{s_i \mid t_i = t\} \cup \{t\}$$

is called a *bush of the second kind of* (S_n, T_n) ; $\{t\}$ is called the *root*, and $\{s_i \mid t_i = t\}$ is called the *crown* of the bush.

Graphically one can depict a bush of the second kind as follows:

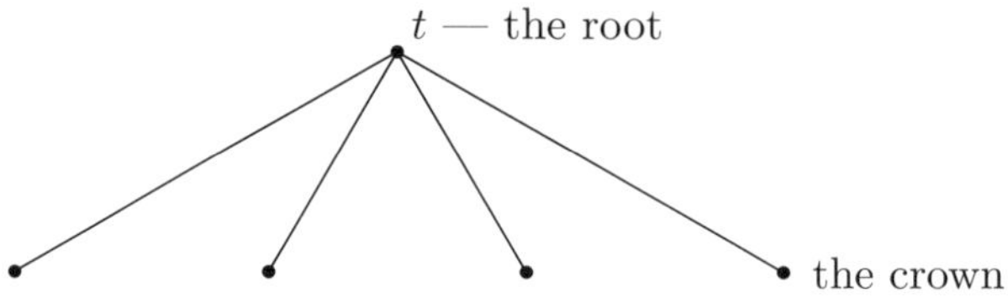

If s belongs to the crown, then there exists a pair s_i, t_i such that $s = s_i$, $t = t_i$. Another pair s_j, t_j can exist such that $s = s_i = s_j$, $t = t_i = t_j$ $(i \neq j)$. There can exist an $\mathcal{A}_m$-special ultrafilter $s_m = s$. The crown of a bush of the second kind can contain only one point.

4.6. Take $s \in S_n$ and assume that s is contained neither in a bush of the first kind nor in a bush of the second kind. Assume that there do not exist pairs s_i, t_i and s_j, t_j such that $s = s_i \neq s_j$, $t_i = t_j$. Then the set

$$\{t_i \mid s_i = s\} \cup \{s\}$$

is called a *bush of the third kind of* (S_n, T_n); $\{s\}$ is called the *root*, and $\{t_i \mid s_i = s\}$ is called the *crown* of the bush. Graphically one can depict a bush of the third kind as follows:

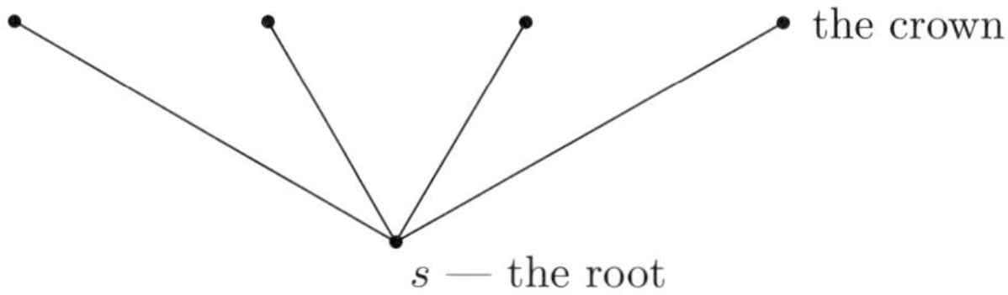

If t belongs to the crown, then there exists a pair s_i, t_i such that $s = s_i$, $t = t_i$. There can exist a pair s_j, t_j such that $s = s_i = s_j$, $t = t_i = t_j$ $(i \neq j)$. There can exist an $\mathcal{A}_m$-special ultrafilter $s_m = s$. The crown of a bush of the third kind should contain more than one point, since otherwise the bush would be of the second kind.

4.7. Take $t \in T_n$ and assume that t is contained neither in a bush of the second kind nor in a bush of the third kind. Put

$$P_1 = \{s_i \mid t_i = t\}, \quad P_2 = \{t_i \neq t \mid s_i \in P_1\}.$$

It is clear that $P_2 \neq \varnothing$ since if $P_2 = \varnothing$, then t is the root of a bush of the second kind. Further, put

$$P_3 = \{s_i \notin P_1 \mid t_i \in P_2\}.$$

It is possible that $P_3 = \varnothing$. If $P_3 \neq \varnothing$, let us continue:

$$P_j = \{t_i \notin P_{j-2} \mid s_i \in P_{j-1}\}$$

for $j \geq 4$ even, and

$$P_j = \{s_i \notin P_{j-2} \mid t_i \in P_{j-1}\}$$

for $j \geq 5$ odd. Our construction will stop at some $P_{j_0} = \varnothing$. The set

$$\{t\} \cup \bigcup_{j=1}^{j_0-1} P_j$$

is called a *bush of the fourth kind of* (S_n, T_n). Let us show the graphical representation of a simplest bush of the fourth kind.

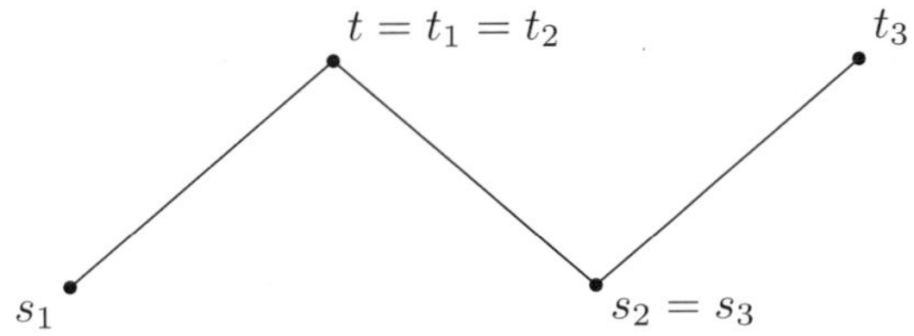

It is possible, for instance, that $s_5 = s_1$, $t_5 = t$ and $s_{10} = s_2$, $t_{10} = t_3$. It is possible, for instance, that s_{11} is an $\mathcal{A}_{11}$-special ultrafilter and $s_{11} = s_2$.

4.8. It is evident that $S_n \cup T_n$ is divided into bushes.

4.9.

(a) If K is a bush of the first kind, then $\#(K) = 1$.
(b) If K is a bush of the second kind, then $\#(K) \geq 2$.
(c) If K is a bush of the third kind, then $\#(K) \geq 3$.
(d) If K is a bush of the fourth kind, then $\#(K) \geq 4$.

4.10. If K is a bush, put

$$\|K\| = \#\big(\{i \mid s_i \in K\}\big).$$

4.11. It is rather easy to check that

$$\#(K) \leq \|K\| + 1.$$

4.12. If a bush K contains an $\mathcal{A}_i$-special ultrafilter s_i, then

$$\#(K) \leq \|K\|.$$

4.13. Put

$$n_0 = \#\big(\{i \mid s_i \text{ is contained in a bush of the fourth kind of } (S_n, T_n)\}\big).$$

Assume that there are l_0 bushes of the fourth kind of (S_n, T_n). By 4.11,

$$n' = \#\big(\{q \mid q \text{ is contained in a bush of the fourth kind of } (S_n, T_n)\}\big) \leq n_0 + l_0.$$

If K is a bush of the fourth kind, $\|K\| \geq 3$. Hence

$$n' \leq n_0 + \frac{n_0}{3} = \frac{4}{3} n_0.$$

Put

$$C = \big\{q \mid q \text{ is the root of a bush of the second or the third kind of } (S_n, T_n), \text{ and this bush does not contain an } \mathcal{A}_i\text{-special ultrafilter } s_i\big\}.$$

By what was said above and by 4.12,

$$\#((S_n \cup T_n) \setminus C) \leq \frac{4}{3} n_0 + n - n_0 \leq \frac{4}{3} n.$$

4.14. Let K be a bush which does not contain an $\mathcal{A}_i$-special ultrafilter s_i. Then K can be "turned over": if $s_i, t_i \in K$, denote s_i by t_i and t_i by s_i. If $\#(K) = 2$, then K was and will remain a bush of the second kind. If $\#(K) \geq 3$, then a bush of the second kind becomes a bush of the third kind, a bush of the third kind becomes a bush of the second kind, and a bush of the fourth kind remains a bush of the fourth kind. The new sets S_n, T_n possess all the properties listed in 4.1. Hence if $Q \in s$ for all $s \in S_n$ (S_n is new) and $Q \notin t$ for all $t \in T_n$ (T_n is new), then $Q \notin \mathcal{A}_i$ for all $i \leq n$.

4.15. PROOF OF THEOREM 2.1. Let $\mathcal{A}_1, \dots, \mathcal{A}_n, \mathcal{A}_{n+1}$, $n \in \mathbb{N}^+$, be a sequence of algebras,

$$\bigcup_{k=1}^{n} \mathcal{A}_k \neq \mathfrak{P}(X),$$

and let there exist more than $\frac{4}{3}n$ pairwise disjoint sets not belonging to $\mathcal{A}_{n+1}$. We must prove that

$$\bigcup_{k=1}^{n+1} \mathcal{A}_k \neq \mathfrak{P}(X).$$

For the algebras $\mathcal{A}_1, \dots, \mathcal{A}_n$ consider the corresponding sets of ultrafilters S_n, T_n. By what was said in 4.13, there exists an ultrafilter $q \in \ker \mathcal{A}_{n+1}$ satisfying one of the following two conditions:

(1) $q \notin S_n \cup T_n$;
(2) q is the root of a bush K of (S_n, T_n), and K does not contain an $\mathcal{A}_i$-special ultrafilter s_i, $i \in [1, n]$.[1]

Our goal is to find the corresponding sets S_{n+1}, T_{n+1} of ultrafilters for the algebras $\mathcal{A}_1, \dots, \mathcal{A}_n, \mathcal{A}_{n+1}$.

Case I. *q is an $\mathcal{A}_{n+1}$-special ultrafilter.* If q is the root of K and K is a bush of the third kind, put $S_{n+1} = S_n$, $T_{n+1} = T_n$. If q is the root of K and K is a bush of the second kind, we turn K over and put $S_{n+1} = S_n$, $T_{n+1} = T_n$. If $q \notin S_n \cup T_n$, put $S_{n+1} = S_n \cup \{q\}$, $T_{n+1} = T_n$. In all these cases put $s_{n+1} = q$.

Case II. *q, r are $\mathcal{A}_{n+1}$-similar ultrafilters, and both of them are contained neither in S_n, nor in T_n.* If one of them is contained in S_n, denote it by s_{n+1}, and denote the other one by t_{n+1}. If neither of them is contained in S_n, then at least one of them does not belong to T_n. Denote it by s_{n+1}, and denote the other one by t_{n+1}. In all these cases put

$$S_{n+1} = S_n \cup \{s_{n+1}\}, \quad T_{n+1} = T_n \cup \{t_{n+1}\}.$$

Case III. *q, r are $\mathcal{A}_{n+1}$-similar ultrafilters, either $q, r \in S_n$, or $q, r \in T_n$.* If $q, r \in S_n$, then q is the root of a bush K of the third kind. Let us turn K over and put $S_{n+1} = S_n$, $T_{n+1} = T_n$. Also put $s_{n+1} = r$, $t_{n+1} = q$. If $q, r \in T_n$, then q is the root of a bush K of the second kind. Let us turn K over and put $S_{n+1} = S_n$, $T_{n+1} = T_n$. Also put $s_{n+1} = q$, $t_{n+1} = r$. □

4.16. REMARK. In the proof of Theorem 2.1, condition (2) can be replaced by a more precise one: q is the root of a bush K of (S_n, T_n), $\|K\| \leq 2$, and $\#(K) = \|K\| + 1$. In connection with this remark see the proof of Lemma 4.23.

[1] If a bush has a root, it is either of the second or of the third kind.

4.17. The following remark is used in the proof of Theorem 2.8.

REMARK. Consider the algebras $\mathcal{A}_1, \dots, \mathcal{A}_n, \mathcal{A}_{n+1}$ from 4.15, and assume that each bush of (S_n, T_n) contains at most one $\mathcal{A}_i$-special ultrafilter s_i, $i \leq n$. Then each bush of (S_{n+1}, T_{n+1}) contains at most one $\mathcal{A}_i$-special ultrafilter s_i, $i \leq n+1$.

4.18. Let us analyze all the cases in the proof of Theorem 2.1 in which we turn the bush K over. In Case I we have turned a bush of the second kind of (S_n, T_n) over and obtained a bush of the second (if $\#(K) = 2$) or the third (if $\#(K) > 2$) kind of (S_{n+1}, T_{n+1}) containing an $\mathcal{A}_{n+1}$-special ultrafilter $s_{n+1} = q$. Let us pass to Case III. Assume that $q, r \in S_n$. If $r = s_i$ and $\{s_i\}$ is a bush of the first kind of (S_n, T_n), then after turning K over it will be contained in a bush of the second kind of (S_{n+1}, T_{n+1}) which contains an $\mathcal{A}_i$-special ultrafilter s_i. If $\{s_i\}$ is not a bush of the first kind of (S_n, T_n), then after turning K over it will be contained in a bush of the fourth kind of (S_{n+1}, T_{n+1}). Now assume that $q, r \in T_n$. Then K is a bush of the second kind of (S_n, T_n) and after turning K over, it will be contained in a bush of the fourth kind of (S_{n+1}, T_{n+1}). Let us summarize all the cases. After turning K over either it is contained in a bush of the second kind containing an $\mathcal{A}_i$-special ultrafilter s_i ($i \leq n+1$), or K becomes a bush of the third kind of (S_{n+1}, T_{n+1}) containing an $\mathcal{A}_{n+1}$-special ultrafilter s_{n+1}, or K is contained in a bush of the fourth kind of (S_{n+1}, T_{n+1}).

Now consider a countable sequence of algebras $\mathcal{A}_1, \dots, \mathcal{A}_k, \dots$ such that for each k there exist more than $\frac{4}{3}(k-1)$ pairwise disjoint sets not belonging to $\mathcal{A}_k$. Let us take sets S_1, T_1 for the algebra $\mathcal{A}_1$, and let us construct, as in the proof of Theorem 2.1, the sets

$$S_2, T_2; \dots; S_k, T_k; \dots .$$

Here it will be more convenient to denote elements of S_k by $s_i^{(k)}$ and elements of T_k by $t_i^{(k)}$. Let i be an arbitrary natural number. From the above it follows that there exists $k_i \geq i$ such that

$$s_i^{(k_i)} = s_i^{(k_i+1)} = \cdots = s_i^{(k)} = \cdots = s_i,$$

and also if the ultrafilter $t_i^{(i)}$ is considered, then

$$t_i^{(k_i)} = t_i^{(k_i+1)} = \cdots = t_i^{(k)} = \cdots = t_i.$$

Put

$$S = \{s_1, \dots, s_i, \dots\},$$

and denote by T the set of ultrafilters of kind t_i. The following is obvious:

(a) either s_i is an $\mathcal{A}_i$-special ultrafilter, or s_i is $\mathcal{A}_i$-similar to t_i;
(b) $S \cap T = \varnothing$;
(c) $S \cup T = \bigcup_k (S_k \cup T_k)$.

Since it is possible that $\overline{S} \cap \overline{T} \neq \varnothing$, one cannot assert that $\bigcup \mathcal{A}_k \neq \mathfrak{P}(X)$.

4.19. Let us present a refinement of Theorem 2.2(1). The idea of its proof will also be used below for proving other results.

THEOREM. *Let $\mathcal{A}_1, \dots, \mathcal{A}_n$ be a finite sequence of algebras, and for each $k \in [1, n]$ let there exist a finite sequence of pairwise disjoint sets $U_1^k, \dots, U_{m_k}^k \notin \mathcal{A}_k$, and let $m_k > \frac{4}{3}(k-1)$. Then for each U_i^k one can take a set $\widehat{U}_i^k \subset U_i^k$ such that $\widehat{U}_i^k \notin \mathcal{A}_k$; we have either $\widehat{U}_i^k = \widehat{U}_j^l$, or $\widehat{U}_i^k \cap \widehat{U}_j^l = \varnothing$; there exists a set Q, which is the union of sets of the type of $\widehat{U}_j^l$, such that $Q \notin \mathcal{A}_k$ for all $k \leq n$.*

PROOF. For each set U_i^k consider either an $\mathcal{A}_k$-special ultrafilter $s_i^k \ni U_i^k$, or $\mathcal{A}_k$-similar ultrafilters s_i^k, t_i^k, $s_i^k \ni U_i^k$, $t_i^k \not\ni U_i^k$. The theorem will be proved if we find the corresponding sets S_n and T_n of ultrafilters such that

$$S_n \subset \bigcup_{l,j} \{s_j^l\}.$$

For each $k \in [1, n]$ consider a set

$$Z_k \subset \bigcup_{l,j} \{s_j^l\}$$

with the following properties:

(a) $\#(Z_k) > \frac{4}{3}(k-1)$;

(b) for each $z \in Z_k$ one of the following three statements is true:

  (b$_1$) z is an $\mathcal{A}_k$-special ultrafilter;

  (b$_2$) z has an $\mathcal{A}_k$-similar ultrafilter in Z_k;

  (b$_3$) for z the statement (b$_2$) is not true, but z has an $\mathcal{A}_k$-similar ultrafilter in $\beta X \setminus \bigcup_{l,j}\{s_j^l\}$.

Define an algebra $\mathcal{A}_k^\circ$ by setting $\ker \mathcal{A}_k^\circ = Z_k$. An ultrafilter $z \in Z_k$ is $\mathcal{A}_k^\circ$-special if and only if z satisfies either the condition (b$_1$) or the condition (b$_3$). Two ultrafilters in Z_k are $\mathcal{A}_k^\circ$-similar if and only if they are $\mathcal{A}_k$-similar. By Theorem 2.2(1), for the algebras $\mathcal{A}_1^\circ, \dots, \mathcal{A}_n^\circ$ one can construct the corresponding sets S_n°, T_n° of ultrafilters (to avoid confusion, we do not denote them by S_n, T_n). Obviously,

$$S_n^\circ \subset \bigcup_{l,j} \{s_j^l\}.$$

Now for the algebras $\mathcal{A}_1, \dots, \mathcal{A}_n$ one naturally constructs the corresponding sets S_n and T_n of ultrafilters, where $S_n = S_n^\circ$. □

4.20. A theorem was formulated at the beginning of 2.34. Let us reformulate it in complete form, and then proceed to the proof.

THEOREM. *Let $\mathcal{A}_1, \dots, \mathcal{A}_n$ be a finite sequence of algebras, and for each $k \in [1, n]$ let there exist more than $\frac{4}{3}(k-1)$ pairwise disjoint sets not belonging to $\mathcal{A}_k$. Then there exist sets $U_1, \dots, U_n$ possessing the following properties*:

(a) *either $U_i = U_j$ or $U_i \cap U_j = \varnothing$;*

(b) *if $U_k \subset Q \subset \bigcup_{i=1}^n U_i$, then $Q \notin \mathcal{A}_k$.*

PROOF. By Theorem 2.2(1), the corresponding sets S_n, T_n of ultrafilters exist. For each $s \in S_n$, let us take a set $U(s) \in s$ such that $\overline{U(s)} \cap T_n = \varnothing$ and $U(s) \cap U(s') = \varnothing$ for $s \neq s'$. It remains to put $U_i = U(s_i)$. □

4.21. REMARK. Theorem 4.20 is presented here because it provides a connection between Theorems 2.2 and 2.27. (However, this has already been mentioned in 2.34.) Here is one more reason. Theorem 4.20 is related to the far from obvious Remark 4.26, which appears in the same Section 2.34 and which precedes the following four sections.

4.22. DEFINITION. Let $s \in S_n$; the *n-index* of s is the number

$$\iota_n(s) = \#(\{i \mid s_i = s\}).$$

4.23. LEMMA. *Let $\mathcal{A}_1, \dots, \mathcal{A}_n, \mathcal{A}_{n+1}$, $n \in \mathbb{N}^+$, be a sequence of algebras; for the algebras $\mathcal{A}_1, \dots, \mathcal{A}_n$ let the corresponding sets S_n, T_n of ultrafilters exist, and let $\#(\ker \mathcal{A}_{n+1}) > \frac{4}{3}n$. Then one can construct sets S_{n+1}, T_{n+1} of ultrafilters such that the following hold:*

(1) *if $\iota_n(s) \geq 4$ and $s \in S_{n+1}$, then $\iota_{n+1} = \iota_n(s)$;*
(2) *if $\iota_n(s) < 4$ and $s \in S_{n+1}$, then $\iota_{n+1} \leq \iota_n(s) + 1$;*
(3) *if $s \in S_{n+1} \setminus S_n$, then $\iota_{n+1}(s) \leq 4$.*

PROOF. We will freely use the properties of bushes from 4.11, 4.12. Put

$$n_0^* = \#\big(\{i \mid s_i \text{ is contained in a bush } K \text{ of } (S_n, T_n) \text{ and } \|K\| \geq 3\}\big),$$

$$n'_* = \#\big(\{q \mid q \text{ is contained in a bush } K \text{ of } (S_n, T_n) \text{ and } \|K\| \geq 3\}\big),$$

$$C^* = \big\{q \mid q \text{ is the root of a bush } K \text{ of } (S_n, T_n),\ \|K\| < 3, \\ \text{and } K \text{ does not contain an } \mathcal{A}_i\text{-special ultrafilter } s_i\ (i \leq n)\big\}.$$

Assume that a bush K of (S_n, T_n) possesses the following properties:

(i) there exists an ultrafilter $s \in K$ such that $\iota_n(s) \geq 4$;
(ii) $\#(K) = \|K\| + 1$;
(iii) there exists an ultrafilter $t \in K \cap T_n$ such that

$$\#\big(\{i \mid t_i = t\}\big) \geq 5.$$

Such a bush is said to be *wide*. If a bush K of (S_n, T_n) satisfies conditions (i), (ii) and does not satisfy condition (iii), then we call it *sparse*. If a bush K of (S_n, T_n) satisfies condition (i) and does not satisfy condition (ii) (we are not concerned by condition (iii)), then we call it *thick*. Put

$$\gamma(K) = \#\big(\{s \in K \cap S_n \mid \iota_n(s) \geq 4\}\big).$$

One easily checks that if K is a wide bush, then

$$\|K\| \geq 3 \cdot \gamma(K) + 5.$$

Put

$$\{K_1^1, \dots, K_{\alpha^1}^1\}: \text{ all wide bushes,}$$
$$\{K_1^2, \dots, K_{\alpha^2}^2\}: \text{ all sparse bushes,}$$
$$\{K_1^3, \dots, K_{\alpha^3}^3\}: \text{ all thick bushes.}$$

Put

$$\gamma^j = \sum_{i=1}^{\alpha^j} \gamma(K_i^j)$$

for all $j \in [1,3]$. We have

$$n'_* - \sum_{i=1}^{\alpha^1} \|K_i^1\| - \sum_{i=1}^{\alpha^2} \|K_i^2\| - \sum_{i=1}^{\alpha^3} \|K_i^3\| - \alpha^1 - \alpha^2$$
$$\leq \frac{4}{3}\left(n_0^* - \sum_{i=1}^{\alpha^1} \|K_i^1\| - \sum_{i=1}^{\alpha^2} \|K_i^2\| - \sum_{i=1}^{\alpha^3} \|K_i^3\|\right).$$

Hence

$$
\begin{aligned}
n'_* &\le \frac{4}{3}\left(n_0^* - \sum_{i=1}^{\alpha^1}\|K_i^1\| - \sum_{i=1}^{\alpha^2}\|K_i^2\| - \sum_{i=1}^{\alpha^3}\|K_i^3\|\right) \\
&\quad + \sum_{i=1}^{\alpha^1}\|K_i^1\| + \sum_{i=1}^{\alpha^2}\|K_i^2\| + \sum_{i=1}^{\alpha^3}\|K_i^3\| + \alpha^1 + \alpha^2 \\
&= \frac{4}{3}n_0^* - \frac{1}{3}\sum_{i=1}^{\alpha^1}\|K_i^1\| - \frac{1}{3}\sum_{i=1}^{\alpha^2}\|K_i^2\| - \frac{1}{3}\sum_{i=1}^{\alpha^3}\|K_i^3\| + \alpha^1 + \alpha^2 \\
&\le \frac{4}{3}n_0^* - \gamma^1 - \gamma^2 - \gamma^3 + \alpha^2 \le \frac{4}{3}n_0^* - \gamma^1 - \gamma^3.
\end{aligned}
$$

Therefore,

$$\#((S_n \cup T_n) \setminus C^*) \le \frac{4}{3}n_0^* - \gamma^1 - \gamma^3 + n - n_0^*.$$

In the construction of S_{n+1} and T_{n+1}, we will use one of the ultrafilters from the set

$$M = \ker \mathcal{A}_{n+1} \setminus ((S_n \cup T_n) \setminus C^*).$$

Put $\nu = \#(M)$. Obviously,

$$\nu > \gamma^1 + \gamma^3.$$

If M contains an $\mathcal{A}_{n+1}$-special ultrafilter, then one constructs S_{n+1}, T_{n+1} in an evident way. In the opposite case, by the statement from 3.8, in the construction of S_{n+1}, T_{n+1} we can use $\mathcal{A}_{n+1}$-similar ultrafilters q and r such that $q \in M$, and if $r \in S_n$, then either $\iota_n(r) < 4$ or $\iota_n(r) \ge 4$ and $r \in K_{i_0}^2$. If $\iota_n(r) \ge 4$, then it is easy to construct S_{n+1} and T_{n+1} after having turned the bush $K_{i_0}^2$ over. In all other cases the argument is exactly the same as in the proof of Theorem 2.1. □

4.24. Example. Let us construct algebras $\mathcal{A}_1, \ldots, \mathcal{A}_9$ and sets of ultrafilters S_8, T_8 (for algebras $\mathcal{A}_1, \ldots, \mathcal{A}_8$) such that $\iota_8(s) \le 3$ for each $s \in S_8$, $\#(\ker \mathcal{A}_9) = 11 > \frac{4}{3}(9-1)$, and there do not exist the corresponding sets of ultrafilters S_9 and T_9 satisfying the condition $\iota_9(s) \le 3$ for each $s \in S_9$.

Construction. Take pairwise distinct ultrafilters s^1, s_4, s_5, s^2, t_1, t_2, t, t_7, t_8, q_1, q_2 and put $s^1 = s_1 = s_2 = s_3$, $s^2 = s_6 = s_7 = s_8$, $t = t_3 = t_4 = t_5 = t_6$; $\ker \mathcal{A}_i = \{s_i, t_i\}$ and s_i, t_i are $\mathcal{A}_i$-similar ultrafilters for $i \in [1, 8]$; $\{s^1, q_1\}$, $\{s^2, q_2\}$, $\{s_4, s_5\}$, $\{t_1, t_2, t, t_7, t_8\}$ are $\mathcal{A}_9$-similar sets.

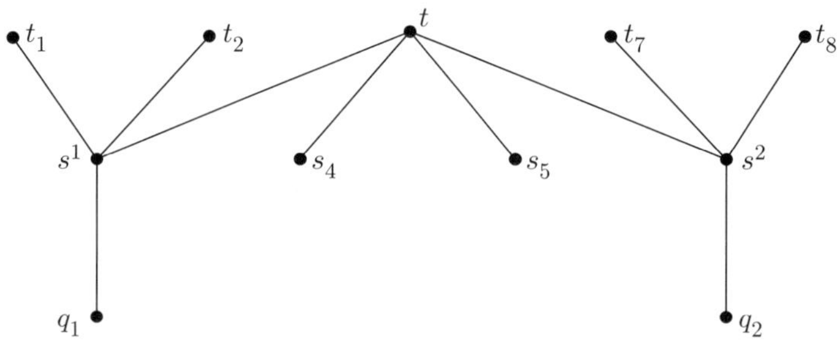

4.25. The following lemma, which is used in Chapter 8 together with Lemma 4.23, is an obvious corollary of Lemma 4.23.

LEMMA. *Let $\mathcal{A}_1, \dots, \mathcal{A}_n$ be a finite sequence of algebras, and let*

$$\#(\ker \mathcal{A}_k) > \frac{4}{3}(k-1) \quad \textit{for all } k \in [1,n].$$

Then one can construct sets S_n, T_n of ultrafilters such that $\iota_n(s) \le 4$ for all $s \in S_n$.

4.26. REMARK. Due to Lemma 4.25, the sets $U_1, \dots, U_n$ in Theorem 4.20 can be constructed so that

$$\#\big(\{i \in [1,n] \mid U_i = U_k\}\big) \le 4$$

for each $k \in [1,n]$.

4.27. REMARK. The only corollary of Lemma 4.25 used later in Chapter 8 is that there exists a constant c independent of n such that $\iota_n(s) \le c$ for all $s \in S_n$. We have shown that one can take $c = 4$. But it is much simpler to show that one can take $c = 6$. Let us return to the proof of Lemma 4.23, and put

$$\{K_1, \dots, K_\alpha\} = \big\{K \mid K \text{ is a bush of } (S_n, T_n), \text{ and there exists an ultrafilter } s \in K \text{ such that } \iota_n(s) \ge 6\big\}.$$

We have

$$n'_* - \sum_{i=1}^{\alpha} \|K_i\| - \alpha \le \frac{4}{3}\left(n_0^* - \sum_{i=1}^{\alpha} \|K_i\|\right).$$

Therefore

$$n'_* \le \frac{4}{3}\left(n_0^* - \sum_{i=1}^{\alpha} \|K_i\|\right) + \sum_{i=1}^{\alpha} \|K_i\| + \alpha = \frac{4}{3}n_0^* - \frac{1}{3}\sum_{i=1}^{\alpha} \|K_i\| + \alpha.$$

Put

$$\gamma = \#\left(\left\{s \in S_n \cap \bigcup_{i=1}^{\alpha} K_i \mid \iota_n(s) \ge 6\right\}\right).$$

Obviously,

$$n'_* \le \frac{4}{3}n_0^* - 2\gamma + \alpha < \frac{4}{3}n_0^* - \gamma.$$

Hence

$$\#((S_n \cup T_n) \setminus C^*) \le \frac{4}{3}n_0^* - \gamma + n - n_0^*, \quad \text{etc.}$$

4.28. Consider three ultrafilters s, τ_1, τ_2, and define algebras $\mathcal{B}_1, \mathcal{B}_2, \mathcal{B}_3$ with $\#(\ker \mathcal{B}_k) > \frac{4}{3}(k-1)$ for all $k \in [1,3]$:

$$\ker \mathcal{B}_1 = \{s\};$$
$$\ker \mathcal{B}_2 = \{s, \tau_1\} \text{ and } X \in \mathcal{B}_2;$$
$$\ker \mathcal{B}_3 = \{s, \tau_1, \tau_2\} \text{ and } \{s, \tau_2\} \text{ is a } \mathcal{B}_2\text{-similar set.}$$

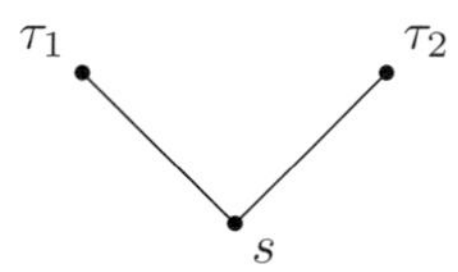

There exists a unique way to construct the sets S_3, T_3:

$$S_3 = \{s\}, \quad T_3 = \{\tau_1, \tau_2\}.$$

It is clear that $\iota_3(s) = 3$.

Denote by π_* the minimal natural number such that if $\mathcal{A}_1, \dots, \mathcal{A}_n$ is an arbitrary finite sequence of algebras with $\#(\ker \mathcal{A}_k) > \frac{4}{3}(k-1)$ for all $k \in [1,n]$, then one can construct the sets S_n, T_n of ultrafilters, and $\iota_n(s) \le \pi_*$ for all $s \in S_n$. By Lemma 4.25 and by the arguments from the present section,

$$3 \le \pi_* \le 4.$$

4.29. Let us return to Theorem 4.20. For each $k \in [1,n]$ consider a finite sequence of pairwise disjoint sets

$$U_1^k, \dots, U_{m_k}^k \notin \mathcal{A}_k,$$

with $m_k > \frac{4}{3}(k-1)$. By the same arguments as in the proof of Theorem 4.19, one can construct the sets U_k in such a way that $U_k \subset U_{\psi_k}^{\xi_k}$ and

$$\#(\{i \in [1,n] \mid U_i = U_k\}) \le \pi_*$$

for each $k \in [1,n]$. In connection with this result it is appropriate to formulate the following theorem.[2]

THEOREM. (1) *Let $\mathcal{A}_1, \dots, \mathcal{A}_n$ be a finite sequence of algebras, and for each k let there exist a finite sequence of pairwise disjoint sets $U_1^k, \dots, U_{m_k}^k \notin \mathcal{A}_k$, and $m_k > 2(k-1)$. Then for each $k \in [1,n]$ there exists a subset $U_k \subset U_{\psi_k}^k$ such that the following hold*:

(i) *either $U_k = U_l$ or $U_k \cap U_l = \varnothing$;*
(ii) *$\#(\{i \mid U_i = U_k\}) \le 2$ for each $k \in [1,n]$;*
(iii) *if $U_k \subset Q \subset \bigcup_{i=1}^n U_i$, then $Q \notin \mathcal{A}_k$.*

(2) *Regarding an improvement of the estimate $2(k-1)$ we know the following. For each $n \in \mathbb{N}^+$ one can construct algebras $\mathcal{A}_1, \dots, \mathcal{A}_n, \mathcal{A}_{n+1}$ and for each $k \le n+1$ pairwise disjoint sets $U_1^k, \dots, U_{m_k}^k \notin \mathcal{A}_k$ such that $m_k > 2(k-1)$ for $k \le n$; $m_{n+1} = 2n$ if n is even, and $m_{n+1} = 2n-1$ if n is odd. On the other hand, one cannot find sets $U_1', \dots, U_n', U_{n+1}'$ satisfying all the conditions of part* (1) *of the theorem with n replaced by $n+1$.*

PROOF. **(1)** As in the proof of Theorem 4.19, consider the ultrafilters s_i^k and t_i^k. Assume that the corresponding sets S_l, T_l of ultrafilters exist for the algebras $\mathcal{A}_1, \dots, \mathcal{A}_l$, $l < n$, and the following hold:

(a) $s_i = s_{\psi_i}^i$ for all $i \le l$;
(b) $\iota_l(s) \le 2$ for all $s \in S_l$.

Put

$$S^{(l+1)} = \{s_1^{l+1}, \dots, s_{m_{l+1}}^{l+1}\}, \quad H = S^{(l+1)} \setminus (S_l \cup T_l).$$

It is clear that

$$\#(H) > \#(S_l \setminus S^{(l+1)}) + \#(\{s \in S_l \cap S^{(l+1)} \mid \iota_l(s) = 2\}).$$

Denote by s_{l+1} some $\mathcal{A}_{l+1}$-special ultrafilter in H. And if H does not contain $\mathcal{A}_{l+1}$-special ultrafilters, then let us take $\mathcal{A}_{l+1}$-similar ultrafilters q, r such that

[2] An infinite analog of the first part of this theorem has already been mentioned in 2.34.

$q \in H$. Using the above and the arguments in 3.8, we will assume that there are two possible cases:

(1) $r \notin S_l$,

(2) $r \in S_l \cap S^{(l+1)}$ and $\iota_l(r) = 1$.

In the first case, put $s_{l+1} = q$, $t_{l+1} = r$. In the second one, put $s_{l+1} = r$, $t_{l+1} = q$. Thus, for the algebras $\mathcal{A}_1, \ldots, \mathcal{A}_l, \mathcal{A}_{l+1}$ we have constructed sets S_{l+1}, T_{l+1} of ultrafilters satisfying conditions (a) and (b) with l replaced by $l+1$. This implies part (1) of our theorem.

(2) Let $l \in \mathbb{N}^+$. Take pairwise distinct points of βX:

$$a_1, \ldots, a_l, b_1, \ldots, b_{2l}, c_1, \ldots, c_l.$$

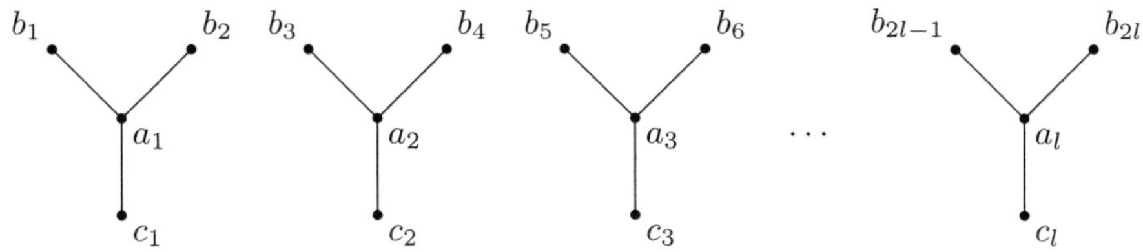

If $n = 1$, then $\ker \mathcal{A}_1 = \{a_1, b_1\}$ and $X \in \mathcal{A}_1$; $\ker \mathcal{A}_2 = \{b_1\}$; $m_1 = m_2 = 1$, and $s_1^1 = a_1$, $s_1^2 = b_1$.

If $n = 2$, then for the algebra $\mathcal{A}_1$ everything remains the same; $\ker \mathcal{A}_2 = \{a_1, b_1, b_2\}$, $\{a_1, b_2\}$ is an $\mathcal{A}_2$-similar set; $\ker \mathcal{A}_3 = \{a_1, b_1, b_2, c_1\}$, and $\{b_1, b_2\}$, $\{a_1, c_1\}$ are $\mathcal{A}_3$-similar sets; $m_2 = 3$, $m_3 = 4$; denote each of the ultrafilters from $\ker \mathcal{A}_2$ in an arbitrary way by the unique symbol s_i^2, $i \in [1, m_2]$; denote each of the ultrafilters from $\ker \mathcal{A}_3$ in an arbitrary way by the unique symbol s_i^3, $i \in [1, m_3]$.

If $n = 3$, then for the algebras $\mathcal{A}_1, \mathcal{A}_2$ everything remains the same; $\ker \mathcal{A}_3 = \{a_1, a_2, b_1, b_2, b_3, c_1\}$, and $\{b_1, b_2\}$, $\{a_1, c_1\}$, $\{a_2, b_3\}$ are $\mathcal{A}_3$-similar sets; $\ker \mathcal{A}_4 = \{a_1, b_1, b_2, b_3, c_1\}$, and $\{b_1, b_2\}$, $\{a_1, c_1\}$ are $\mathcal{A}_4$-similar sets (hence b_3 is an $\mathcal{A}_4$-similar ultrafilter); $m_3 = m_4 = 5$; denote each of the ultrafilters from $\ker \mathcal{A}_3 \setminus \{b_3\}$ in an arbitrary way by the unique symbol s_i^3, $i \in [1, m_3]$; denote each of the ultrafilters from $\ker \mathcal{A}_4$ in an arbitrary way by the unique symbol s_i^4, $i \in [1, m_4]$.

If $n = 4$, then everything for the algebras $\mathcal{A}_1, \mathcal{A}_2, \mathcal{A}_3$ remains the same; $\ker \mathcal{A}_4 = \{a_1, a_2, b_1, b_2, b_3, b_4, c_1\}$, and $\{b_1, b_2, b_3\}$, $\{a_1, c_1\}$, $\{a_2, b_4\}$ are $\mathcal{A}_4$-similar sets; $\ker \mathcal{A}_5 = \{a_1, a_2, b_1, b_2, b_3, b_4, c_1, c_2\}$, and $\{b_1, b_2, b_3, b_4\}$, $\{a_1, c_1\}$, $\{a_2, c_2\}$ are $\mathcal{A}_5$-similar sets; $m_4 = 7$, $m_5 = 8$; each of the ultrafilters from $\ker \mathcal{A}_4$ is denoted in an arbitrary way by the unique symbol s_i^4, $i \in [1, m_4]$; each of the ultrafilters from $\ker \mathcal{A}_5$ is denoted in an arbitrary way by the unique symbol s_i^5, $i \in [1, m_5]$.

Continuing our construction in a natural way, let us put $n = 2l - 1$, $l > 2$; everything for the algebras $\mathcal{A}_1, \ldots, \mathcal{A}_{2l-2}$ remains the same;

$$\ker \mathcal{A}_{2l-1} = \{a_1, \ldots, a_l, b_1, \ldots, b_{2l-1}, c_1, \ldots, c_{l-1}\},$$

and $\{b_1, \ldots, b_{2l-2}\}$, $\{a_1, c_1\}$, $\ldots$, $\{a_{l-1}, c_{l-1}\}$, $\{a_l, b_{2l-1}\}$ are $\mathcal{A}_{2l-1}$-similar sets;

$$\ker \mathcal{A}_{2l} = \{a_1, \ldots, a_{l-1}, b_1, \ldots, b_{2l-1}, c_1, \ldots, c_{l-1}\},$$

and $\{b_1, \ldots, b_{2l-2}\}$, $\{a_1, c_1\}$, $\ldots$, $\{a_{l-1}, c_{l-1}\}$ are $\mathcal{A}_{2l}$-similar sets (hence, b_{2l-1} is an $\mathcal{A}_{2l}$-special ultrafilter); $m_{2l-1} = m_{2l} = 4l - 3$; each of the ultrafilters from $\ker \mathcal{A}_{2l-1} \setminus \{b_{2l-1}\}$ is denoted in an arbitrary way by the unique symbol s_i^{2l-1}, $i \in [1, m_{2l-1}]$; each of the ultrafilters from $\ker \mathcal{A}_{2l}$ is denoted in an arbitrary way by the unique symbol s_i^{2l}, $i \in [1, m_{2l}]$.

Finally, put $n = 2l$, $l > 2$; everything for the algebras $\mathcal{A}_1, \dots, \mathcal{A}_{2l-1}$ remains the same;

$$\ker \mathcal{A}_{2l} = \{a_1, \dots, a_l, b_1, \dots, b_{2l}, c_1, \dots, c_{l-1}\},$$

and $\{b_1, \dots, b_{2l-1}\}$, $\{a_1, c_1\}$, ... , $\{a_{l-1}, c_{l-1}\}$, $\{a_l, b_{2l}\}$ are $\mathcal{A}_{2l}$-similar sets;

$$\ker \mathcal{A}_{2l+1} = \{a_1, \dots, a_l, b_1, \dots, b_{2l}, c_1, \dots, c_l\},$$

and $\{b_1, \dots, b_{2l}\}$, $\{a_1, c_1\}, \dots$, $\{a_l, c_l\}$ are $\mathcal{A}_{2l+1}$-similar sets; $m_{2l} = 4l-1$, $m_{2l+1} = 4l$; each of the ultrafilters from $\ker \mathcal{A}_{2l}$ is denoted in an arbitrary way by the unique symbol s_i^{2l}, $i \in [1, m_{2l}]$; each of the ultrafilters from $\ker \mathcal{A}_{2l+1}$ is denoted in an arbitrary way by the unique symbol s_i^{2l+1}, $i \in [1, m_{2l+1}]$.

Concluding the construction, we agree that $s_i^k \ni U_i^k$. □

4.30. In this chapter it is appropriate to formulate and prove a lemma which will be used in the proof of Theorem 2.8.

LEMMA. *Let $\mathcal{A}_1, \dots, \mathcal{A}_n, \mathcal{A}_{n+1}$, $n \in \mathbb{N}^+$, be a sequence of algebras, and let $\#(\ker \mathcal{A}_{n+1}) > \frac{4}{3}n$. Let*

$$\bigcup_{k=1}^{n} \mathcal{A}_k \neq \mathfrak{P}(X),$$

and let S_n, T_n be corresponding sets of ultrafilters. Let K be a bush of (S_n, T_n) consisting of r points. Then there exist pairwise distinct ultrafilters $q_1, \dots, q_p \in \ker \mathcal{A}_{n+1}$ such that $p > r/3 - \rho$ (where $\rho = 4/3$ if K is of the fourth kind, and $\rho = 1/3$ otherwise), and for each $j \in [1, p]$ one of the two conditions holds:

(1) *$q_j \notin S_n \cup T_n$;*
(2) *q_j is the root of a bush K_j of (S_n, T_n), and K_j does not contain an $\mathcal{A}_i$-special ultrafilter s_i, $i \leq n$.*[3]

PROOF.[4] We will use the notation and formulas from the previous sections of this chapter. Thus,

$$\#((S_n \cup T_n) \setminus C) \leq n' + (n - n_0) \leq \frac{4}{3} n_0 + (n - n_0).$$

Since $\#(\ker \mathcal{A}_{n+1}) > \frac{4}{3}n$, we have

$$p > \frac{4}{3}n - \#((S_n \cup T_n) \setminus C) \geq \frac{n - n_0}{3}.$$

Case I. *If K is a bush of the first, the second, or the third kind,* then we have $n - n_0 \geq r - 1$ and

$$p > \frac{r}{3} - \frac{1}{3}.$$

Case II. *If K is a bush of the fourth kind,* then using the arguments from 4.13, we obtain:

$$\#(\{q \mid q \text{ is contained in a bush of the fourth kind of } (S_n, T_n) \text{ and } q \notin K\})$$
$$\leq n_0 - \|K\| + l_0 - 1 \leq \frac{4}{3}(n_0 - \|K\|).$$

Hence

$$n' \leq r + \frac{4}{3}(n_0 - \|K\|).$$

[3]If a bush has a root, then it can be either of the second or of the third kind.

[4]It is based on the same ideas as the proof of Lemma 4.23.

Since $\|K\| \geq r-1$, we have

$$\#((S_n \cup T_n) \setminus C) \leq r + \frac{4}{3}(n_0 - r + 1) + (n - n_0).$$

Therefore

$$p > \frac{4}{3}n - \#((S_n \cup T_n) \setminus C) \geq \frac{n-n_0}{3} + \frac{r}{3} - \frac{4}{3}. \qquad \square$$

4.31. Remark to Lemma 4.30. Actually the proof of Theorem 2.8 uses only the fact that if $\#(K)$ is "large", then p can also be taken "large".

4.32. Let us finally pass to the

Proof of Theorem 2.2(2). Let us take $m \in \mathbb{N}^+$ and construct algebras $\mathcal{A}_1, \dots, \mathcal{A}_{3m}$, each of which contains X.[5] To this end, consider pairwise distinct points of βX:

$$a_1^1 \ a_2^1 \ b_1^1 \ b_2^1$$
$$\dots\dots\dots\dots$$
$$a_1^m \ a_2^m \ b_1^m \ b_2^m.$$

Let $j \in [1, m]$. Let us list all $\mathcal{A}_{3j-2}$-similar sets: $\{a_1^j, b_1^j\}$, and for each natural number $i < j$, $\{a_1^i, a_2^i\}$, $\{b_1^i, b_2^i\}$. Let us list all $\mathcal{A}_{3j-1}$-similar sets: $\{a_2^j, b_1^j\}$, and for each natural number $i < j$, $\{a_1^i, a_2^i\}$, $\{b_1^i, b_2^i\}$. Let us list all $\mathcal{A}_{3j}$-similar sets: $\{a_1^j, a_2^j, b_2^j\}$, and for each natural number $i < j$, $\{a_1^i, a_2^i\}$, $\{b_1^i, b_2^i\}$.

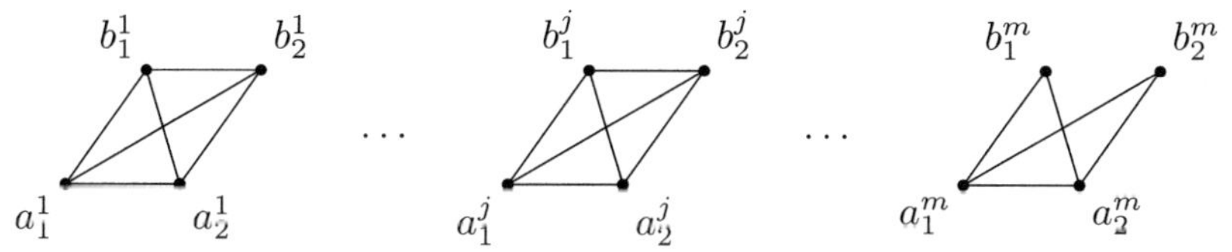

Let us construct an algebra $\mathcal{A}_{3m+1}^*$ by listing all $\mathcal{A}_{3m+1}^*$-similar sets: $\{a_1^i, a_2^i\}$, $\{b_1^i, b_2^i\}$ for each $i \leq m$. For each $k \leq 3m$,

$$\#(\ker \mathcal{A}_k) > \frac{4}{3}(k-1),$$

and $\#(\ker \mathcal{A}_{3m+1}^*) = 4m$. Let us prove that

$$\mathcal{A}_{3m+1}^* \cup \bigcup_{k=1}^{3m} \mathcal{A}_k = \mathfrak{P}(X).$$

Suppose the contrary, i.e., assume that there exists

$$Q \notin \mathcal{A}_1, \dots, \mathcal{A}_{3m}, \mathcal{A}_{3m+1}^*.$$

Then either

$$\overline{Q} \supset \{a_1^1, a_2^1\}, \quad \overline{Q} \cap \{b_1^1, b_2^1\} = \varnothing, \quad \text{or}$$
$$\overline{Q} \supset \{b_1^1, b_2^1\}, \quad \overline{Q} \cap \{a_1^1, a_2^1\} = \varnothing.$$

[5]The proof of our theorem also contains the construction of other algebras. All of them contain X. Recall that if $X \in \mathcal{A}$, then $\ker \mathcal{A}$ is the union of $\mathcal{A}$-similar sets.

This implies that either

$$\overline{Q} \supset \{a_1^2, a_2^2\}, \quad \overline{Q} \cap \{b_1^2, b_2^2\} = \varnothing, \quad \text{or}$$
$$\overline{Q} \supset \{b_1^2, b_2^2\}, \quad \overline{Q} \cap \{a_1^2, a_2^2\} = \varnothing,$$

etc. This implies that $Q \in \mathcal{A}^*_{3m+1}$, a contradiction.

Now we construct the algebras $\widehat{\mathcal{A}}_{3m+1}, \widehat{\mathcal{A}}_{3m+2}$. Take

$$q \in \beta X \setminus \ker \mathcal{A}^*_{3m+1}.$$

All $\mathcal{A}^*_{3m+1}$-similar sets except $\{a_1^m, a_2^m\}$ are $\widehat{\mathcal{A}}_{3m+1}$-similar. There exists one more $\widehat{\mathcal{A}}_{3m+1}$-similar set, namely $\{a_1^m, a_2^m, q\}$. All $\mathcal{A}^*_{3m+1}$-similar sets except $\{b_1^m, b_2^m\}$ are $\widehat{\mathcal{A}}_{3m+2}$-similar. There exists one more $\widehat{\mathcal{A}}_{3m+2}$-similar set, namely $\{b_1^m, b_2^m, q\}$. Obviously,

$$\#(\ker \widehat{\mathcal{A}}_{3m+1}) = \#(\ker \widehat{\mathcal{A}}_{3m+2}) = 4m + 1.$$

Let us prove that

$$\widehat{\mathcal{A}}_{3m+1} \cup \widehat{\mathcal{A}}_{3m+2} \cup \bigcup_{k=1}^{3m} \mathcal{A}_k = \mathfrak{P}(X).$$

Suppose the contrary, i.e., assume that there exists

$$Q \notin \mathcal{A}_1, \dots, \mathcal{A}_{3m}, \widehat{\mathcal{A}}_{3m+1}, \widehat{\mathcal{A}}_{3m+2}.$$

We have already proved that for each $j \in [1, m]$ either

$$\overline{Q} \supset \{a_1^j, a_2^j\}, \quad \overline{Q} \cap \{b_1^j, b_2^j\} = \varnothing, \quad \text{or}$$
$$\overline{Q} \supset \{b_1^j, b_2^j\}, \quad \overline{Q} \cap \{a_1^j, a_2^j\} = \varnothing.$$

For example, assume that

$$\overline{Q} \supset \{a_1^m, a_2^m\}.$$

Since $Q \notin \widehat{\mathcal{A}}_{3m+2}$, we have $Q \in q$ and

$$\overline{Q} \supset \{a_1^m, a_2^m, q\}.$$

This implies that $Q \in \widehat{\mathcal{A}}_{3m+1}$, a contradiction.

Let us now construct the algebras $\mathcal{A}'_{3m+1}, \mathcal{A}'_{3m+2}, \mathcal{A}'_{3m+3}$. Take pairwise distinct ultrafilters

$$q_1, q_2, q_3 \in \beta X \setminus \ker \mathcal{A}^*_{3m+1}.$$

All $\mathcal{A}^*_{3m+1}$-similar sets are $\mathcal{A}'_{3m+i}$-similar for each $i \in [1, 3]$. For each algebra $\mathcal{A}'_{3m+i}$ there is one more $\mathcal{A}'_{3m+i}$-similar set, namely

$$\{q_1, q_2, q_3\} \setminus \{q_i\}.$$

We will also consider three algebras for $m = 0$: $\mathcal{A}'_1, \mathcal{A}'_2, \mathcal{A}'_3$; the algebra $\mathcal{A}'_i$ has one $\mathcal{A}'_i$-similar set, namely

$$\{q_1, q_2, q_3\} \setminus \{q_i\}.$$ [6]

Obviously,

$$\#(\ker \mathcal{A}'_{3m+1}) = \#(\ker \mathcal{A}'_{3m+2}) = \#(\ker \mathcal{A}'_{3m+3}) = 4m + 2.$$

---

[6] These algebras are just the algebras from 1.13. The algebra $\mathcal{A}'_i$ from 1.13 has one $\mathcal{A}'_i$-similar set, namely $X \setminus \{x_i\}$.

It is rather easy to check that

$$\mathcal{A}'_{3m+1} \cup \mathcal{A}'_{3m+2} \cup \mathcal{A}'_{3m+3} \cup \bigcup_{k=1}^{3m} \mathcal{A}_k = \mathfrak{P}(X).^7$$

Let us summarize the arguments. Having constructed the algebras $\mathcal{A}_1, \dots, \mathcal{A}_{3m}, \mathcal{A}^*_{3m+1}$, we have proved the theorem for $n \equiv 0 \pmod 3$. Having constructed the algebras $\mathcal{A}_1, \dots, \mathcal{A}_{3m}, \widehat{\mathcal{A}}_{3m+1}, \widehat{\mathcal{A}}_{3m+2}$, we have proved the theorem for $n \equiv 1 \pmod 3$. Having constructed the algebras

$$\mathcal{A}_1, \dots, \mathcal{A}_{3m}, \mathcal{A}'_{3m+1}, \mathcal{A}'_{3m+2}, \mathcal{A}'_{3m+3},$$

we have proved the theorem for $n \equiv 2 \pmod 3$. □

4.33. Although the purpose of this chapter is to study finite sequences of algebras, it makes sense to present at this point an interesting example, which shows that Theorem 2.2(1) is not true if one considers countable sequences of algebras. The reason for putting it here is simple: this example uses the construction from the proof of Theorem 2.2(2) given in the previous section.

EXAMPLE. There exists a countable sequence of algebras $\mathcal{A}_1, \dots, \mathcal{A}_k, \dots$ possessing the following properties:

(1) for each k there exist more than $\frac{4}{3}(k-1)$ pairwise disjoint sets not belonging to $\mathcal{A}_k$;
(2) $\bigcup \mathcal{A}_k = \mathfrak{P}(X)$;
(3) $\bigcup_{k \neq n} \mathcal{A}_k \neq \mathfrak{P}(X)$ for each $n \in \mathbb{N}^+$.[8]

CONSTRUCTION. Let $\#(X) = \aleph_0$. Let us divide X into four pairwise disjoint countable sets:

$$A_1 = \{a_1^2, a_1^3, \dots, a_1^m, \dots\}, \quad A_2 = \{a_2^2, a_2^3, \dots, a_2^m, \dots\},$$
$$B_1 = \{b_1^2, b_1^3, \dots, b_1^m, \dots\}, \quad B_2 = \{b_2^2, b_2^3, \dots, b_2^m, \dots\}.$$

Consider the Stone–Čech compactifications $\beta A_1, \beta A_2, \beta B_1, \beta B_2$, each of which is similar to $\beta\mathbb{N}$ and all of them are similar to one another.[9] The similarities between them are constructed in the natural way, and the points $a_1^m, a_2^m, b_1^m, b_2^m$ are similar to one another. Let us take similar points

$$a_1^1 \in \beta A_1 \setminus A_1, \quad b_1^1 \in \beta A_2 \setminus A_2, \quad a_2^1 \in \beta B_1 \setminus B_1, \quad b_2^1 \in \beta B_2 \setminus B_2.$$

For each $m \in \mathbb{N}^+$ consider the pairwise distinct points:

$$\begin{matrix} a_1^1 & a_2^1 & b_1^1 & b_2^1 \\ \dots & \dots & \dots & \dots \\ a_1^m & a_2^m & b_1^m & b_2^m, \end{matrix}$$

and let us construct the algebras $\mathcal{A}_1, \dots, \mathcal{A}_{3m}$ as in the previous section. Thus, a countable sequence of algebras $\mathcal{A}_1, \dots, \mathcal{A}_k, \dots$ is constructed, and for the algebra $\mathcal{A}_k$ there exist more than $\frac{4}{3}(k-1)$ pairwise disjoint sets not belonging to it.

---

[7] This equality is true also for $m = 0$, i.e., $\mathcal{A}'_1 \cup \mathcal{A}'_2 \cup \mathcal{A}'_3 = \mathfrak{P}(X)$.

[8] Property (3) is an interesting one.

[9] As usual, by $\beta\mathbb{N}$ we denote the Stone–Čech compactification of the space $\mathbb{N}$ with discrete topology.

Let us prove that there does not exist Q such that $Q \notin \mathcal{A}_k$ for all k. Assume the contrary, i.e., let such a Q exist. Then for each $m \in \mathbb{N}^+$ either

$$\overline{Q} \cap \{a_1^m, a_2^m, b_1^m, b_2^m\} = \{a_1^m, a_2^m\}, \quad \text{or}$$
$$\overline{Q} \cap \{a_1^m, a_2^m, b_1^m, b_2^m\} = \{b_1^m, b_2^m\}.$$

If $\overline{Q} \supset \{a_1^1, a_2^1\}$, then there exists m such that $\overline{Q} \supset \{a_1^m, b_1^m\}$. If $\overline{Q} \supset \{b_1^1, b_2^1\}$, then there exists m such that $\overline{Q} \supset \{a_2^m, b_2^m\}$, a contradiction.

Let us prove that $\bigcup_{k \neq n} \mathcal{A}_k \neq \mathfrak{P}(X)$ for each $n \in \mathbb{N}^+$. For each $m \in \mathbb{N}^+$ let us construct sets $Q_{3m-2}, Q_{3m-1}, Q_{3m}$ such that if $l < m$, then

$$\overline{Q}_n \cap \{a_1^l, a_2^l, b_1^l, b_2^l\} = \{a_1^l, a_2^l\}$$

for each $n \in [3m-2, 3m]$; in addition, we have

$$\{a_1^m, a_2^m, b_1^m, b_2^m\} \cap \overline{Q}_{3m-2} = \{a_2^m\},$$
$$\{a_1^m, a_2^m, b_1^m, b_2^m\} \cap \overline{Q}_{3m-1} = \{a_1^m\},$$
$$\{a_1^m, a_2^m, b_1^m, b_2^m\} \cap \overline{Q}_{3m} = \{b_1^m\}.$$

One easily checks that if $n \in [3m-2, 3m]$ and $k \neq n$, then $Q_n \notin \mathcal{A}_k$. □

4.34. REMARK. In Example 4.33 for each of the algebras $\mathcal{A}_k$ there exist at most a finite number of pairwise disjoint sets not belonging to it. Hence Theorem 2.4(1) is a stronger result.

4.35. DEFINITION. If there exists a σ-two-valued measure on X such that all subsets of X are measurable, then $\#(X)$ is said to be a σ-measurable cardinal. A cardinal which is not σ-measurable is said to be σ-nonmeasurable.

4.36. REMARK. Some authors (see [**E**], [**G-Je**], [**Ku-Mo**]) used the term "measurable cardinal" for a σ-measurable cardinal. However, in modern set theory the following definition is standard. A cardinal $\varkappa > \aleph_0$ is said to be *measurable* if on a set X, $\#(X) = \varkappa$, one can define a $\varkappa$-additive two-valued measure such that all subsets of X are measurable. (A measure μ is said to be *$\varkappa$-additive* if for any family $\{M_\lambda\}_{\lambda \in \Lambda}$, $\#(\Lambda) < \varkappa$, of μ-measurable sets the set $\bigcup_{\lambda \in \Lambda} M_\lambda$ is μ-measurable, and if $\{M_\lambda\}_{\lambda \in \Lambda}$ is also a family of pairwise disjoint sets, then $\mu(\bigcup_{\lambda \in \Lambda} M_\lambda) = \sum_{\lambda \in \Lambda} \mu(M_\lambda)$.) To avoid confusion, we have introduced the notion of σ-measurable cardinal which seems reasonable, since in its definition one assumes the existence of a σ-two-valued measure. Denote by $\varkappa_0$ the minimal measurable cardinal. It is clear that a cardinal $\varkappa$ is σ-measurable if and only if $\varkappa \geq \varkappa_0$.

4.37. Let us quote [**E**]: "It is evident that the class $\mathcal{N}$ of all σ-nonmeasurable cardinals contains the cardinal $\aleph_0$. One can prove that if $\varkappa \in \mathcal{N}$, then $\mathcal{N}$ contains: all cardinals less than $\varkappa$, the sum of any family $\{\varkappa_s\}_{s \in S}$ of cardinals from $\mathcal{N}$ where $\#(S) \leq \varkappa$ (i.e., the cardinality of the set $\bigcup_{s \in S} A_s$, where $\#(A_s) = \varkappa_s$ and $A_s \cap A_{s'} = \varnothing$ for $s \neq s'$), and also the cardinal $2^\varkappa$. One can also prove that the first strongly inaccessible cardinal belongs to $\mathcal{N}$ (a cardinal $\aleph_\alpha$ is called *strongly inaccessible* if $\alpha > 0$, the ordinal ω_α is regular, and the inequality $\varkappa < \aleph_\alpha$ implies $2^\varkappa < \aleph_\alpha$). It is known that the assumption of σ-nonmeasurability of all cardinals (and even the assumption that there does not exist a strongly inaccessible cardinal) is compatible with the usual axioms of set theory. It is also known that the assumption of the existence of a σ-measurable cardinal is stronger than the assumption of the self-consistency of set theory." (See [**G-Je**], [**Ke-T**], [**Ku-Mo**], [**U1**].)

4.38. DEFINITION. A point $a \in \beta X$ is said to be *regular* if there exists a countable sequence of sets $M_1, \dots, M_k, \dots$ such that $a \notin \overline{M_k}$ for all k, but $a \in \overline{\bigcup M_k}$. A point of βX is said to be *irregular* if it is not regular.

4.39. Since a point of βX is an ultrafilter on X and, vice versa, an ultrafilter on X is a point of βX, we will use the notion of *regular ultrafilter* along with the notion of regular point, and use the notion of *irregular ultrafilter* along with the notion of irregular point.

4.40. All points of X are irregular. A point a is regular if and only if there exists a continuous real function f on βX such that $f(a) = 0$ and $f(x) > 0$ for all $x \in X$. If $\#(X)$ is a σ-measurable cardinal and μ is a σ-two-valued measure defined on all subsets of X, then $a = \{M \mid \mu(M) = 1\}$ is an irregular ultrafilter and $a \in \beta X \setminus X$. On the other hand, a point $a \in \beta X \setminus X$ is irregular if and only if the equality $\mu(M) = 1$ for all $M \in a$ defines a σ-two-valued measure defined on all subsets of X. So, the cardinal $\#(X)$ is σ-measurable if and only if $\beta X \setminus X$ contains irregular points.[10]

4.41. Assume that $\#(X)$ is a σ-measurable cardinal and all points $a_i^j, b_i^j, q, q_1, q_2, q_3$ from the proof of Theorem 2.2(2) are irregular and belong to $\beta X \setminus X$. Then each algebra $\mathcal{A}$ from the proof of Theorem 2.2(2) is the algebra of all measurable sets of some σ-measure μ. Also, $\mu(M) = 0$ if and only if $\mathfrak{P}(M) \subset \mathcal{A}$, $\mu(\{x\}) = 0$ for each $x \in X$, there exist only a finite number of pairwise disjoint μ-nonmeasurable sets, and one can put $\mu(X) = 1$.[11]

4.42. An algebra and a measure from the previous section can exist only under the assumption that $\#(X)$ is a σ-measurable cardinal. This follows from the following simple statement.

STATEMENT. *If $\#(X)$ is a σ-nonmeasurable cardinal, $\mathcal{A}$ is an almost σ-algebra, $\mathcal{A} \neq \mathfrak{P}(X)$, and $\{x\} \in \mathcal{A}$ for all $x \in X$, then there exist $\aleph_0$ pairwise disjoint sets not belonging to $\mathcal{A}$.*

PROOF. It is clear that $\varnothing \neq \ker \mathcal{A} \subset \beta X \setminus X$. If $\#(\ker \mathcal{A}) < \aleph_0$, then each point of $\ker \mathcal{A}$ is irregular. Hence $\ker \mathcal{A}$ is an infinite set. □

---

[10] Here one should point out the paper **[M]** and the notion of Q-space or functionally closed space (there are many other names for these spaces) introduced by Hewitt in **[He]**. By definition, any functionally closed space is a completely regular topological space. In **[Mr1]** the following result is obtained: a completely regular space Y is functionally closed if and only if for any point $y_0 \in \beta Y \setminus Y$ there exists a continuous real function f on βY such that $f(y_0) = 0$ and $f(y) > 0$ for $y \in Y$.

[11] Let $\mathcal{A}$ be our algebra, and let $\mathcal{A}^*$ be the Boolean algebra $\mathcal{A}/I_{\mathcal{A}}$, where $I_{\mathcal{A}}$ is the ideal of all sets M such that $\mathfrak{P}(M) \subset \mathcal{A}$. (The definition of an ideal is given in Chapter 6; the quotient of an algebra by an ideal is defined in the natural way.) On $\mathcal{A}^*$ one naturally defines a measure μ which is finite, strictly positive (i.e., takes the value 0 at the zero element only), and σ-additive. It is clear that here we have $\#(\mathcal{A}^*) < \aleph_0$. The following problem was also studied: under which conditions does there exist a strictly positive finite measure on a Boolean algebra (and when can such a measure be made σ-additive)? See the book **[Sik]** and papers **[Ho-Hor]**, **[Hor-T]**, **[Kel]**, **[Ma]**.

4.43. In connection with our considerations, it makes sense to make the following remark, which does not pretend to be new.

REMARK. If x is a point of a topological space, then, as usual, the minimal cardinal number of kind $\#(B(x))$, where $B(x)$ is a base of x, is called the *character* of x and denoted by $\chi(x)$. The character of a regular point a can be large and is always greater than $\aleph_0$; this is true despite the existence of a countable sequence of subsets $L_1, \ldots, L_n, \ldots$ of X such that $\bigcap L_n = \varnothing$ and $\overline{L}_n \ni a$ for each n. Let $x \in \beta X \setminus X$, and let $\varkappa$ be the minimal cardinal of the type of $\#(M)$, where $M \subset X$ and $\overline{M} \ni x$. If $\{M_\xi\}$ is a family consisting of $\varkappa$ subsets of X and $\overline{M}_\xi \ni x$ for each ξ, then one can find $A \subset X$ such that $A \cap M_\xi \neq \varnothing$ and $(X \setminus A) \cap M_\xi \neq \varnothing$ for each ξ. Therefore $\chi(x) > \varkappa$.

4.44. REMARK. In Chapter 8 finite sequences of algebras $\mathcal{A}_1, \ldots, \mathcal{A}_n$ are studied such that $\#(\ker \mathcal{A}_k) > \frac{3}{2}(k-1)$ for all $k \in [1, n]$. The statements and proofs of the results given there are similar to the corresponding results of the present chapter.

CHAPTER 5

Countable Sequences of Algebras (1). Proof of Theorem 2.4

5.1. DEFINITION. An algebra $\mathcal{A}$ is said to be *simple* if there exists $Z \subset \beta X$ such that:

(1) $\#(Z) \leq \aleph_0$;
(2) if $Z \neq \varnothing$, then all points of Z are irregular;
(3) $\ker \mathcal{A} \subset \overline{Z}$.

With this definition, the algebra $\mathfrak{P}(X)$ is simple.

Although this chapter contains the proof of one of the main results (Theorem 2.4), the central result of the chapter is Theorem 5.17; due to this theorem, the crucial point in studying countable sequences of almost σ-algebras is the consideration of simple algebras.

5.2. The following theorem will be used below.

THEOREM. *Let $\mathcal{A}_1, \dots, \mathcal{A}_k, \dots$ be a countable sequence of algebras, and let there exist a matrix*

$$\begin{pmatrix} U_1^1 & & & \\ U_1^2 & U_2^2 & & \\ \dots & \dots & \dots & \dots \\ U_1^k & \dots & U_k^k & \\ \dots & \dots & \dots & \dots \end{pmatrix}$$

of pairwise disjoint sets such that $U_i^k \notin \mathcal{A}_k$. Let $W \subset X \setminus \bigcup_{k,i} U_i^k$. Then there exists a set U such that either $U = \varnothing$ or

$$U = \bigcup_i U_{\beta_i}^{\alpha_i}, \quad \textit{where } \alpha_1 < \alpha_2 < \dots < \alpha_i < \dots \textit{ is a finite or countable sequence,}$$

and $W \cup U \notin \mathcal{A}_k$ for all k.[1]

PROOF. Since

$$(W \cup U_1^1) \setminus W = U_1^1 \notin \mathcal{A}_1,$$

among the two sets $W, W \cup U_1^1$ there is at least one not belonging to $\mathcal{A}_1$. Denote it by V_1. Assume that there exists a finite or infinite sequence of sets

$$U_{\beta_1^1}^{\alpha_1^1}, \dots, U_{\beta_m^1}^{\alpha_m^1}, \dots \quad (1 < \alpha_1^1 < \dots < \alpha_m^1 < \dots)$$

[1] Our theorem can be strengthened. Its statement is true if the matrix of pairwise disjoint sets U_i^k satisfies the conditions of Theorem 2.21(1). The proof of the theorem in this form is almost the same as the proof of Theorem 2.21(1). Hence Theorem 5.2 is a particular case of Theorem 2.21(1). Nevertheless, it makes sense to present this theorem here. The reason is that the proof of Theorem 2.21(1) is more complicated and is based on the Main Statement 3.10, whereas the proof of our theorem is simple and quite elementary.

such that

$$\mathcal{A}_1 \ni V_1 \cup \bigcup_m U_{\beta_m^1}^{\alpha_m^1}.$$

We call all the sets $U_{\beta_1^1}^{\alpha_1^1}, \dots, U_{\beta_m^1}^{\alpha_m^1}, \dots$ *marked.* Now consider the sets U_1^2, U_2^2 from the second row of the matrix. At least one of them, say U_2^2, is not marked. Since

$$(V_1 \cup U_2^2) \setminus V_1 = U_2^2 \notin \mathcal{A}_2,$$

among the two sets $V_1, V_1 \cup U_2^2$ there is at least one not belonging to $\mathcal{A}_2$. Denote it by V_2. Assume that there exists a finite or infinite sequence of sets

$$U_{\beta_1^2}^{\alpha_1^2}, \dots, U_{\beta_m^2}^{\alpha_m^2}, \dots, \quad \text{where } 2 < \alpha_1^2 < \dots < \alpha_m^2 < \cdots,$$

such that

$$\mathcal{A}_2 \ni V_2 \cup \bigcup_m U_{\beta_m^2}^{\alpha_m^2}.$$

Then we call all the sets $U_{\beta_1^2}^{\alpha_1^2}, \dots, U_{\beta_m^2}^{\alpha_m^2}, \dots$ marked.

Continuing our construction inductively, for $k > 2$ consider a set

$$V_{k-1} \subset W \cup \bigcup_{i,l<k} U_i^l$$

and the k-th row $U_1^k, \dots, U_k^k$ of the matrix, which contains a nonmarked set U_j^k. Since

$$(V_{k-1} \cup U_j^k) \setminus V_{k-1} = U_j^k \notin \mathcal{A}_k,$$

among the two sets $V_{k-1}, V_{k-1} \cup U_j^k$ there is at least one not belonging to $\mathcal{A}_k$. Denote it by V_k. Assume that there exists a finite or infinite sequence of sets

$$U_{\beta_1^k}^{\alpha_1^k}, \dots, U_{\beta_m^k}^{\alpha_m^k}, \dots \quad (k < \alpha_1^k < \dots < \alpha_m^k < \cdots)$$

such that

$$\mathcal{A}_k \ni V_k \cup \bigcup_m U_{\beta_m^k}^{\alpha_m^k}.$$

Then we call all the sets $U_{\beta_1^k}^{\alpha_1^k}, \dots, U_{\beta_m^k}^{\alpha_m^k}, \dots$ marked.

Consider all the rows of the matrix, and let us construct a sequence of sets

$$V_1 \subset V_2 \subset \dots \subset V_k \subset \cdots.$$

Put

$$U = \bigcup_k V_k \setminus W.$$

It is clear that either $U \neq \varnothing$ or $U = \bigcup_i U_{\beta_i}^{\alpha_i}$, where $\alpha_1 < \alpha_2 < \dots < \alpha_i < \cdots$ is a finite or countable sequence. Let us prove that $W \cup U \notin \mathcal{A}_k$. This is clear if a sequence of sets $U_{\beta_1^k}^{\alpha_1^k}, \dots, U_{\beta_m^k}^{\alpha_m^k}, \dots$ does not exist. Assume that such a sequence exists and $W \cup U \in \mathcal{A}_k$. Then

$$(W \cup U) \cap (V_k \cup \bigcup_m U_{\beta_m^k}^{\alpha_m^k}) = V_k \in \mathcal{A}_k,$$

a contradiction. □

5.3. The following lemma will be used below.

LEMMA. *If*

$$\begin{pmatrix} U_1^1 & \dots & U_{n_1}^1 \\ \dots & \dots & \dots \\ U_1^k & \dots & U_{n_k}^k \\ \dots & \dots & \dots \end{pmatrix}$$

is an infinite matrix of pairwise disjoint subsets of X, $n_k \to \infty$, $Z \subset \beta X$, $\#(Z) \le \aleph_0$, *and* $Z \cap \overline{U_i^k} = \varnothing$ *for all* U_i^k, *then there exists a sequence* $U_{i_1}^1, \dots, U_{i_k}^k, \dots$ *for which* $Z \cap \overline{\bigcup U_{i_k}^k} = \varnothing$.[2]

PROOF. Let $Z = \{z_1, \dots, z_m, \dots\}$. For each point z_m let us take a sequence of sets

$$U_{r_1^m}^{l_1^m}, \dots, U_{r_j^m}^{l_j^m}, \dots, \quad l_1^m < \dots < l_j^m < \dots$$

(if it exists, of course), such that

$$z_m \in \overline{\bigcup_j U_{r_j^m}^{l_j^m}}.$$

These sequences of sets can be chosen in such a way that in the k-th row of the matrix there exists a set $U_{i_k}^k$ which does not belong to any of the sequences. Since X has the discrete topology, we have

$$Z \cap \overline{\bigcup U_{i_k}^k} = \varnothing. \qquad \square$$

5.4. The next theorem (see 5.11), one of the most important in this chapter, is preceded by three definitions and four more sections. Let us recall that a subset of a topological space is called *separable* if it contains an at most countable everywhere dense set.

DEFINITION. A subset C of a topological space Y is said to be *externally separable* if there exists an at most countable set $Z \subset Y$ such that $C \subset \overline{Z}$. A subset of Y which is not externally separable is said to be *externally nonseparable.*

5.5. DEFINITION. Let C be a subset of a topological space Y and $y_0 \in Y$. If for each neighborhood $U(y_0)$ the set $U(y_0) \cap C$ is externally nonseparable, then y_0 is called a *point of inseparability* of C.

5.6. If Y is a compact set, then a point of inseparability of C exists if and only if C is an externally nonseparable set.

5.7. Below by externally separable and externally nonseparable sets we mean subsets of βX; points of inseparability always belong to βX.

5.8. DEFINITION. An algebra $\mathcal{A}$ is said to be *separable* (*nonseparable*) if $\ker \mathcal{A}$ is an externally separable (resp. nonseparable) set.

5.9. An algebra $\mathcal{A}$ is separable if and only if there exists a not more than countable set of ultrafilters $\{a_\lambda\}$ such that if $M \notin a_\lambda$ for all λ, then $M \in \mathcal{A}$.

[2] Since X has the discrete topology, the equality $Z \cap \overline{\bigcup U_{i_k}^k} = \varnothing$ clearly implies $\overline{Z} \cap \overline{\bigcup U_{i_k}^k} = \varnothing$.

5.10. If there exist $\aleph_1$ pairwise disjoint sets not belonging to an algebra, then, obviously, this algebra is nonseparable. It is also obvious that if an algebra is nonseparable, then there exist $\aleph_0$ pairwise disjoint sets not belonging to it. Let us give an example of a nonseparable σ-algebra such that there do not exist $\aleph_1$ pairwise disjoint sets not belonging to it. Using the result of Solovay [**So**] mentioned in 1.4, consider a σ-additive extension ν of the Lebesgue measure to all subsets of $(0,1)$. Consider the following σ-two-valued measure μ on $(0,1)$: $\mu(M) = 0$ if and only if $\nu(M) = 0$. The required algebra is the algebra $\mathcal{A}_\mu$. A similar example can be constructed in a clear way using Shelah's result [**S5**] mentioned in 6.12.

5.11. THEOREM. *Let $\mathcal{A}_1, \dots, \mathcal{A}_k, \dots$ be a countable sequence of algebras, and let $\bigcup\{\mathcal{A}_k \mid \mathcal{A}_k$ is a separable algebra$\} \neq \mathfrak{P}(X)$. Then $\bigcup \mathcal{A}_k \neq \mathfrak{P}(X)$.*

PROOF. Assume, without loss of generality, that there is a countable subsequence $\mathcal{A}_{k_1}, \dots, \mathcal{A}_{k_n}, \dots$ of all nonseparable algebras from our sequence $\mathcal{A}_1, \dots, \mathcal{A}_k, \dots$. Let y_n be a point of inseparability of $\ker \mathcal{A}_{k_n}$. Take $Z \subset \beta X$ such that $\#(Z) \leq \aleph_0$, Z contains all the points y_n, and $\ker \mathcal{A}_k \subset \overline{Z}$ if the algebra $\mathcal{A}_k$ is separable. Take a set $U_1^1 \notin \mathcal{A}_{k_1}$ such that $Z \cap \overline{U_1^1} = \varnothing$. Since $y_2 \in Z$, one can take pairwise disjoint sets

$$U_1^2, U_2^2, U_3^2, U_4^2 \notin \mathcal{A}_{k_2}$$

such that

$$Z \cap \overline{\bigcup_{i=1}^{4} U_i^2} = \varnothing \quad \text{and} \quad U_1^1 \cap \bigcup_{i=1}^{4} U_i^2 = \varnothing.$$

One constructs in the obvious way an infinite matrix of pairwise disjoint sets

$$\begin{pmatrix} U_1^1 & & \\ \dots & \dots & \dots \\ U_1^n & \dots & U_{n^2}^n \\ \dots & \dots & \dots \end{pmatrix}$$

such that $U_i^n \notin \mathcal{A}_{k_n}$ and $Z \cap \overline{U_i^n} = \varnothing$ for all U_i^n. Put

$$V_i^n = \bigcup_{j=i(n-1)+1}^{in} U_j^n$$

for each $i \in [1, n]$. By Lemma 5.3, there exists a sequence $V_{i_1}^1, \dots, V_{i_n}^n, \dots$ such that

$$Z \cap \overline{\bigcup V_{i_n}^n} = \varnothing.$$

Assume that $i_n = 1$ for all n. Thus, we have an infinite matrix of pairwise disjoint sets

$$\mathfrak{U} = \begin{pmatrix} U_1^1 & & \\ \dots & \dots & \dots \\ U_1^n & \dots & U_n^n \\ \dots & \dots & \dots \end{pmatrix}$$

such that $U_i^n \notin \mathcal{A}_{k_n}$ and

$$Z \cap \overline{\bigcup\{U_i^n \mid U_i^n \in \mathfrak{U}\}} = \varnothing.$$

By the assumption of the theorem, there exists a set W such that $W \notin \mathcal{A}_k$ for all separable algebras $\mathcal{A}_k$. Since $\ker \mathcal{A}_k \subset \overline{Z}$ for all separable algebras $\mathcal{A}_k$, one can assume that

$$W \cap \bigcup \{U_i^n \mid U_i^n \in \mathfrak{U}\} = \varnothing.$$

By Theorem 5.2 there exists a set U such that $W \cup U \notin \mathcal{A}_{k_n}$ for all n and $Z \cap \overline{U} = \varnothing$. Let $\mathcal{A}_k$ be a separable algebra. Since $\ker \mathcal{A}_k \subset \overline{Z}$, we have $W \cup U \notin \mathcal{A}_k$. □

5.12. In the beginning of this chapter, the definition of a simple algebra was given. It is evident that a simple algebra is separable. It is also evident that the converse is not true. (Here we are talking about algebras without any conditions. The following question is nontrivial: is a separable almost σ-algebra simple (see 5.21)?) Now it is natural to give the following definition, after which we proceed to the arguments necessary for the proof of Theorem 5.17.

DEFINITION. A separable algebra is said to be *strictly separable* if it is not simple.[3]

5.13. Consider a strictly separable almost σ-algebra $\mathcal{A}$.[4] There exists a subset $Z \subset \beta X$ such that $\#(Z) = \aleph_0$ and $\ker \mathcal{A} \subset \overline{Z}$. Put

$$Z' = \{z \in Z \mid z \text{ is an irregular point}\},$$
$$Z'' = \{z \in Z \mid z \text{ is a regular point}\}.$$

Obviously, $\#(Z'') = \aleph_0$, $\overline{Z''} \supset \ker \mathcal{A} \setminus \overline{Z'} \neq \varnothing$, $Z'' \subset \beta X \setminus X$, and for each point $z \in Z''$ there exists a continuous function f on βX such that $f(z) = 0$ and $f(x) > 0$ if $x \in X$. Let $U \subset X$ satisfy $Z' \cap \overline{U} = \varnothing$, $\overline{U} \cap \ker \mathcal{A} \neq \varnothing$, and

$$Z_U = Z \cap \overline{U} = Z'' \cap \overline{U}.$$

Since $\mathcal{A}$ is an almost σ-algebra, let us define on X a σ-two-valued measure $\mu_{\mathcal{A}}^U$ by setting $\mu_{\mathcal{A}}^U(M) = 0$ if and only if $\mathfrak{P}(M \cap U) \subset \mathcal{A}$. Obviously, $\mu_{\mathcal{A}}^U(M) = 0$ if and only if

$$\overline{M \cap U} \cap \ker \mathcal{A} = \varnothing.$$

Now let us consider a countable sequence of strictly separable almost σ-algebras $\mathcal{A}_1, \dots, \mathcal{A}_k, \dots$. For each algebra $\mathcal{A}_k$ consider the corresponding sets $Z^k, U_k, Z_{U_k}^k$, and the σ-two-valued measure $\mu_{\mathcal{A}_k}^{U_k}$. Put $E = \bigcup_k Z_{U_k}^k$. Obviously, $\#(E) = \aleph_0$. Let $E = \{e_1, \dots, e_n, \dots\}$. For each point e_n consider a continuous function f_n on βX such that $f_n(e_n) = 0$ and $f_n(x) > 0$ if $x \in X$. Consider a one-to-one correspondence between the cardinal $2^{\aleph_0}$ and the set of all countable sequences of positive real numbers. If $\iota \in 2^{\aleph_0}$ and ι corresponds to the sequence $\alpha_1, \dots, \alpha_n, \dots$, then put

$$P_\iota = \{x \in X \mid f_1(x) \geq \alpha_1, \dots, f_n(x) \geq \alpha_n, \dots\}.$$

We know that $2^{\aleph_0}$ is a completely ordered set. If ι_0 is the first element of $2^{\aleph_0}$, then put $Q_{\iota_0} = P_{\iota_0}$; if $\iota_0 < \iota \in 2^{\aleph_0}$, then put $Q_\iota = P_\iota \setminus \bigcup_{\lambda < \iota} P_\lambda$. For each $k \in \mathbb{N}^+$ define a σ-two-valued measure μ_k on $2^{\aleph_0}$: $\mu_k(M) = 0$ if and only if

$$\mu_{\mathcal{A}_k}^{U_k}\left(\bigcup_{\iota \in M} Q_\iota\right) = 0.$$

[3] Since the algebra $\mathfrak{P}(X)$ is simple, it is not strictly separable.

[4] Note that an almost σ-algebra is separable if and only if the set $\overline{\ker \mathcal{A}}$ is separable.

By the Gitik–Shelah theorem (formally by the Gitik–Shelah theorem and Theorem 1.19) there exist pairwise disjoint sets

$$B_1, \ldots, B_k, \ldots \subset 2^{\aleph_0}$$

such that B_k is a μ_k-nonmeasurable set. Put

$$B_k^* = U_k \cap \bigcup_{\iota \in B_k} Q_\iota.$$

Clearly, $B_1^*, \ldots, B_k^*, \ldots$ are pairwise disjoint subsets of X. Besides that, the set $\overline{B_k^*} \cap \ker \mathcal{A}_k \neq \varnothing$ and contains only regular points.[5] Hence

$$\#(\overline{B_k^*} \cap \ker \mathcal{A}_k) \geq \aleph_0.$$

It is obvious that if $\mathfrak{C}$ is a certain set consisting of irregular points of βX and $\#(\mathfrak{C}) \leq \aleph_0$, then the sets $U_1, \ldots, U_k, \ldots$ can be chosen in such a way that $\mathfrak{C} \cap \overline{U_k} = \varnothing$ for all k. It is also obvious that B_k^* contains $\aleph_0$ pairwise disjoint sets each of which does not belong to $\mathcal{A}_k$.

5.14. The results presented in this and the next two sections are not related to the proof of Theorem 5.17. However, it is natural to present them here.

We continue the discussion of properties of kernels started in 3.20. Using the notions introduced in the previous section, let us prove that

$$\#(\overline{B_k^*} \cap \ker \mathcal{A}_k) = \varkappa = 2^{\mathfrak{c}}.$$

Since $\overline{B_k^*} \cap \ker \mathcal{A}_k$ is an externally separable set, $\varkappa \leq 2^{\mathfrak{c}}$. Consider a discrete set $M \subset \overline{B_k^*} \cap \ker \mathcal{A}_k$ such that $\#(M) = \aleph_0$. Assume, without loss of generality, that M does not contain $\mathcal{A}_k$-special ultrafilters. There exist sets of ultrafilters M', M'' such that $M' \subset M$, $\#(M') = \aleph_0$, $\overline{M'} \cap \overline{M''} = \varnothing$, and for each ultrafilter from M' there exists an $\mathcal{A}_k$-similar ultrafilter in M''. Clearly, $\overline{M'} \subset \ker \mathcal{A}_k$. But $\#(\overline{M'}) = \#(\beta\mathbb{N}) = 2^{\mathfrak{c}}$. (The fact that $\#(\overline{M'}) = \#(\beta\mathbb{N})$ is obvious; the equality $\#(\beta\mathbb{N}) = 2^{\mathfrak{c}}$ is a well-known fact in general topology appearing in several textbooks (see also [**Po**], [**Mr2**]).) So we have proved that $\varkappa = 2^{\mathfrak{c}}$. This statement does not make the proof of Theorem 5.17 easier, but it is used below in the proof of other theorems. These considerations imply two statements:

(1) there does not exist an algebra $\mathcal{A}$ such that

$$\aleph_0 \leq \#(\ker \mathcal{A}) < 2^{\mathfrak{c}};$$

(2) $\ker \mathcal{A}$ is a sequentially compact subset of βX. (Recall the notion introduced by Fréchet in [**Fre**]: a topological space is called *sequentially compact* if from any of its countable open coverings one can choose a finite subcovering.)

5.15. The above considerations of this chapter easily imply the following result, which should be compared with Theorem 7.25.

STATEMENT. *Let $\mathcal{A}_1, \ldots, \mathcal{A}_k, \ldots$ be a countable sequence of almost σ-algebras, and for each k let there exist k pairwise disjoint sets not belonging to $\mathcal{A}_k$. Then there exists a countable sequence of pairwise disjoint sets $U_1, \ldots, U_k, \ldots$ such that $U_k \notin \mathcal{A}_k$.*

[5] In fact we only need to know that among its points at least one is regular.

5.16. Statement 5.15 and Statement 3.18 imply the following interesting result.

STATEMENT. *Let $\mathcal{A}_1, \dots, \mathcal{A}_k, \dots$ be a countable sequence of almost σ-algebras, and for each k let there exist k pairwise disjoint sets not belonging to $\mathcal{A}_k$. Then there exists a countable sequence of pairwise disjoint sets $U_1, \dots, U_k, \dots$ $(U_k \notin \mathcal{A}_k)$, and for any real number $\rho < \frac{2}{3}$ there exist sets $\mathbb{N}_* \subset \mathbb{N}^* \subset \mathbb{N}^+$ possessing the following properties:*

(i) $p(\mathbb{N}^*) \geq \frac{2}{3}$;
(ii) $p_a(\mathbb{N}^*) > \rho$;
(iii) *if $k \in \mathbb{N}^*$, then $\mathcal{A}_k \not\ni \bigcup_{i \in \mathbb{N}_*} U_i$.*

5.17. THEOREM. *Let $\mathcal{A}_1, \dots, \mathcal{A}_k, \dots$ be a countable sequence of almost σ-algebras, and let $\bigcup\{\mathcal{A}_k | \mathcal{A}_k$ is a simple algebra$\} \neq \mathfrak{P}(X)$. Then $\bigcup \mathcal{A}_k \neq \mathfrak{P}(X)$.*

PROOF. By Theorem 5.11, one can assume that all the algebras $\mathcal{A}_k$ are separable. Assume, without loss of generality, that all strictly separable algebras from our sequence $\mathcal{A}_1, \dots, \mathcal{A}_k, \dots$ form a countable subsequence $\mathcal{A}_{k_1}, \dots, \mathcal{A}_{k_n}, \dots$. Let $\mathfrak{C}$ be a set of irregular points, $\#(\mathfrak{C}) \leq \aleph_0$, and $\ker \mathcal{A}_k \subset \overline{\mathfrak{C}}$ if $\mathcal{A}_k$ is a simple algebra. By the arguments from 5.13, there exists a matrix of pairwise disjoint sets

$$\begin{pmatrix} U_1^1 & & \\ \dots & \dots & \dots \\ U_1^n & \dots & U_n^n \\ \dots & \dots & \dots \end{pmatrix}$$

such that $U_i^n \notin \mathcal{A}_{k_n}$ and

$$\mathfrak{C} \cap \overline{\bigcup_{n,i} U_i^n} = \varnothing.$$

Arguing as at the end of the proof of Theorem 5.11 and using Theorem 5.2, we obtain the proof of our theorem. $\square$

5.18. The strength of Theorem 5.17 is demonstrated by the following obvious corollary.

THEOREM. *Let $\mathcal{A}_1, \dots, \mathcal{A}_k, \dots$ be a countable sequence of almost σ-algebras, each of which is not equal to $\mathfrak{P}(X)$ and contains all one-point sets. Assume that $\#(X)$ is a σ-nonmeasurable cardinal. Then $\bigcup \mathcal{A}_k \neq \mathfrak{P}(X)$.*

5.19. REMARK. Theorem 5.18 is a particular case of Theorem 2.8, since for the algebra $\mathcal{A}_k$ from Theorem 5.18 there exist $\aleph_0$ pairwise disjoint sets not belonging to $\mathcal{A}_k$.

5.20. Suppose that in the assumptions of Theorem 5.18, $\#(X) = \mathfrak{c}$. Then the proof of Theorem 5.18 follows immediately from the Gitik–Shelah theorem and Theorem 5.2. Indeed, for each k let us define a σ-two-valued measure μ_k as follows: $\mu_k(M) = 0$ if and only if $\mathfrak{P}(M) \subset \mathcal{A}_k$. By the Gitik–Shelah theorem, there exists a countable sequence of pairwise disjoint sets $M_1, \dots, M_k, \dots$ such that M_k is a μ_k-nonmeasurable set. This implies that one can construct a matrix of pairwise disjoint sets

$$\begin{pmatrix} U_1^1 & & \\ \dots & \dots & \dots \\ U_1^k & \dots & U_k^k \\ \dots & \dots & \dots \end{pmatrix}$$

such that $U_i^k \notin \mathcal{A}_k$. By Theorem 5.2, Theorem 5.18 is proved (under the assumption that $\#(X) = \mathfrak{c}$). □

5.21. In 5.13 we considered σ-two-valued measures μ_n on the interval $(0,1)$. It is clear that the kernel of the algebra $\mathcal{A}_{\mu_n}$ is a closed separable subset of $\beta X \setminus X$. (By X we mean here, of course, the interval $(0,1)$ with the discrete topology.) Therefore a strictly separable almost σ-algebra exists if and only if there exists a σ-two-valued measure on a set of cardinality continuum such that the algebra of all measurable sets is separable. If $\mathfrak{c} = \aleph_1$, then, by Theorem 1.6, there does not exist a strictly separable almost σ-algebra. In any model in which there does not exist a strictly separable almost σ-algebra, Theorem 5.11, in whose proof the Gitik–Shelah theorem is not used, is a generalization of the Gitik–Shelah theorem. In such a model Theorem 5.17 is an obvious corollary of Theorem 5.11. But it is known that there exists a model in which one can construct a σ-two-valued measure $\hat{\mu}$ on a set of cardinality continuum such that the algebra of all $\hat{\mu}$-measurable sets is separable (see [**Ba-D**] and the Appendix).[6]

5.22. Considering σ-algebras instead of almost σ-algebras leads to simplifications.

STATEMENT. *Let $\mathcal{A}_1, \dots, \mathcal{A}_k, \dots$ be a countable sequence of σ-algebras, and let $W, U_1, \dots, U_k, \dots$ be a countable sequence of pairwise disjoint sets; for each k let there exist a regular point $s_k \in \overline{U}_k \cap \ker \mathcal{A}_k$. Then there exists a countable sequence of sets $V_1, \dots, V_k, \dots$ such that:*

(i) *$V_k \subset U_k$;*
(ii) *if $\mathfrak{Q} \subset \{W, V_1, \dots, V_k, \dots\}$ and $V_k \in \mathfrak{Q}$, then $\mathcal{A}_k \not\ni \bigcup \mathfrak{Q}$.*

PROOF. The proof is based on the main property of a σ-algebra $\mathcal{A}$: if we have $\{L_n\}_{n \in \mathbb{N}^+} \subset \mathcal{A}$, then $\mathcal{A} \ni \bigcap_n L_n$. For each k consider a countable sequence of sets

$$U_k^1 \supset U_k^2 \supset \cdots \supset U_k^n \supset \cdots$$

such that

$$U_k^1 \subset U_k, \quad \bigcap_n U_k^n = \varnothing, \quad U_k^n \notin \mathcal{A}_k.$$

(Such sets exist, because there exists a regular point $s_k \in \overline{U}_k \cap \ker \mathcal{A}_k$.) Let us choose i_1 so that $W \cup U_1^{i_1} \notin \mathcal{A}_1$. Put $V_1 = U_1^{i_1}$. Take n_1 so that if

$$\mathfrak{Q} \subset \{W, U_2^{j_2}, \dots, U_k^{j_k}, \dots\}$$

and $n_1 \le j_k$ for all $k > 1$, then $\mathcal{A}_1 \not\ni V_1 \cup \bigcup \mathfrak{Q}$. Take $i_2 \ge n_1$ so that

$$\mathcal{A}_2 \not\ni W \cup V_1 \cup U_2^{i_2}, V_1 \cup U_2^{i_2}, W \cup U^{i_2}.$$

Put $V_2 = U_2^{i_2}$. Take $n_2 \ge n_1$ so that if

$$\mathfrak{Q} \subset \{W, V_1, U_3^{j_3}, \dots, U_k^{j_k}, \dots\}$$

and $n_2 \le j_k$ for all $k > 2$, then $\mathcal{A}_2 \not\ni V_2 \cup \bigcup \mathfrak{Q}$. Continuing our construction inductively in the natural way, we find all the required sets V_k. □

---

[6] After the arguments in 5.14 it is clear that $\#(\ker \mathcal{A}_{\hat{\mu}}) = 2^{\mathfrak{c}}$. In fact, such a measure is supported on a set of cardinality $< \mathfrak{c}$ (this will be shown in Chapter 6). We have mentioned the existence of such a σ-two-valued measure in 3.20.

5.23. If in Theorem 5.17 the algebras $\mathcal{A}_k$ are σ-algebras, then in its proof one can use Statement 5.22 instead of the simple but not completely trivial Theorem 5.2. However, in the proof of Theorem 5.17 we used Theorem 5.11, and in the proof of Theorem 5.11 we used Theorem 5.2.

Let $U \subset X$ and $\#(\overline{U} \cap \ker \mathcal{A}) \geq 2^{\mathfrak{c}}$. Then, according to 5.14, there exists a regular ultrafilter $a \in \ker \mathcal{A}$ with $a \ni U$. Therefore, if the $\mathcal{A}_k$'s are σ-algebras, in the proof of Theorem 5.11 one can use Statement 5.22 instead of Theorem 5.2.

5.24. Assume that in Theorem 5.18 $\#(X) = \mathfrak{c}$, and $\mathcal{A}_k$ are σ-algebras. As in 5.20, let us consider σ-two-valued measures μ_k, and construct, by the Gitik–Shelah theorem, a countable sequence of pairwise disjoint sets $U_1, \dots, U_k, \dots$ such that $U_k \notin \mathcal{A}_k$. Further, arguing as in the proof of Statement 5.22 (but here one can manage without ultrafilters and kernels of algebras), let us construct a countable sequence of pairwise disjoint sets $V_1, \dots, V_k, \dots$ possessing the following property:

$$\text{if } \mathfrak{Q} \subset \{V_1, \dots, V_k, \dots\} \text{ and } V_k \in \mathfrak{Q}, \text{ then } \mathcal{A}_k \not\ni \bigcup \mathfrak{Q}.$$

5.25. REMARK. Below (see Theorem 17.1) it is shown that the result of the previous section can be improved. But we want to emphasize how simply the result of the previous section follows from the Gitik–Shelah theorem.

5.26. Sections 5.26, 5.27, 5.28 are devoted to certain very interesting problems. First let us repeat a well-known definition.

DEFINITION. An uncountable subset M of the set of real numbers is called a *Q-set* if any subset of M is an F_σ-set in the induced topology on M.[7]

Do Q-sets exist? (This question was first asked by Hausdorff; we should mention that the notion of a Q-set was also introduced by Hausdorff.) If $2^{\aleph_0} < 2^{\aleph_1}$, then, obviously, they do not exist. However, even in models with $2^{\aleph_0} = 2^{\aleph_1}$, Q-sets do not necessarily exist (see [**Bu**]). The first paper in which the consistency of the existence of Q-sets is shown was [**Mar-So**]. Early papers on Q-sets are [**Hau2**], [**Si2**], [**Ro**]. Among later papers, let us mention [**F**], which is mentioned in Chapter 6. Incidentally, in Chapter 6 the following well-known statement is proved: the Lebesgue measure of a Q-set equals 0.

5.27. Let $X = (0, 1)$, and let $M \subset (0, 1)$ be a Q-set; let μ be a σ-two-valued measure such that $\mu(M) = 1$. Define an algebra $\mathcal{A}$: its generators are all μ-measurable sets and all intervals $[\alpha, \beta] \subset (0, 1)$. If M' is a μ-nonmeasurable set, then M' contains a subset not belonging to $\mathcal{A}$. It is clear that $\mathcal{A}$ contains all one-element sets. It is also clear that $\mathcal{A}$ is an almost σ-algebra but not a σ-algebra, and that the minimal σ-algebra $\overline{\mathcal{A}} \supset \mathcal{A}$ is $\mathfrak{P}(X)$.

5.28. In 5.21 it was mentioned that in some models one can construct a σ-two-valued measure μ on $(0, 1)$ such that the algebra $\mathcal{A}_\mu$ is separable. Denote by $\varkappa_\mu$ the minimal cardinal such that there exists a set M of cardinality $\varkappa_\mu$ with $\mu(M) = 1$. One can assume that if $M \subset (0, 1)$ and $\#(M) < \varkappa_\mu$, then $\mu(M) = 0$. By Theorem 1.6, $\aleph_1 < \varkappa_\mu$. In footnote 5.6 we have already mentioned that $\varkappa_\mu < \mathfrak{c}$. In the Appendix Saharon Shelah constructed, in a certain model, a σ-two-valued measure $\widetilde{\mu}$ on $(0, 1)$ such that:

(1) the algebra $\mathcal{A}_{\widetilde{\mu}}$ is separable;

[7]Recall that an F_σ-set is a union of a countable family of closed sets.

(2) if $M \subset (0,1)$ and $\aleph_0 < \#(M) < \varkappa_{\widetilde{\mu}}$, then M is a Q-set.

It would be interesting to construct a σ-two-valued measure $\widetilde{\mu}$ on $(0,1)$ that satisfies condition (1) and the following condition:

(2*) there exists a Q-set $\widetilde{M}$ such that $\widetilde{\mu}(\widetilde{M}) = 1$.

5.29. Let us now proceed to the proof of Theorem 2.4. First let us recall a well-known definition.

DEFINITION. A map φ from a topological space Y onto a topological space Y' is called *irreducible* if for any closed set $F \subset Y$ $(F \neq Y)$ we have $\varphi(F) \neq Y'$.

5.30. The following theorem is a well-known fact from general topology (see e.g. [**E**]).

THEOREM. *Let K be a separable compact set. Then there exists a compact set $A \subset \beta\mathbb{N}$ and a continuous irreducible map from A onto K.*

PROOF. Let $S \subset K$, $\#(S) \leq \aleph_0$, and $\overline{S} = K$. Let us consider only the nontrivial case when $\#(S) = \aleph_0$. Let φ be a one-to-one map from $\mathbb{N}$ onto S, and let $x \in K$. Put

$$\mathfrak{f}_x = \{M \subset \mathbb{N} \mid x \in \overline{\varphi(M)}\}.$$

Let us extend the map φ to $\beta\mathbb{N}$ so that $\beta\mathbb{N}$ is mapped onto K: all ultrafilters on $\mathbb{N}$ containing the filter $\mathfrak{f}_x$ are mapped to x. Let U be an open subset of K, and let $\{V_\alpha\}$ be an open base of U such that $\overline{V_\alpha} \subset U$ for each α. Then

$$\varphi^{-1}(U) = \bigcup_\alpha \{a \text{ is an ultrafilter on } \mathbb{N} \mid a \ni \varphi^{-1}(V_\alpha \cap S)\}.$$

So $\varphi^{-1}(U)$ is a union of open sets, hence is itself an open set. We have proved that the map $\varphi\colon \beta\mathbb{N} \to K$ is continuous. The required compact set A, such that φ is an irreducible map from A onto K, exists due to the classical Kuratowski–Zorn lemma.[8] □

5.31. Let us recall one more well-known definition.

DEFINITION. A family $\{U_\alpha\}$ of open subsets of a topological space is called a *π-base* if for any open subset $U \neq \varnothing$ there exists $U_\alpha \subset U$, $U_\alpha \neq \varnothing$.

5.32. PROOF OF THEOREM 2.4(1). Let K be a compact set without isolated points admitting a countable π-base, and let $K \neq \varnothing$. Of course, K is a separable compact set. For K one can take the interval $[0,1]$ with the natural topology. By Theorem 5.30 there exists a corresponding compact set $A \subset \beta\mathbb{N}$. Since K does not have isolated points, A also does not have isolated points and $A \subset \beta\mathbb{N} \setminus \mathbb{N}$. Since K has a countable π-base, A also has a countable π-base $U_1, \dots, U_k, \dots$. For each $k \in \mathbb{N}^+$ let us take a discrete subset $D_k \subset U_k$ with $\#(D_k) = \aleph_0$. Put $E_k = \overline{D_k} \setminus D_k$. Define the two-valued measure μ_k on $\mathbb{N}$ as follows: $\mu_k(M) = 0$ if and only if $\overline{M} \cap E_k = \varnothing$.[9] It is clear that each subset of $\mathbb{N}$ is μ_k-measurable

---

[8]Recall the Kuratowski–Zorn lemma, which is equivalent to the Axiom of Choice. If a set P has an ordering $\geq$ and for each linearly ordered subset L of P there exists an element l_0 such that $l \geq l_0$ for all $l \in L$, then P has a minimal element. In our case P is the set of all closed subsets $F \subset \beta\mathbb{N}$ such that $\varphi(F) = K$. The ordering $\geq$ is defined as follows: $F_1 \geq F_2$ if and only if $F_1 \supseteq F_2$.

[9]In the proof of Theorem 2.4(1) $X = \mathbb{N}$, and it is natural that if $M \subset \mathbb{N}$, then $\overline{M}$ is the closure of M in $\beta\mathbb{N}$.

for some k. For each k consider a family $\{Q_\alpha^k\}_\alpha$ consisting of $\mathfrak{c}$ pairwise disjoint nonempty open subsets of E_k (see Remark 5.34). Let $M_\alpha^k \subset \mathbb{N}$ and

$$\varnothing \neq \overline{M_\alpha^k} \cap E_k \subset Q_\alpha^k.$$

The sets M_α^k satisfy all the conditions of Theorem 2.4(1). □

5.33. REMARK. The facts that A does not have isolated points and has a countable π-base $U_1, \dots, U_k, \dots$ immediately imply the existence of two-valued measures $\nu_1, \dots, \nu_k, \dots$ on $\mathbb{N}$ such that each subset of $\mathbb{N}$ is ν_k-measurable for some k, and for each k there exist $\aleph_0$ pairwise disjoint ν_k-nonmeasurable sets. Indeed, let us define ν_k as follows: $\nu_k(M) = 0$ if and only if $\overline{M} \cap U_k = \varnothing$.

5.34. REMARK. In the proof of Theorem 2.4(1) in 5.32 we used a well-known result presented in textbooks on general topology: there exist $\mathfrak{c}$ pairwise disjoint nonempty open subsets of $\beta\mathbb{N} \setminus \mathbb{N}$. In [**E**] it is said that a construction which implies this result is contained in the paper by Sierpiński [**Si1**]. The result itself is contained in the papers [**N-K**], [**Kat**]. For completeness, let us outline the construction from Katětov's paper [**Kat**]. Let us represent

$$\mathbb{N} = \{r \text{ is a rational number, } 0 < r < 1\}.$$

For each irrational number $t \in (0,1)$ choose a sequence

$$\mathbb{N}_t = \{r_1^{(t)}, \dots, r_n^{(t)}, \dots\} \subset \mathbb{N}$$

such that $r_n^{(t)} \to t$. The family

$$\{\overline{\mathbb{N}}_t \setminus \mathbb{N}\}$$

of open sets in $\beta\mathbb{N} \setminus \mathbb{N}$ satisfies all the required conditions.

5.35. In the assumptions of Theorems 2.4(2), 2.4(3), all algebras are nonseparable. Hence these theorems follow from Theorem 5.11.

5.36. Here it would be appropriate to construct measures μ', μ'' from footnote 2.3. Let us describe only the construction of the measure μ'. It is convenient to give the construction of the measure μ'' in Chapter 6. Let us take for μ' an arbitrary σ-two-valued measure defined on X, where $\#(X) = \aleph_1$, all considerations relating to a model for which $2^{\aleph_1} = \mathfrak{c}$.[10] By Theorem 1.6, there exist $\aleph_1$ pairwise disjoint μ'-nonmeasurable sets.

It is clear that an algebra satisfying the condition of Theorem 2.4(2) but not of Theorem 2.4(3) can exist only in a model where $2^{\aleph_1} = \mathfrak{c}$. It is clear that one can construct μ' without appealing to Theorem 1.6. Namely, let $\#(X) = \aleph_1$ and $\mu'(M) = 0$ if and only if $\#(M) \leq \aleph_0$ (clearly, $2^{\aleph_1} = \mathfrak{c}$).

5.37. Recall a well-known definition which first appeared in the paper [**St2**] by M. Stone.

DEFINITION. A topological space is said to be *extremally disconnected* if the closure of each of its open subsets is an open subset.

5.38. LEMMA. *Let $\mathcal{A}_1, \dots, \mathcal{A}_k, \dots$ be an at most countable sequence of separable almost σ-algebras. Then $\overline{\bigcup_k \ker \mathcal{A}_k}$ is an extremally disconnected compact set.*

[10] Such models were mentioned in 5.26.

PROOF. Let $\mathcal{A}$ be a separable almost σ-algebra. Let us prove that $\overline{\ker \mathcal{A}}$ is an extremally disconnected compact set. Let U be an open subset of $\overline{\ker \mathcal{A}}$, $\overline{\ker \mathcal{A}} \backslash \overline{U} \neq \varnothing$, and let $\xi_1, \ldots, \xi_n, \ldots$ be a countable everywhere dense subset of U. Let

$$V_n \subset X, \quad \xi_n \in \overline{V}_n, \quad \overline{V}_n \cap \overline{\ker \mathcal{A}} \subset U.$$

Put $V = \bigcup_n V_n$. Let

$$a \in \overline{W} \cap (\overline{\ker \mathcal{A}} \setminus \overline{U}),$$

where $W \subset X$ and $\overline{W} \cap \overline{U} = \varnothing$. Obviously, $\mathfrak{P}(W \cap V_n) \subset \mathcal{A}$. Since $\mathcal{A}$ is an almost σ-algebra, $\mathfrak{P}(W \cap V) \subset \mathcal{A}$. Therefore

$$\ker \mathcal{A} \cap \overline{W \cap V} = \varnothing, \quad a \notin \overline{V}.$$

Hence

$$\overline{V} \cap \overline{\ker \mathcal{A}} = \overline{U},$$

and $\overline{U}$ is an open subset of $\overline{\ker \mathcal{A}}$. Thus, $\overline{\ker \mathcal{A}}$ is an extremally disconnected compact set. After these arguments, the proof of the lemma in the general case involves no difficulties. □

5.39. REMARK. It is clear that βX is an extremally disconnected compact set. However, a closed subset $F \subset \beta X$ is not necessarily an extremally disconnected compact set. For example, $\beta\mathbb{N} \setminus \mathbb{N}$ is not an extremally disconnected compact set. This statement can be obtained using the result given in 5.34 on the existence of $\mathfrak{c}$ pairwise disjoint nonempty open subsets of $\beta\mathbb{N} \setminus \mathbb{N}$ (see [**G-Je**], [**E**]).

5.40. The following lemma is a well-known fact of general topology.

LEMMA. *Any separable extremally disconnected compact set can be homeomorphically embedded into $\beta\mathbb{N}$.*

PROOF. Let K be such a compact set. Using Theorem 5.30, consider a compact set $A \subset \beta\mathbb{N}$ and a continuous irreducible map φ from A onto K. Assume that $a, b \in A$, $a \neq b$, $\varphi(a) = \varphi(b)$. Let U, V be open subsets of A, $a \in U$, $b \in V$, $U \cap V = \varnothing$. Put

$$U' = K \setminus \varphi(A \setminus U), \quad V' = K \setminus \varphi(A \setminus V).$$

Clearly, $U' \cap V' = \varnothing$ and $\overline{U'} \cap \overline{V'} \ni \varphi(a)$. But this contradicts the extremal disconnectedness of K. Hence, φ is a one-to-one map from A onto K. Therefore, K is homeomorphic to A. □

5.41. Consider an at most countable sequence of almost σ-algebras $\mathcal{A}_1, \ldots, \mathcal{A}_k, \ldots$. Let $\mathcal{A}_{k_1}, \ldots, \mathcal{A}_{k_n}, \ldots$ be the subsequence of all separable algebras. By Lemmas 5.38 and 5.40 there exists a map φ which embeds $\overline{\bigcup_n \ker \mathcal{A}_{k_n}}$ homeomorphically into $\beta\mathbb{N}$. For each algebra $\mathcal{A}_{k_n}$ let us construct an algebra $\mathcal{B}_n$ on $\mathbb{N}$. A set M belongs to $\mathcal{B}_n$ if and only if for some $M' \in \mathcal{A}_{k_n}$ we have

$$\varphi(\overline{M'} \cap \ker \mathcal{A}_{k_n}) = \overline{M} \cap \varphi(\ker \mathcal{A}_{k_n}).$$

The sequence of algebras $\mathcal{A}_{k_1}, \ldots, \mathcal{A}_{k_n}, \ldots$ can be said to be similar to the sequence of algebras $\mathcal{B}_1, \ldots, \mathcal{B}_n, \ldots$. By Theorem 5.11, $\bigcup \mathcal{A}_k \neq \mathfrak{P}(X)$ if and only if we have $\bigcup \mathcal{B}_n \neq \mathfrak{P}(\mathbb{N})$. One can hardly obtain meaningful corollaries from this curious result. The reason is that in general $\mathcal{B}_n$ are not almost σ-algebras. On the other hand, let $\mathcal{A}_{l_1}, \ldots, \mathcal{A}_{l_n}, \ldots$ be the subsequence of all simple algebras of our sequence $\mathcal{A}_1, \ldots, \mathcal{A}_k, \ldots$. It is clear that the sequence of algebras $\mathcal{A}_{l_1}, \ldots, \mathcal{A}_{l_n}, \ldots$ is similar to a sequence of almost σ-algebras $\mathcal{B}'_1, \ldots, \mathcal{B}'_n, \ldots$ on $\mathbb{N}$ (here, of course, one need not use the map φ). By Theorem 5.17, $\bigcup \mathcal{A}_k \neq \mathfrak{P}(X)$ if and only if $\bigcup \mathcal{B}'_n \neq \mathfrak{P}(\mathbb{N})$.

5.42. For the proof of Theorem 2.4(1) I needed to know whether one can construct a nonempty subset of $\beta\mathbb{N}$ having no isolated points and admitting a countable π-base. I obtained the answer in Moscow in the winter of 1994 from Vyacheslav Malykhin. The answer is positive and such a set is an absolute of the interval $[0, 1]$, homeomorphically embedded into $\beta\mathbb{N}$. Recall that each compact set K is the continuous image of an extremally disconnected compact set A under a continuous map; A is called an absolute of K. Any two absolutes of K are homeomorphic. In the proof of Theorem 5.30 an absolute A of a separable compact set K is constructed. The first paper on absolutes is the well-known paper [**Gl**] by Gleason. (However, in [**Gl**] an absolute is called a projective space, in the fashion of homological algebra.)

CHAPTER 6

Proof of the Gitik–Shelah Theorem, and More from Set Theory[1]

6.1. Let $I \subset \mathfrak{P}(X)$; I is called an *ideal* (the notion of an ideal is dual to the notion of a filter) if:

(a) $M_1 \subset M_2 \in I$ implies $M_1 \in I$;
(b) $M_1, M_2 \in I$ implies $M_1 \cup M_2 \in I$;
(c) $\varnothing \in I$; $X \notin I$.

I is said to be *simple* if $\{x\} \in I$ for all $x \in X$. If $\{M \mid \#(M) < \#(X)\} \subset I$, then I is said to be *homogeneous*. I is said to be *ρ-additive* (ρ is a cardinal) if for each family $\{M_\lambda\}_{\lambda\in\Lambda} \subset I$, where $\#(\Lambda) < \rho$, we have $\bigcup_{\lambda\in\Lambda} M_\lambda \in I$; an $\aleph_1$-additive ideal is said to be *σ-additive*. If I is simple and σ-additive, then one naturally defines a σ-two-valued measure μ as follows: I is the system of all sets whose μ-measure is 0. Below *by an ideal we will always mean a simple and σ-additive ideal.* Let us introduce the following notation:

$$I^+ = \{M \mid M \notin I\}, \quad I^c = \{M \mid X \setminus M \in I\}.$$

Consider the algebra $I \cup I^c$, i.e., the algebra $\mathcal{A}_\mu$. If this algebra is separable, then I is said to be *separable.* Another name for a separable ideal is *σ-centered ideal*; see the Appendix. If I is not separable, then it is said to be *nonseparable.* If there exists a countable family

$$M_1, \dots, M_k, \dots \subset I^+$$

such that for any $Q \in I^+$ there exists a set M_k such that $M_k \setminus Q \in I$, then I is said to be *countably dense.* For I to be countably dense it is necessary and sufficient that the kernel of the algebra $I \cup I^c$, *which we denote by* $\ker(I)$, have a countable π-base. It is evident that any countably dense ideal is separable.

Take the minimal cardinal $\varkappa$ for which there exists a family $\{M_\alpha \in I \mid \alpha < \varkappa\}$ and $I^+ \ni \bigcup_{\alpha<\varkappa} M_\alpha$. Put

$$M'_0 = M_0, \quad M'_\alpha = M_\alpha \setminus \bigcup_{\beta<\alpha} M_\beta \quad (0 < \alpha < \varkappa).$$

Let us consider M'_α as elements of $\varkappa$.[2] An ideal I is naturally defined on $\varkappa$; I is $\varkappa$-additive, and hence homogeneous. Since I is σ-additive, $\varkappa \geq \aleph_1$.

[1] Recall what was mentioned in 1.18: we will give a proof of the Gitik–Shelah theorem following [**Ka**]. All essential points related to the proof of this theorem are known to the experts. The exception is the proof of Theorem 6.2. It appears likely that a purely mathematical proof of the result of 6.33 is published for the first time here.

[2] Instead of considering as elements of $\varkappa$ all ordinals less than $\varkappa$.

6.2. THEOREM.[3] *Let ρ be a cardinal such that $\rho \geq \aleph_1$. The following statements are equivalent:*

(a) *Any ideal defined on ρ is not countably dense.*
(b) *Any $\varkappa$-additive ideal defined on $\varkappa \leq \rho$ is not countably dense.*
(c) *For any countable sequence of σ-two-valued measures defined on ρ there exists a set nonmeasurable with respect to each of these measures.*

PROOF. (a) $\Longrightarrow$ (b). Obvious.

(b) $\Longrightarrow$ (c). Let $\mu_1, \ldots, \mu_k, \ldots$ be a countable sequence of σ-two-valued measures defined on ρ. Denote by I_k the ideal corresponding to the measure μ_k. As above, consider the ideal I_1 on $\varkappa$, where I_1 is a $\varkappa$-additive not countably dense ideal. Let us suppose that the construction of I_1 on $\varkappa$ involves, as above, a family of sets $\{M_\alpha \in I_1 \mid \alpha < \varkappa\}$. Assume without loss of generality that $\mu_k(M_\alpha) = 0$ for all $\alpha < \varkappa$ and $k > 1$, and

$$\mu_k\Big(\bigcup_{\alpha<\varkappa} M_\alpha\Big) \neq 0$$

for all $k > 1$. Therefore one can consider the ideals I_k on $\varkappa$ for all $k \in \mathbb{N}^+$, and these ideals considered on $\varkappa$ have kernels $\ker(I_k)$.

Case 1. *The ideal I_1 on $\varkappa$ is separable, i.e., the compact set* $\ker(I_1)$ *is separable.* The separability of $\ker(I_1)$ and σ-additivity of I_k imply that the set $\ker(I_1)\cap\ker(I_k)$ is open in $\ker(I_1)$. Since $\ker(I_1)$ does not have a countable π-base, there exists a μ_1-nonmeasurable set U_1 such that $\rho \setminus U_1$ is not a set of μ_k-measure zero for all $k > 1$.

Case 2. *The ideal I_1 on $\varkappa$ is nonseparable, i.e., the compact set* $\ker(I_1)$ *is nonseparable.* Let us take a point $z_k \in \ker(I_k)$ for each $k > 1$, and denote the set of these points by Z. Let us take a set $U^* \subset \varkappa$ such that $U^* \notin I_1, I_1^c$, and $\overline{U^*}\cap Z = \varnothing$. Put

$$U_1 = \bigcup\{M'_\alpha \mid M'_\alpha \in U^*\}.$$

It is clear that U_1 is a μ_1-nonmeasurable set and $\rho \setminus U_1$ is not a set of μ_k-measure zero for all $k > 1$.

Now the arguments are similar to those used in the proof of Theorem 1.19.

(c) $\Longrightarrow$ (a). Assume that a countably dense ideal I on ρ is given. Let us take a countable family $M_1, \ldots, M_k, \ldots \in I^+$ such that for any $Q \in I^+$ there exists M_k such that $M_k \setminus Q \in I$. For each M_k let us define a σ-two-valued measure μ_k on ρ: $\mu_k(M) = 0$ if and only if $M \cap M_k \in I$. Obviously, each subset of ρ is measurable with respect to some measure μ_k, a contradiction. □

6.3. By Theorem 6.2, for the proof of the Gitik–Shelah theorem it suffices to prove the following theorem.

THEOREM. *Any $\varkappa$-additive ideal defined on $\varkappa \leq \mathfrak{c}$ is not countably dense.*

6.4. Before passing to the proof of Theorem 6.3, we make the following remark.

REMARK. Ulam's result given in 1.5 in its general form sounds as follows: let $\rho^+ \geq \aleph_1$ (as usual, ρ^+ denotes the cardinal following the cardinal ρ). Then there exists a matrix $\{M_\xi^\alpha\}$ of subsets of ρ^+ having ρ rows and ρ^+ columns (α is the

[3]This theorem belongs to Taylor (see [**Ta**]) and is used in [**Gi-S1**]. In [**Gi-S1**], the paper [**S1**] is also cited in connection with this theorem.

index of a row, ξ is the index of a column) such that $M^\alpha_{\xi_1} \cap M^\alpha_{\xi_2} = \varnothing$ if $\xi_1 \neq \xi_2$, and $\#\big(\rho^+ \setminus \bigcup_\alpha M^\alpha_\xi\big) \leq \rho$. The sets M^α_ξ are constructed in the following way. For each $\beta < \rho^+$ let us take (if this is possible) a one-to-one map f_β from the set $\{\xi \mid \xi \leq \beta\}$ onto the set $\{\alpha \mid \alpha < \rho\}$, and put

$$M^\alpha_\xi = \{\xi \leq \beta < \rho^+ \mid f_\beta(\xi) = \alpha\}.$$

Let us assume that a $\varkappa$-additive separable ideal is defined on $\varkappa$. By Ulam's result, $\varkappa \neq \rho^+$ (if $\varkappa = \rho^+$, then by Ulam's result there exist ρ^+ pairwise disjoint sets from I^+ which are elements of one row of the matrix $\{M^\alpha_\xi\}$). Besides, $\varkappa \neq \sum_{\nu<\eta} \varkappa_\nu$, where $\varkappa_\nu < \varkappa$ and $\eta < \varkappa$. That is, $\varkappa$ is a limit regular cardinal and $\varkappa > \aleph_0$. One also usually calls such cardinals *weakly inaccessible*. The existence of weakly inaccessible cardinals cannot be proved in ZFC (see [**J**]). It is clear that a weakly inaccessible cardinal $\varkappa$ must be "very large" because $\varkappa = \aleph_\varkappa$ (this is mentioned e.g. in [**A**]). Hence, in the models where the Gitik–Shelah theorem is a nontrivial statement, the cardinal $\mathfrak{c}$ is "very large". From the point of view of modern set theory, the situation in which the cardinal $\mathfrak{c}$ is "very large" is not special. On the contrary, such a situation is rather natural (regarding this point, see the arguments of P. Cohen himself at the end of his book [**Co2**]). Furthermore, $\varkappa < \mathfrak{c}$, since a homogeneous ideal on $\mathfrak{c}$ is nonseparable (this is mentioned later in this chapter). Let us sum up our arguments. It is enough to prove Theorem 6.3 only for weakly inaccessible cardinals $\varkappa < \mathfrak{c}$. But this restriction will not help us. And we will be proving Theorem 6.3 for all $\varkappa$, i.e., $\aleph_1 \leq \varkappa \leq \mathfrak{c}$. In 5.21, models were mentioned in which there exists a cardinal $\varkappa < \mathfrak{c}$ such that one can construct a $\varkappa$-additive separable ideal on $\varkappa$. Let us mention the following known fact: Using Ulam's general result one can assume that in Theorems 1.6 and 1.8 the cardinality of the set on which the measures are given is less than the first weakly inaccessible cardinal.

6.5. Consider a set A on which an ideal I is defined, and a set Y on which an ideal J is defined. Let $D \subset A \times Y$, $\alpha \in A$, $y \in Y$. Put

$$D_\alpha = \{y \in Y \mid (\alpha, y) \in D\}, \quad D^y = \{\alpha \in A \mid (\alpha, y) \in D\}.$$

Define two ideals $I \times^{\mid} J$ and $I \times^{-} J$ on the set $A \times Y$ as follows:

$$D \in I \times^{\mid} J \text{ if and only if } \{\alpha \in A \mid D_\alpha \in J^+\} \in I;$$
$$D \in I \times^{-} J \text{ if and only if } \{y \in Y \mid D^y \in I^+\} \in J.$$

Our notation seems reasonable. Since D_α is a "vertical section of D", we use the notation $\times^{\mid}$, and since D^y is a "horizontal section of D", we use the notation $\times^{-}$.

Assume that Theorem 6.3 is not true; below we denote by I a $\varkappa$-additive countably dense ideal defined on a cardinal $\varkappa \leq \mathfrak{c}$.

6.6. Consider the set 2^ω with the usual topology, and the ideal $\mathbb{K}$ on 2^ω of sets of the first category.[4]

LEMMA. $I \times^{\mid} \mathbb{K} \subset I \times^{-} \mathbb{K}$.

---

[4] 2^ω is the product of a countable number of two-point sets; $M \in \mathbb{K}$ if and only if $M \subset \bigcup_{i=1}^\infty F_i$, F_i is a closed set, and $2^\omega \setminus F_i$ is an everywhere dense set.

PROOF. Let $D \subset \varkappa \times 2^\omega$, and for each $\alpha < \varkappa$ let D_α be an everywhere dense open subset of 2^ω. It suffices to prove that in this case the set

$$\{y \in 2^\omega \mid D^y \in I^c\}$$

is everywhere dense and G_δ.[5] Let $\{U_1, \dots, U_n, \dots\}$ be a countable open base of 2^ω, and let

$$P_n = \{\alpha < \varkappa \mid U_n \subset D_\alpha\}.$$

It is clear that

$$\varkappa = \bigcup_n P_n, \quad D = \bigcup_n P_n \times U_n, \quad D^y = \bigcup_n \{P_n \mid y \in U_n\}.$$

Since the ideal I is countably dense, one can take a countable family

$$M_1, \dots, M_l, \dots \subset I^+$$

with the following property: if $Q \subset \varkappa$ and $Q \cap M_l \in I^+$ for each l, then $Q \in I^c$.[6] Put

$$H_l = \bigcup_n \{U_n \mid M_l \cap P_n \in I^+\}.$$

Since the ideal I is countably dense,

$$\bigcap_l H_l = \{y \in 2^\omega \mid D^y \in I^c\}.$$

Clearly, $\bigcap_l H_l$ is a G_δ set. It remains to prove that H_l is everywhere dense in 2^ω. Let U be an open subset of 2^ω. Let us prove that $U \cap H_l \neq \varnothing$. Since D_α is an everywhere dense open subset of 2^ω,

$$\varkappa = \{\alpha < \varkappa \mid D_\alpha \cap U \neq \varnothing\} = \bigcup_n \{\alpha < \varkappa \mid U_n \subset D_\alpha \cap U\} = \bigcup_n \{P_n \mid U_n \subset U\}.$$

Let us now choose n so that $M_l \cap P_n \in I^+$ and $U_n \subset U$. It is clear that we have $U_n \subset U \cap H_l \neq \varnothing$. □

6.7. It is needless to repeat that by Ulam's result from 1.5, Theorem 6.3 is true if $\mathfrak{c} = \aleph_1$. Let us show that in this case Theorem 6.3 is a corollary of Lemma 6.6. So let $\varkappa = \#(2^\omega) = \aleph_1$. Let f be a one-to-one map from the set $\{\alpha \mid \alpha < \varkappa\}$ to 2^ω. For each $\alpha < \varkappa$ put

$$O_\alpha = \{f(\beta) \mid \beta \leq \alpha\};$$

thus we define the corresponding set $O \subset \varkappa \times 2^\omega$. Since the at most countable set O_α belongs to $\mathbb{K}$, Lemma 6.6 implies that $O \in I \times^- \mathbb{K}$. But this contradicts the fact that

$$O^y = \{\alpha < \varkappa \mid \alpha \geq f^{-1}(y)\} \in I^c$$

for any $y \in 2^\omega$. Hence the ideal I cannot be countably dense. □

The above construction of the set O has been known for a long time. Thus, in the book [**Sh-Gu**] the same construction is given as an example for understanding the Fubini theorem; this construction is attributed to Tolstov. It is clear that the spirit of the Fubini theorem is present in Lemma 6.6, which plays the key role in the proof of Theorem 6.3.

---

[5] Let us recall that a set G_δ is an intersection of a countable family of open sets.

[6] Of course, the converse statement also holds. Let a countable family $M_1, \dots, M_l, \dots \subset \varkappa$ have the following property: $Q \in I^c$ if and only if $Q \cap M_l \in I^+$ for all l. Then for each $Q \in I^+$ there exists l such that $M_l \setminus Q \in I$.

6.8. Let us interrupt the exposition of the proof of Theorem 6.3. The reason is that in the proof of the following theorem given in this section, the ingredients of the proof of Lemma 6.6 and the construction of the set O from the previous section are crucial, and it uses the Fubini theorem. Besides, the theorem itself seems rather interesting to me. Its direct corollaries are the Banach–Kuratowski result from 1.2 and the Lusin theorem from [**Lu**], which form the subject of the next four sections.

THEOREM. *Let a nontrivial σ-measure μ possess the following properties:*

(1) *there exists a family of sets $\{M_\alpha\}_{\alpha\in A}$, A being linearly ordered, such that $M_\alpha \subset M_\beta$ for $\alpha < \beta$, and $\bigcup M_\alpha$ is not a set of μ-measure zero;*
(2) *there exists a countable family $\mathcal{U}$ of μ-measurable sets such that $M_\alpha \subset \bigcup \mathcal{U}$ and*

$$\inf\{\mu(\bigcup \mathcal{V}) \mid \mathcal{V} \subset \mathcal{U} \text{ and } M_\alpha \subset \bigcup \mathcal{V}\} = 0$$

for each α.[7]

Then there exist μ-nonmeasurable sets.

PROOF. Let us prove that there exists a μ-nonmeasurable subset of the set $M = \bigcup M_\alpha$. Assume the contrary. We know that $\mu(M) < \infty$ (see footnote 1.4). Let $0 < \varepsilon < \mu(M)$. For each $t \in M$ put

$$D_t = \bigcap \{M_\alpha \mid t \in M_\alpha\}$$

and choose $\mathcal{V}_t \subset \mathcal{U}$ so that $D_t \subset \bigcup \mathcal{V}_t$ and $\mu(\bigcup \mathcal{V}_t) \le \varepsilon$. As usual, the measure is defined on X. Put

$$\Pi = \{(t,x) \in M \times X \mid x \in \bigcup \mathcal{V}_t\},$$
$$\mathcal{U} = \{U_1, \dots, U_n, \dots\},$$
$$P_n = \{t \in M \mid U_n \in \mathcal{V}_t\}.$$

By assumption, P_n is a μ-measurable set. It is clear that $\Pi = \bigcup P_n \times U_n$, and hence Π is a $(\mu \times \mu)$-measurable set. To estimate its measure, let us use the Fubini theorem. Since the measure of each "vertical section" $\bigcup \mathcal{V}_t$ is not greater than ε,

$$(\mu \times \mu)(\Pi) \le \varepsilon \cdot \mu(M).$$

Since the measure of each "horizontal section" Π^x for $x \in M$ is equal to $\mu(M)$,

$$(\mu \times \mu)(\Pi) \ge \mu(M)^2,$$

a contradiction. □

6.9. Suppose that on $\aleph_1$ a nontrivial σ-measure is given such that all one-point sets are of measure 0, and all subsets of $\aleph_1$ are measurable. Consider the countable sets

$$M_\alpha = \{\beta \le \alpha \mid \alpha \in \aleph_1\}.^8$$

Let $\aleph_1 \subset (0,1)$. Put

$$\mathcal{U} = \{(a,b) \cap \aleph_1 \mid 0 < a < b < 1; \text{ the numbers } a,b \text{ are rational}\}.$$

The families $\{M_\alpha\}_{\alpha\in\aleph_1}$ and $\mathcal{U}$ satisfy all the conditions of Theorem 6.8. Hence there exist nonmeasurable sets. By the contradiction thus obtained, the Banach–Kuratowski result from 1.2 is proved. □

---

[7] Obviously, $\mu(M_\alpha) = 0$.

[8] Since below we consider $\aleph_1$ as a subset of the interval $(0,1)$, let us emphasize that here by $<$ we denote the usual complete ordering of elements of $\aleph_1$.

6.10. Denote by $\mathfrak{m}$ the Lebesgue measure on $(0,1)$. Denote the external measure by $\mathfrak{m}^*$, and the internal measure by $\mathfrak{m}_*$.

LUSIN THEOREM. *Each* $Q \subset (0,1)$ *can be divided into disjoint subsets* Q_1 *and* Q_2 *such that* $\mathfrak{m}^*(Q_1) = \mathfrak{m}^*(Q_2) = \mathfrak{m}^*(Q)$.

PROOF. Consider the nontrivial case where $\mathfrak{m}^*(Q) > 0$. Introduce a complete ordering $\prec$ of elements of Q. Let γ be the minimal element such that $\mathfrak{m}^*(M_\gamma) > 0$, where $M_\gamma = \{\alpha \mid \alpha \prec \gamma\}$. Denote by ν the measure induced by $\mathfrak{m}$ on M_γ. The two families of sets

$$\{M_\alpha\}_{\alpha \prec \gamma}, \quad \text{where } M_\alpha = \{\beta \mid \beta \prec \alpha\},$$
$$\mathcal{U} = \{(a,b) \cap M_\gamma \mid 0 < a < b < 1;\ a, b \text{ are rational numbers}\}$$

satisfy all conditions of Theorem 6.8. Let Q_1' be a ν-nonmeasurable set such that $\mathfrak{m}_*(Q_1') = 0$. Let us take a ν-measurable set $Q' \supset Q_1'$ such that $\nu(Q') = \mathfrak{m}^*(Q_1')$. Put $Q_2' = Q' \setminus Q_1'$. It is clear that $\mathfrak{m}^*(Q') = \mathfrak{m}^*(Q_1') = \mathfrak{m}^*(Q_2')$. If we have $\mathfrak{m}^*(Q') = \mathfrak{m}^*(Q)$, the proof is over. Otherwise, using standard arguments, we get the required result. □

6.11. REMARK. Theorem 6.8 is presented in [**F**] under the name Lusin Theorem. Theorem 6.8 implies the statement mentioned in 5.26: the Lebesgue measure of a Q-set equals 0. To prove this statement, one can use the arguments in the previous section.[9]

6.12. Consider a finite sequence of sets $A_1, \dots, A_n \subset (0,1)$. By the Lusin theorem, there exist pairwise disjoint sets $B_1, \dots, B_n$ such that $B_i \subset A_i$ and $\mathfrak{m}^*(B_i) = \mathfrak{m}^*(A_i)$. In [**Gi-S2**] the following generalization is obtained by the forcing method: for each countable sequence of sets $A_1, \dots, A_n, \dots \subset (0,1)$ there exist pairwise disjoint sets $B_1, \dots, B_n, \dots$ such that

$$B_n \subset A_n, \quad \mathfrak{m}^*(B_n) = \mathfrak{m}^*(A_n).$$[10]

If $\mathfrak{c} = \aleph_1$, then this generalization of the Lusin theorem is clear, because the existence of Ulam's matrix (see 1.5) implies rather easily that any subset of $(0,1)$ of nonzero measure[11] can be divided into $\aleph_1$ pairwise disjoint nonzero measure subsets.[12] (Obviously, this statement is equivalent to the following one: if $\mathfrak{c} = \aleph_1$, then any set $M \subset (0,1)$ can be divided into $\aleph_1$ pairwise disjoint subsets M_α such that $\mathfrak{m}^*(M_\alpha) = \mathfrak{m}^*(M)$.) Some time ago Moti Gitik informed me that the following problem is open: Is it true in the general case that any set of nonzero measure in $(0,1)$ can be divided into $\aleph_1$ pairwise disjoint nonzero measure subsets? In [**S5**] Saharon Shelah solved this problem negatively. In some models there exists a nonzero measure subset of $(0,1)$ which cannot be so divided.

---

[9] The same proof of this statement is given in [**F**].

[10] We will also meet a generalized Lusin theorem at the end of this chapter.

[11] By a zero measure set in $(0,1)$ we mean a set whose Lebesgue measure equals 0. All other sets from $(0,1)$ are called nonzero measure sets.

[12] The last statement holds also under a more general assumption: $\mathfrak{c}$ is smaller than the first weakly inaccessible cardinal (see 6.4). Further, Moti Gitik informed me that the following fact is clear: one can construct a model in which $(0,1)$ is the union of $\aleph_1$ zero measure sets, and $\mathfrak{c}$ is a weakly inaccessible cardinal. It is obvious that in such a model any nonzero measure set from $(0,1)$ can be divided into $\aleph_1$ pairwise disjoint nonzero measure sets.

6.13. REMARK. For sets of the second category there are results similar to the above results on nonzero measure sets. Obviously, any set of the second category on the real line can be divided into $\aleph_0$ pairwise disjoint sets of the second category. By the existence of Ulam's matrix, if $\mathfrak{c} = \aleph_1$, then one can find $\aleph_1$ such sets (in [**Ko**] the paper [**U2**] and also [**Ku**] and [**Si3**] are cited regarding this point). Clearly, the last statement is valid also under a more general assumption: $\mathfrak{c}$ is less than the first weakly inaccessible cardinal. Can one construct a model in which the real line is the union of $\aleph_1$ sets of the first category, and $\mathfrak{c}$ is a weakly inaccessible cardinal? I obtained the positive answer to this question from Martin Goldstern and Saharon Shelah in [**Gol**]. There two such models are constructed. It is clear that in such models any set of the second category can be divided into $\aleph_1$ pairwise disjoint sets of the second category. In [**Ko**] a result for sets of the second category is obtained similar to the result from [**S5**] for nonzero measure sets.

6.14. Let us return to the proof of Theorem 6.3. For the proof of the following lemma, one must use the arguments from 6.7 regarding the construction of the set O. Recall that here and below I is a $\varkappa$-additive countably dense ideal defined on $\varkappa$. The ideal J is defined on Y.

LEMMA. *Let $I \ltimes J \subset I \times^{-} J$ and $\{z_\beta \mid \beta < \varkappa\} \subset Y$. If we have $A \subset \varkappa$ and $\{z_\beta \mid \beta \in A\} \in J^+$, then $\{z_\beta \mid \beta \in A \cap \alpha\} \in J^+$ for some $\alpha < \varkappa$.*

6.15. LEMMA. *If $\#(Y) < \varkappa$, then $I \times^{-} J \subset I \ltimes J$.*

PROOF. Let $D \subset \varkappa \times Y$ and $D^y \in I$ for each $y \in Y$. It is enough to prove that $D \in I \ltimes J$. Since $\#(Y) < \varkappa$, we have $I \ni \bigcup_{y\in Y} D^y$. It is also obvious that

$$\{\alpha < \varkappa \mid D_\alpha \in J^+\} \subset \{\alpha < \varkappa \mid D_\alpha \neq \varnothing\} = \bigcup_{y\in Y} D^y \in I. \qquad \square$$

6.16. Let us pass to functions whose set of values is $\mathfrak{c}$.

DEFINITION. A family $\mathcal{F}$ of functions defined on Y is said to be *J-disjoint* if for each two distinct functions $f_1, f_2 \in \mathcal{F}$ we have

$$\{y \in Y \mid f_1(y) = f_2(y)\} \in J.$$

6.17. DEFINITION. A family $\mathcal{F}$ of functions defined on Y is said to be $(J, \varkappa, \varkappa)$-*concentrated* if there exists a subset $Z \subset Y$ such that

$$\#(Z) \leq \varkappa, \quad \{y \in Z \mid f(y) < \varkappa\} \in J^+ \quad \text{for all } f \in \mathcal{F}.$$

6.18. LEMMA. *Assume that $I \ltimes J \subset I \times^{-} J$ and a J-disjoint and $(J, \varkappa, \varkappa)$-concentrated family $\mathcal{F}$ of functions on Y is given. Then $\#(\mathcal{F}) \leq \varkappa$.*

PROOF. Suppose the contrary: $\#(\mathcal{F}) > \varkappa$. Let us take $Z \subset Y$ so that $\#(Z) \leq \varkappa$ and

$$\{y \in Z \mid f(y) < \varkappa\} \in J^+$$

for all $f \in \mathcal{F}$. Assume that $\#(Z) = \varkappa$. Consider a one-to-one map from $\varkappa$ onto $Z\colon \beta \to z_\beta$. Let A_f be the preimage under this map of the set $\{y \in Z \mid f(y) < \varkappa\}$, where $f \in \mathcal{F}$. Using Lemma 6.14, let us find $\alpha < \varkappa$ and $\mathcal{F}' \subset \mathcal{F}$ so that $\#(\mathcal{F}') > \varkappa$ and for each $f \in \mathcal{F}'$ we have

$$\{z_\beta \mid \beta \in A_f \cap \alpha\} \in J^+.$$

This implies that we can assume the inequality $\#(Z) < \varkappa$. Choose $f_0 \in \mathcal{F}$. Since $\#(Z) < \varkappa$, there exists $\eta_0 < \varkappa$ such that

$$\{y \in Z \mid f_0(y) < \eta_0\} = \{y \in Z \mid f_0(y) < \varkappa\} \in J^+.$$

For this reason and from the fact that $\#(\mathcal{F}) > \varkappa$, we see that there exist $\eta < \varkappa$ and a family $\mathcal{F}^* \subset \mathcal{F}$ such that $\#(\mathcal{F}^*) = \varkappa$ and

$$\{y \in Z \mid f(y) < \eta\} \in J^+$$

for all $f \in \mathcal{F}^*$. Consider a one-to-one map from $\varkappa$ onto $\mathcal{F}^*: \alpha \to f_\alpha$. Put

$$D_\alpha = \{y \in Z \mid f_\alpha(y) < \eta\},$$

and let us consider $D \subset \varkappa \times Z$. The ideal J induces an ideal J_Z on Z. Obviously, $D \notin I \ltimes J_Z$, and by Lemma 6.15, $D \notin I \times^- J_Z$. Hence

$$W = \{y \in Z \mid \{\alpha < \varkappa \mid f_\alpha(y) < \eta\} \in I^+\} \in J^+.$$

Since $\eta < \varkappa$ and the ideal I is $\varkappa$-additive, for each $y \in W$ there exists $h(y) < \eta$ such that

$$\{\alpha < \varkappa \mid f_\alpha(y) = h(y)\} \in I^+.$$

The ideal J induces an ideal J_W on W, and, obviously, $I \ltimes J_W \subset I \times^- J_W$. For each $y \in W$ put

$$E^y = \{\alpha < \varkappa \mid f_\alpha(y) = h(y)\},$$

and let us consider $E \subset \varkappa \times W$. Since $E \notin I \times^- J_W$, let us find $\alpha' < \varkappa$ and $W' \in W \cap J^+$ such that $f_{\alpha'}(y) = h(y)$ for all $y \in W'$. Similarly, let us consider the ideals $J_{W'}, I \ltimes J_{W'}, I \times^- J_{W'}$, and find $\alpha'' < \varkappa$ and $W'' \in W' \cap J^+$ such that $f_{\alpha''}(y) = h(y)$ for all $y \in W''$, and also $\alpha'' \neq \alpha'$. So $f_{\alpha'}$ and $f_{\alpha''}$ are distinct functions from $\mathcal{F}$ which coincide on $W'' \in J^+$, a contradiction. □

6.19. Let us summarize our arguments. We have proved the following statement:

STATEMENT. *Assume that a $\mathbb{K}$-disjoint and $(\mathbb{K}, \varkappa, \varkappa)$-concentrated family $\mathcal{F}$ of functions on 2^ω is given. If there exists a $\varkappa$-additive and countably dense ideal I on $\varkappa$, then $\#(\mathcal{F}) \leq \varkappa$.*

Now our goal is to show that under the assumption of the existence of such an ideal I, there exists a $\mathbb{K}$-disjoint and $(\mathbb{K}, \varkappa, \varkappa)$-concentrated family of functions $\mathcal{F}$ and $\#(\mathcal{F}) > \varkappa$, i.e., to get a contradiction implying Theorem 6.3.

6.20. As usual, an element of the set 2^ω is an infinite sequence of zeros and ones $(a_1, \ldots, a_r, \ldots)$. Let $(s_1, \ldots, s_k)$ be a finite sequence of zeros and ones. The sets

$$\Pi(s_1, \ldots, s_k) = \{(a_1, \ldots, a_r, \ldots) \in 2^\omega \mid a_i = s_i \text{ for all } i \leq k\}$$

form an open base of 2^ω.

6.21. Since the ideal I is countably dense, let us take a countable family $M_1, \dots, M_l, \dots \subset I^+$ possessing the following property: for any $Q \in I^+$ there exists M_l such that $M_l \setminus Q \in I$. Let n_1 be the minimal natural number such that $M_{n_1} \notin I^c$. Let us assign to M_{n_1} the set $\Pi(0)$, and to the set $\varkappa \setminus M_{n_1}$ let us assign the set $\Pi(1)$. If the set M_{n_1+1} coincides up to a set from I with one of the three sets $\varkappa, M_{n_1}, \varkappa \setminus M_{n_1}$, then let us pass to the consideration of the set M_{n_1+2}. Otherwise there are three possibilities:

(1) $M_{n_1+1} \subset M_{n_1}$;
(2) $M_{n_1+1} \subset \varkappa \setminus M_{n_1}$;
(3) neither (1) nor (2) holds.

Clearly, the inclusions in (1) and (2) are inclusions up to a set from I. In case (1), let us assign to $M_{n_1} \cap M_{n_1+1}$ the set $\Pi(0,0)$, and to the set $M_{n_1} \setminus M_{n_1+1}$ let us assign the set $\Pi(0,1)$. In case (2), let us assign to $(\varkappa \setminus M_{n_1}) \cap M_{n_1+1}$ the set $\Pi(1,0)$, and to the set $(\varkappa \setminus M_{n_1}) \setminus M_{n_1+1}$ let us assign the set $\Pi(1,1)$. In case (3), the assignments are the following:

$$M_{n_1} \setminus M_{n_1+1} \to \Pi(0,0), \quad M_{n_1} \cap M_{n_1+1} \to \Pi(0,1),$$
$$(\varkappa \setminus M_{n_1}) \cap M_{n_1+1} \to \Pi(1,0), \quad (\varkappa \setminus M_{n_1}) \setminus M_{n_1+1} \to \Pi(1,1).$$

Let us continue our construction in the natural way: each set $\Pi(s_1, \dots, s_k)$ is assigned to some set from I^+ which we denote by $M(s_1, \dots, s_k)$. The sets $M(s_1, \dots, s_k)$ possess the following properties:

(a) $M(s_1, \dots, s_k) \cap M(s'_1, \dots, s'_k) = \varnothing$ if $(s_1, \dots, s_k) \neq (s'_1, \dots, s'_k)$;
(b) $I^c \ni M^{(k)} = \bigcup M(s_1, \dots, s_k)$.

Put $M = \bigcap_k M^{(k)}$. Obviously, $M \in I^c$.

6.22. Define a map $e \colon M \to 2^\omega$ as follows. Let $(a_1, \dots, a_r, \dots) \in 2^\omega$. Put

$$M(a_1, \dots, a_r, \dots) = \bigcap_r M(a_1, \dots, a_r).$$

If $M(a_1, \dots, a_r, \dots) \neq \varnothing$, put

$$e[M(a_1, \dots, a_r, \dots)] = (a_1, \dots, a_r, \dots).$$

6.23. LEMMA. *If $M' \in M \cap I^+$, then $e[M'] \in \mathbb{K}^+$.*

PROOF. Assume that $e[M'] \in \mathbb{K}$. Then $e[M'] \subset \bigcup_{i=1}^\infty F_i$ and each F_i is a closed nowhere dense set. Assume that $e^{-1}[F_{i_0}] \in I^+$. Then there exists $M(s_1, \dots, s_k)$ such that

$$M(s_1, \dots, s_k) \setminus e^{-1}[F_{i_0}] \in I.$$

There exists a subset

$$\Pi(s'_1, \dots, s'_{k'}) \subset \Pi(s_1, \dots, s_k)$$

such that

$$F_{i_0} \cap \Pi(s'_1, \dots, s'_{k'}) = \varnothing.$$

Hence

$$M(s'_1, \dots, s'_{k'}) \subset M(s_1, \dots, s_k), \quad e^{-1}[F_{i_0}] \cap M(s'_1, \dots, s'_{k'}) = \varnothing.$$

Therefore

$$M(s_1, \dots, s_k) \setminus e^{-1}[F_{i_0}] \in I^+,$$

because $M(s'_1, \dots, s'_{k'}) \in I^+$. The contradiction obtained above implies that $e^{-1}[F_{i_0}] \in I$. But this means that $M' \in I$, a contradiction. □

6.24. LEMMA. *If a closed set $F \subset 2^\omega$ and $e^{-1}[F] \in I$, then $F \in \mathbb{K}$.*

PROOF. Assume that $F \in \mathbb{K}^+$. Then F contains an open set and there exists a subset $\Pi(s_1, \dots, s_k) \subset F$. Hence $M(s_1, \dots, s_k) \in I$. But $M(s_1, \dots, s_k) \in I^+$. □

6.25. Denote by $\mathbb{C}$ a Cantor perfect set isomorphic to 2^ω. (We consider a Cantor perfect set only for convenience, to avoid confusion between two identical sets 2^ω.) The sets $\Pi(s_1, \dots, s_k)$ that lie in $\mathbb{C}$ will be denoted by $\Pi'(s_1, \dots, s_k)$.

LEMMA. *For an arbitrary map $\varphi\colon \varkappa \to \mathbb{C}$ there exist sets $M_\varphi \in M \cap I^c$, $L_\varphi \in \mathbb{K}^c$, and a continuous function $f_\varphi\colon L_\varphi \to \mathbb{C}$ such that $e[M_\varphi] \subset L_\varphi$ and the diagram*

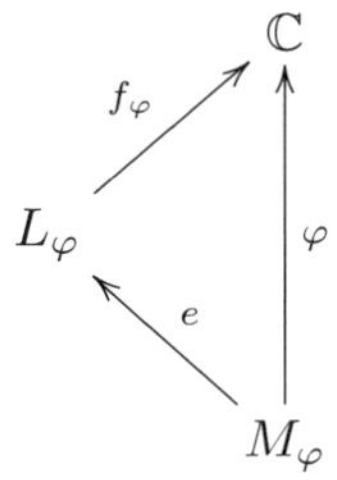

is commutative.

PROOF. Let

$$\varphi^{-1}[\Pi'(s_1, \dots, s_k)] = A \in I^+.$$

Put

$$M^\varphi(s_1, \dots, s_k) = \bigcup \{A \cap M(a_1, \dots, a_r) \mid M(a_1, \dots, a_r) \setminus A \in I\},$$
$$\Pi^\varphi(s_1, \dots, s_k) = \bigcup \{\Pi(a_1, \dots, a_r) \mid M(a_1, \dots, a_r) \setminus A \in I\},$$
$$M_k^\varphi = \bigcup M^\varphi(s_1, \dots, s_k), \quad L_k^\varphi = \bigcup \Pi^\varphi(s_1, \dots, s_k),$$
$$M_\varphi = \bigcap_k M_k^\varphi, \quad L_\varphi = \bigcap_k L_k^\varphi.$$

Let $y \in L_\varphi$. There exists a uniquely defined sequence of sets

$$\Pi^\varphi(s_1^1), \Pi^\varphi(s_1^2, s_2^2), \dots, \Pi^\varphi(s_1^k, \dots, s_k^k), \dots,$$

each of which contains y. Obviously, $\bigcap_k \Pi'(s_1^k, \dots, s_k^k)$ contains one point $p \in \mathbb{C}$. Put $p = f_\varphi[y]$. □

6.26. Below by f_φ we will mean some function $2^\omega \to \mathbb{C}$ which coincides on L_φ with the function f_φ defined in the previous lemma. Clearly, we cannot assert (and actually we do not need this) that f_φ is continuous on 2^ω.

6.27. Let us imagine that $\mathbb{C}$ is the set of all ordinals $< \mathfrak{c}$.

LEMMA. *Let Φ be an I-disjoint family of functions mapping $\varkappa$ to $\varkappa \subset \mathbb{C}$. Then the family of functions $\mathcal{F} = \{f_\varphi \mid \varphi \in \Phi\}$ is $\mathbb{K}$-disjoint and $(\mathbb{K}, \varkappa, \varkappa)$-concentrated.*

PROOF. Consider two distinct functions $f_{\varphi_1}, f_{\varphi_2} \in \mathcal{F}$. Since these functions are continuous on $L_{\varphi_1} \cap L_{\varphi_2}$, the set

$$F = \{y \in L_{\varphi_1} \cap L_{\varphi_2} \mid f_{\varphi_1}(y) = f_{\varphi_2}(y)\}$$

is closed in $L_{\varphi_1} \cap L_{\varphi_2}$. Let $\overline{F}$ be the closure of F in 2^ω. By Lemma 6.25,

$$e^{-1}[\overline{F}] \subset \{\alpha < \varkappa \mid \varphi_1(\alpha) = \varphi_2(\alpha)\} \cup (\varkappa \setminus (M_{\varphi_1} \cap M_{\varphi_2})) \in I.$$

By Lemma 6.24, $F \in \mathbb{K}$. Since $L_{\varphi_1} \cap L_{\varphi_2} \in \mathbb{K}^c$, we have

$$\{y \in 2^\omega \mid f_{\varphi_1}(y) = f_{\varphi_2}(y)\} \in \mathbb{K}.$$

We have proved that the family $\mathcal{F}$ is $\mathbb{K}$-disjoint. Obviously, $\#(e[M]) \leq \varkappa$.[13] By Lemma 6.25, for each function $f_\varphi \in \mathcal{F}$ we have

$$\{y \in e[M] \mid f_\varphi(y) < \varkappa\} \supset e[M_\varphi].$$

But since $M_\varphi \in I^c$, then, by Lemma 6.23, $e[M_\varphi] \in \mathbb{K}^+$, and so

$$\{y \in e[M] \mid f_\varphi(y) < \varkappa\} \in \mathbb{K}^+.$$

We have proved that the family of functions $\mathcal{F}$ is $(\mathbb{K}, \varkappa, \varkappa)$-concentrated. □

6.28. LEMMA. *Let Υ be a homogeneous ideal on a cardinal ρ. Then any maximal Υ-disjoint family of functions mapping ρ to itself has cardinality $> \rho$.*[14]

PROOF. Let $\Phi = \{f_\alpha \mid \alpha < \rho_0\}$ be a maximal Υ-disjoint family of functions mapping ρ to itself, and let $\rho_0 \leq \rho$. Consider a function $f^*\colon \rho \to \rho$ which takes at $\beta < \rho$ some value from the set $\rho \setminus \{f_\alpha(\beta) \mid \alpha \leq \beta\}$. Clearly, $f^* \notin \Phi$ and $\Phi \cup \{f^*\}$ is a Υ-disjoint family of functions mapping ρ to itself, a contradiction. □

6.29. In Lemma 6.27 families of functions Φ and $\mathcal{F}$ are considered. Obviously, $\#(\Phi) = \#(\mathcal{F})$. Together with Lemmas 6.27 and 6.28, this implies that one can construct a $\mathbb{K}$-disjoint and $(\mathbb{K}, \varkappa, \varkappa)$-concentrated family of functions $\mathcal{F}$ such that $\#(\mathcal{F}) > \varkappa$. We have obtained the contradiction which was our goal in 6.19. *Theorem 6.3 is proved.*

6.30. Let $\#(X) = \mathfrak{c}$, and let Υ be a homogeneous ideal on X. By Lemma 6.28, a maximal family of Υ-disjoint functions mapping X to $[0, 1]$ has cardinality $> \mathfrak{c}$. Since, by assumption, the ideal Υ is σ-additive, the cardinality of the set of all continuous real functions on $\ker(\Upsilon)$ is $> \mathfrak{c}$. Therefore the ideal Υ is nonseparable.

Let $\mu_1, \ldots, \mu_k, \ldots$ be a countable sequence of σ-two-valued homogeneous measures defined on $\mathfrak{c}$ (the homogeneity condition means that if $\mu_k(M) \neq 0$, then $\#(M) = \mathfrak{c}$, i.e., the ideal I_k corresponding to the measure μ_k is homogeneous). Let us show how simply the Gitik–Shelah theorem is proved in this case. Indeed, we have just shown that the ideals I_k are nonseparable. It remains to refer to the simple arguments presented in Case 2 in the proof of Theorem 6.2. If $\mathfrak{c} = \aleph_1$, then any σ-two-valued measure on $\mathfrak{c}$ is homogeneous. We come to one more proof of the Gitik–Shelah theorem in the case $\mathfrak{c} = \aleph_1$.

6.31. If $\mathfrak{c} \leq \#(X) <$ (the first measurable cardinal), then on X there does not exist a homogeneous countably dense ideal. This is shown in [**T**] for $\#(X) > \mathfrak{c}$. For $\#(X) = \mathfrak{c}$, this had been shown in [**Kr-P**], [**Fr-Gut**], [**J-Pr**], and also by Baumgartner, Kunin, Taylor, and possibly others before the Gitik–Shelah theorem appeared (all this is discussed in [**Gi-S1**]). The results of this monograph already presented (including the results of Chapter 5) imply a general theorem:

THEOREM. *Any simple σ-additive ideal defined on a σ-nonmeasurable cardinal is not countably dense.*[15]

---

[13] We know that the ideal I is $\varkappa$-additive and $e^{-1}[y] \in I$ for all $y \in 2^\omega$. Hence $\#(e[M]) = \varkappa$.

[14] Here we do not assume that Υ is σ-additive.

[15] It is clear that this theorem is an easy corollary of Theorem 6.3 and Tarski's result from [**T**] cited above.

6.32. Let us return to Solovay's result from 1.4. Let μ be a σ-additive extension of the Lebesgue measure to all subsets of $(0,1)$. One can assume that if $\{M_\lambda\}_{\lambda\in\Lambda}$ is a family of pairwise disjoint subsets of $(0,1)$ and $\#(\Lambda) < \mathfrak{c}$, then

$$\mu\left(\bigcup_{\lambda\in\Lambda} M_\lambda\right) = \sum_{\lambda\in\Lambda} \mu(M_\lambda).$$

Let us now construct the measure μ'' from footnote 2.3. By definition, $\mu''(M) = 0$ if and only if $\mu(M) = 0$. Let $X = (0,1)$, and let I be the ideal corresponding to the measure μ''. Let us take a family $\{M_\alpha\}_{\alpha\in A}$ of μ''-nonmeasurable sets possessing the following properties:

(1) $(M_\alpha \setminus M_\beta) \cup (M_\beta \setminus M_\alpha)$ is a μ''-nonmeasurable set if $\alpha \neq \beta$;
(2) for any two distinct points $x, y \in \ker(I)$ there exists $\alpha \in A$ such that $x \in \overline{M}_\alpha$ and $y \notin \overline{M}_\alpha$.

By Lemma 6.28 the cardinality of the family of all continuous real functions on $\ker(I)$ is

$$\#(C(\ker(I))) > \mathfrak{c}.$$

Hence by the Weierstrass–Stone theorem,[16] $\#(A) > \mathfrak{c}$. Obviously, there do not exist $\aleph_1$ pairwise disjoint μ''-nonmeasurable sets. □

6.33. PROOF OF THE GENERALIZED LUSIN THEOREM IN THE HOMOGENEOUS CASE. Let us first state the theorem:

THEOREM. *Assume that the Lebesgue measure on the interval* $(0,1)$ *is homogeneous: if* $M \subset (0,1)$ *and* $\#(M) < \mathfrak{c}$, *then* $\mathfrak{m}(M) = 0$. *Let* $A_0, A_1, \dots, A_k, \dots \subset (0,1)$ *be a countable sequence of sets. Then there exist pairwise disjoint sets* $B_0, B_1, \dots, B_k, \dots$ *such that* $B_k \subset A_k$ *and* $\mathfrak{m}^*(A_k) = \mathfrak{m}^*(B_k)$ *for all* $k \in \mathbb{N}$.

PROOF. Assume, without loss of generality, that $\mathfrak{m}^*(A_k) > 0$ for all $k \in \mathbb{N}$. The Lebesgue measure induces the measure μ_k on A_k. It suffices to prove that there exists M such that $\mu_0^*(M) > 0$[17] and

$$\{S \subset M \mid S \text{ is } \mu_k\text{-measurable and } \mu_k(S) > 0\} = \varnothing$$

for each $k \in \mathbb{N}^+$. Suppose the contrary. Consider an arbitrary real function f defined on A_0. For any real number α we have

$$\{x \in A_0 \mid f(x) < \alpha\} = \bigcup_{k=1}^{\infty} M_k^{f,\alpha},$$

where $M_k^{f,\alpha}$ is a μ_k-measurable set. Therefore the cardinality of the set of all functions f (of course, f is defined up to values on a set of measure 0) is $\mathfrak{c}$. By the construction from 6.28, the cardinality is greater than $\mathfrak{c}$. □

---

[16]The following generalization of the classical Weierstrass theorems was given by Stone in [**St1**]. Assume that on a compact space $\mathcal{K}$ a family M of continuous real functions is given, which contains all constants and which is an algebraic ring with the usual operations of sum and product, and for any two points $x_1 \neq x_2$ there exists a function $f \in M$ such that $f(x_1) \neq f(x_2)$. Then each continuous real function on $\mathcal{K}$ is the limit of a uniformly convergent sequence of functions from M.

[17]By μ_0^* we denote the external measure of the measure μ_0.

CHAPTER 7

Proof of Theorems 1.17, 2.7, 2.8

7.1. DEFINITION. The set $\{a \in \ker \mathcal{A} \mid a \text{ is an irregular point}\}$ is called the *spectrum* of an algebra $\mathcal{A}$ and is denoted $\operatorname{sp} \mathcal{A}$.

7.2. If $\mathcal{A}$ is a separable algebra, then $\#(\operatorname{sp} \mathcal{A}) \leq \aleph_0$.

7.3. LEMMA. *If $\mathcal{A}$ is a simple almost σ-algebra, then* $\ker \mathcal{A} \subset \overline{\operatorname{sp} \mathcal{A}}$.

PROOF. Suppose the contrary. Then there exist a subset $U \subset X$ and a countable sequence of irregular points $H = \{h_1, \dots, h_k, \dots\}$ such that

$$\overline{U} \supset \overline{H} \supset \overline{U} \cap \ker \mathcal{A} \neq \varnothing \quad \text{and} \quad H \cap \ker \mathcal{A} = \varnothing.$$

Put $H_k = H \setminus \{h_k\}$. Obviously, $\overline{H}_k \supset \overline{U} \cap \ker \mathcal{A}$, $h_k \notin \overline{H}_k$. Hence one can take $M_k \in h_k$ such that $\mathfrak{P}(M_k) \subset \mathcal{A}$. Since the algebra $\mathcal{A}$ is almost σ-additive, $\mathfrak{P}(\bigcup M_k) \subset \mathcal{A}$. But this contradicts the fact that

$$\overline{\bigcup M_k} \supset \overline{H} \supset \overline{U} \cap \ker \mathcal{A} \neq \varnothing. \qquad \square$$

7.4. LEMMA. *Let $\mathcal{A}$ be a σ-algebra, $Q_1, \dots, Q_k, \dots$ a countable sequence of pairwise disjoint sets, and a, b $\mathcal{A}$-similar ultrafilters; $Q_1 \in a$, $b \ni \bigcup_k Q_k$, and $Q_k \notin b$ for all k. Then there exist $\mathcal{A}$-similar ultrafilters a', b' such that $Q_1 \in a'$, $Q_k \in b'$ and $k > 1$.*

PROOF. Let $D \in \mathcal{A}, a, b$ and $Q'_k = Q_k \cap D$. Assume that if $Q'_1 \in a'$ and $Q'_k \in b'$ $(k > 1)$, then a', b' are not $\mathcal{A}$-similar ultrafilters. Then for each $k > 1$ there exists $U_k \in \mathcal{A}$ such that $Q'_k \subset U_k$ and $Q'_1 \cap U_k = \varnothing$. Since $\mathcal{A}$ is a σ-algebra, $\mathcal{A} \ni U = \bigcup_{k>1} U_k$. But $U \notin a$, $U \in b$; hence $U \notin \mathcal{A}$, a contradiction. $\square$

7.5. Lemma 7.4 is used in the following chapters. The present section contains a simple corollary of Lemma 7.4 which is used in this chapter.

LEMMA. *Suppose $\mathcal{A}$ is a simple σ-algebra, and $a \in \operatorname{sp} \mathcal{A}$ is not an $\mathcal{A}$-special ultrafilter. Then*

$$\{b \in \operatorname{sp} \mathcal{A} \mid a \text{ is } \mathcal{A}\text{-similar to } b\} \neq \varnothing.$$

7.6. DEFINITION. An algebra is said to be *ω-saturated* if there does not exist an infinite number of pairwise disjoint sets not belonging to it.[1]

7.7. An algebra $\mathcal{A}$ is ω-saturated if and only if $\#(\ker \mathcal{A}) < \aleph_0$. An ω-saturated almost σ-algebra $\mathcal{A}$ is a σ-algebra and $\ker \mathcal{A} = \operatorname{sp} \mathcal{A}$. Obviously, any ω-saturated σ-algebra is simple.

7.8. DEFINITION. A simple but not ω-saturated algebra is called *strictly simple.*

[1]The notion of ω-saturation is discussed e.g. in [**Grz**], [**Kr-P**], [**Ta**].

7.9. DEFINITION. An algebra $\mathcal{A}$ *exceeds* an algebra $\mathcal{B}$ (notation: $\mathcal{A} > \mathcal{B}$ or $\mathcal{B} < \mathcal{A}$) if each $\mathcal{A}$-special ultrafilter is $\mathcal{B}$-special and each pair of $\mathcal{A}$-similar ultrafilters is a pair of $\mathcal{B}$-similar ultrafilters.

7.10. Here are some simple properties of the relation $>$:

(a) if $\mathcal{A} > \mathcal{B}$, then $\ker \mathcal{A} \subset \ker \mathcal{B}$ and $\operatorname{sp} \mathcal{A} \subset \operatorname{sp} \mathcal{B}$;

(b) if $\{\mathcal{A}_\lambda\}_{\lambda \in \Lambda}$ and $\{\mathcal{B}_\lambda\}_{\lambda \in \Lambda}$ are two families of algebras such that $\mathcal{A}_\lambda > \mathcal{B}_\lambda$ for all λ and $\bigcup_{\lambda \in \Lambda} \mathcal{A}_\lambda \neq \mathfrak{P}(X)$, then $\bigcup_{\lambda \in \Lambda} \mathcal{B}_\lambda \neq \mathfrak{P}(X)$;[2]

(c) let the kernel of the algebra $\mathcal{B}$ contain no more than one $\mathcal{B}$-special ultrafilter; in this case $\mathcal{A} > \mathcal{B}$ if and only if $\mathcal{A} \supset \mathcal{B}$.[3]

7.11. The following lemma is a corollary of Lemma 7.5.

LEMMA. *Let $\mathcal{A}$ be a strictly simple σ-algebra, and let $n \in \mathbb{N}^+$. Then there exists an ω-saturated σ-algebra $\mathcal{A}(n) > \mathcal{A}$ such that $\#(\ker \mathcal{A}(n)) > n$.*

7.12. LEMMA. *Let $\mathcal{A}_1, \dots, \mathcal{A}_k, \dots$ be a countable sequence of ω-saturated σ-algebras, and for each k let $\bigcup_{i=1}^k \mathcal{A}_i \neq \mathfrak{P}(X)$. Then $\bigcup \mathcal{A}_k \neq \mathfrak{P}(X)$.*

PROOF. Assume for simplicity that $X \in \mathcal{A}_k$ for all k. As in Chapter 4, for each finite sequence of algebras $\mathcal{A}_1, \dots, \mathcal{A}_k$ consider two sets of ultrafilters S_k, T_k. Let

$$S_k = \{s_1^k, \dots, s_k^k\}, \quad T_k = \{t_1^k, \dots, t_k^k\}.$$

Since $\ker \mathcal{A}_1$ is a finite set, there exists an infinite sequence of natural numbers

$$\mathbb{N}^1 = \{n_1^1 < n_2^1 < \dots < n_m^1 < \dots\}$$

such that

$$s_1^{n_1^1} = s_1^{n_2^1} = \dots = s_1^{n_m^1} = \dots \quad \text{and} \quad t_1^{n_1^1} = t_1^{n_2^1} = \dots = t_1^{n_m^1} = \dots .$$

Since $\ker \mathcal{A}_2$ is a finite set, there exists an infinite sequence of natural numbers

$$\mathbb{N}^2 = \{n_1^2 < n_2^2 < \dots < n_m^2 < \dots\}$$

such that

$$\mathbb{N}^2 \subset \mathbb{N}^1, \quad s_2^{n_1^2} = s_2^{n_2^2} = \dots = s_2^{n_m^2} = \dots \quad \text{and} \quad t_2^{n_1^2} = t_2^{n_2^2} = \dots = t_2^{n_m^2} = \dots .$$

Further, consider the sequences $\mathbb{N}^3, \mathbb{N}^4, \dots$; we have

$$\mathbb{N}^1 \supset \mathbb{N}^2 \supset \mathbb{N}^3 \supset \dots \supset \mathbb{N}^k \supset \dots .$$

Put

$$S = \{s_1^{n_1^1}, s_2^{n_1^2}, \dots, s_k^{n_1^k}, \dots\}, \quad T = \{t_1^{n_1^1}, t_2^{n_1^2}, \dots, t_k^{n_1^k}, \dots\}.$$

The ultrafilters $s_k^{n_1^k}, t_k^{n_1^k}$ are $\mathcal{A}_k$-similar and $S \cap T = \varnothing$. Since all points in S and T are irregular, $\overline{S} \cap \overline{T} = \varnothing$. It remains to apply the Main Statement 3.10. □

7.13. PROOF OF THEOREM 1.17. By Theorem 5.17 we can assume that all σ-algebras $\mathcal{A}_k$ are simple. By Lemma 7.11 we can assume that they all are ω-saturated. Now our theorem is true by Lemma 7.12. □

---

[2] Clearly it is useless here to impose our usual condition on the cardinality of Λ: $\#(\Lambda) \leq \aleph_0$.

[3] In 3.23 we considered algebras $\mathcal{A}$ with at most one $\mathcal{A}$-special ultrafilter.

Obviously, if $\mathcal{A} > \mathcal{B}$, then $\mathcal{A} \supset \mathcal{B}$. In general the converse is not true. Indeed, assume that $\ker \mathcal{B}$ consists of two $\mathcal{B}$-special ultrafilters q_1, q_2; $\ker \mathcal{A} = \{q_1, q_2\}$, and q_1, q_2 are $\mathcal{A}$-similar ultrafilters. It is clear that $\mathcal{A} \supset \mathcal{B}$ but $\mathcal{A}$ does not exceed $\mathcal{B}$.

7.14. REMARK. Since we are speaking about simple σ-algebras, let us note a property of such an algebra $\mathcal{A}$ when $X \in \mathcal{A}$: $\mathcal{A}$ is the algebra of all measurable sets for some σ-measure μ with $\mu(X) = 1$. Indeed, let us divide $\operatorname{sp} \mathcal{A} \neq \varnothing$ into pairwise disjoint sets $V_1, \ldots, V_n, \ldots$; $V_n \neq \varnothing$, any two ultrafilters from V_n are $\mathcal{A}$-similar, and any ultrafilter from V_n does not have an $\mathcal{A}$-similar ultrafilter from V_m if $n \neq m$. Clearly, $\overline{V}_n$ is an $\mathcal{A}$-similar set. (If there are $\aleph_0$ sets V_n, then, in addition to the sets $\overline{V}_n$, there are other $\mathcal{A}$-similar sets.) Let us assign to each V_n a real number m_n; $0 < m_n \leq 1$, $\sum m_n = 1$. A set M is μ-measurable if and only if for each n either $\overline{M} \cap V_n = \varnothing$, or $\overline{M} \supset V_n$, and

$$\mu(M) = \sum \{m_n \mid V_n \subset \overline{M}\}.$$

If $\operatorname{sp} \mathcal{A} = \varnothing$, i.e., $\mathcal{A} = \mathfrak{P}(X)$, then let us take an arbitrary point $x_0 \in X$ (we always have $X \neq \varnothing$) and put $\mu(\{x_0\}) = 1$.

Let us make a concluding remark, which, however, is hardly very useful. In solving the "general" question mentioned in the Introduction, when one considers σ-algebras with "unit", Theorem 5.17 shows that it suffices to deal with σ-measures taking the value 1 on X.

7.15. After what was pointed out above, it is easy to understand what the difficulty in the proof of Theorem 2.8 is. The difficulty is that for strictly simple almost σ-algebras, Lemma 7.11 does not hold. Let us give an appropriate counterexample.

EXAMPLE. There exists an almost σ-algebra $\mathcal{B} \neq \mathfrak{P}(X)$ on $X = \mathbb{N}$ such that if an ω-saturated σ-algebra $\mathcal{A}$ satisfies $\mathcal{A} > \mathcal{B}$, then $\mathcal{A} = \mathfrak{P}(X)$.[4]

CONSTRUCTION. Consider a one-to-one map φ from the set $\mathbb{N}$ onto the set

$$R^* = \left\{ r \text{ is a rational number} \mid 0 < r < 1 \text{ and } r \text{ is not of the form } \frac{p}{2^n} \right\},$$

where $p, n \in \mathbb{N}^+$. The algebra $\mathcal{B}$ is the set of finite unions of sets of the type

$$\varphi^{-1}\left(\left\{ r \in R^* \mid \frac{p-1}{2^n} < r < \frac{p}{2^n} \right\}\right). \qquad \square$$

7.16. REMARK. One can naturally define a measure μ on $\mathbb{N}$; $\mu(\mathbb{N}) = 1$, and the algebra $\mathcal{B}$ from the previous section is the algebra of all μ-measurable sets. Indeed, if $p \leq 2^n$, put

$$\mu\left(\varphi^{-1}\left(\left\{ r \in R^* \mid \frac{p-1}{2^n} < r < \frac{p}{2^n} \right\}\right)\right) = \frac{1}{2^n}.$$

Of course, the measure μ is not σ-additive.[5]

---

[4]Obviously, any algebra on $\mathbb{N}$ is simple.

[5]It is clear that $\ker \mathcal{B}$ is a separable set (moreover, one can easily show that $\ker \mathcal{B} = \beta\mathbb{N}$). Suppose $\mathcal{A}$ is an arbitrary algebra such that $X \in \mathcal{A}$ and $\overline{\ker \mathcal{A}}$ is a separable set. Let $\mathcal{A}^*$ be the Boolean algebra $\mathcal{A}/I_{\mathcal{A}}$, where $I_{\mathcal{A}}$ is the ideal of all sets M such that $\mathfrak{P}(M) \subset \mathcal{A}$. (Such objects already appeared in footnote 4.11.) Clearly, one can define a finite strictly positive measure on $\mathcal{A}^*$ which is not necessarily σ-additive. This follows also from the general Kelley theorem [**Kel**].

7.17. PROOF OF THEOREM 2.7. Using Theorem 5.17, we can assume that all algebras $\mathcal{A}_k$ are simple. For each algebra $\mathcal{A}_k$ one of the following two cases is possible:

(1) there exists an ω-saturated σ-algebra $\mathcal{A}'_k > \mathcal{A}_k$ such that

$$\#(\ker \mathcal{A}'_k) \geq \psi(k);$$

(2) such an algebra $\mathcal{A}'_k$ does not exist.

In case (1),

$$\ker \mathcal{A}'_k \subset Z = \bigcup_{l=1}^{\infty} \operatorname{sp} \mathcal{A}_l.$$

In case (2), we will say that the algebra $\mathcal{A}_k$ *is not reducible to an ω-saturated algebra*. If the algebra $\mathcal{A}_k$ is not reducible to an ω-saturated one, then there exists an infinite set $Z_k \subset \operatorname{sp} \mathcal{A}_k$ such that each ultrafilter $z \in Z_k$ has an $\mathcal{A}_k$-similar ultrafilter $c(z) \in \overline{Z} \setminus Z$.

There are three possible cases:

I. There exists at most a finite number of algebras $\mathcal{A}'_k$.

II. There exists an infinite number of algebras $\mathcal{A}'_k$ and an infinite number of algebras which are not reducible to ω-saturated ones.

III. There exists at most a finite number of algebras which are not reducible to ω-saturated ones.

Case I. Assume that we have n algebras $\mathcal{A}'_k$ and $n > 0$. As in Chapter 4, consider the sets of ultrafilters S_n and T_n corresponding to these algebras. We know that $\#(S_n \cup T_n) \leq 2n$ and $S_n \cup T_n \subset Z$. Let $\mathcal{A}_{l_1}, \dots, \mathcal{A}_{l_m}, \dots$ be all the algebras which are not reducible to ω-saturated ones, and let

$$B = \begin{pmatrix} b_1^1 & & \\ \dots & \dots & \dots \\ b_1^m & \dots & b_m^m \\ \dots & \dots & \dots \end{pmatrix}$$

be a matrix of pairwise distinct points of Z such that $B \cap T_n = \varnothing$, $b_i^m \in \operatorname{sp} \mathcal{A}_{l_m}$, and there exists an ultrafilter $c_i^m \in \overline{Z} \setminus Z$ which is $\mathcal{A}_{l_m}$-similar to b_i^m. Let $C = \bigcup \{c_i^m\}$. Since the points b_i^m are irregular, by Lemma 5.3 there exists a sequence

$$B^* = \{b_{p_1}^1, \dots, b_{p_m}^m, \dots\}$$

such that $\overline{B^*} \cap C = \varnothing$. Let $Q \subset X$ and $\overline{Q} \cap Z = S_n \cup B^*$. Then $Q \notin \bigcup \mathcal{A}'_k \cup \bigcup \mathcal{A}_{l_m}$.

Case II. Let $\mathcal{A}'_{n_1}, \dots, \mathcal{A}'_{n_k}, \dots$, $n_1 < n_2 < \dots < n_k < \dots$, be all the algebras $\mathcal{A}'_k$, and let $\mathcal{A}_{l_1}, \dots, \mathcal{A}_{l_m}, \dots$ be all the algebras which are not reducible to ω-saturated ones. Obviously, $n_k - k$ is a nondecreasing sequence tending to ∞ as $k \to \infty$. Consider a sequence of natural numbers

$$\{n_{k_1} < n_{k_2} < \dots < n_{k_m} < \dots\} \subset \{n_1 < n_2 < \dots < n_k < \dots\}$$

such that

$$\psi(n_{k_m}) - \psi(k_m) \geq m(m+1).$$

Let

$$B = \begin{pmatrix} b_1^1 & & \\ \dots & \dots & \dots \\ b_1^m & \dots & b_m^m \\ \dots & \dots & \dots \end{pmatrix}$$

be a matrix of pairwise distinct points of Z such that $b_i^m \in \operatorname{sp} \mathcal{A}_{l_m}$; $b_i^m \notin \ker \mathcal{A}'_{n_k}$ if $k < k_m$, and there exists an ultrafilter $c_i^m \in \overline{Z} \setminus Z$ which is $\mathcal{A}_{l_m}$-similar to b_i^m. Let $k_m \leq k < k_{m+1}$; consider the algebra $\mathcal{A}'_{n_k}$ and the matrix

$$B^m = \begin{pmatrix} b_1^1 & & \\ \dots & \dots & \dots \\ b_1^m & \dots & b_m^m \end{pmatrix}$$

which has $m(m+1)/2$ elements. Let us construct an algebra $\widetilde{\mathcal{A}}_k > \mathcal{A}'_{n_k}$ in the following way:

(a) an ultrafilter a is $\widetilde{\mathcal{A}}_k$-special if and only if it is $\mathcal{A}'_{n_k}$-special and $a \notin B^m$;

(b) two ultrafilters a, b are $\widetilde{\mathcal{A}}_k$-similar if and only if they are $\mathcal{A}'_{n_k}$-similar and $a, b \notin B^m$.

Let us find out when $a \in \ker \mathcal{A}'_{n_k} \setminus \ker \widetilde{\mathcal{A}}_k$. Obviously, either $a \in B^m$, or $a \notin B^m$ and a has $\mathcal{A}'_{n_k}$-similar ultrafilters, but only in B^m. Clearly,

$$\#(\ker \mathcal{A}'_{n_k} \setminus \ker \widetilde{\mathcal{A}}_k) \leq m(m+1).$$

Since $\#(\ker \mathcal{A}'_{n_k}) \geq \psi(n_k)$ and $\psi(n_k) - m(m+1) \geq \psi(k)$ (because $k \geq k_m$), we obtain:

$$\#(\ker \widetilde{\mathcal{A}}_k) \geq \psi(k).$$

If $k < k_1$, then put $\widetilde{\mathcal{A}}_k = \mathcal{A}'_{n_k}$; obviously, in this case also $\#(\ker \widetilde{\mathcal{A}}_k) \geq \psi(k)$. It is important that $\ker \widetilde{\mathcal{A}}_k \cap B = \varnothing$ for all k. As in the proof of Lemma 7.12, let us construct sets of ultrafilters S and T correspondingly to the algebras $\widetilde{\mathcal{A}}_1, \dots, \widetilde{\mathcal{A}}_k, \dots$. The sets S, T, B are pairwise disjoint. As in Case I, consider the set $C = \bigcup \{c_i^m\}$ and a sequence of ultrafilters B^*. Let

$$B' = \{b_{p_m}^m \in B^* \mid c_{p_m}^m \notin \overline{S}\}.$$

Let $Q \subset X$ and $Q \cap Z = S \cup B'$. Then $Q \notin \bigcup \mathcal{A}'_k \cup \bigcup \mathcal{A}_{l_m}$.

By Lemma 7.12, the theorem is also true in Case III. □

7.18. Proof of Theorem 2.8. Let us return to the proof of Theorem 2.7, setting

$$\psi(k) = \left[\frac{4}{3}(k-1)\right] + 1.$$

One needs to consider only Case III. Consider only the nontrivial situation where among the algebras $\mathcal{A}_k$ there exist algebras which are not reducible to ω-saturated ones. So let

$$\mathcal{A}_{l_1}, \dots, \mathcal{A}_{l_m}$$

$(l_1 < \dots < l_m)$ be all the algebras which are not reducible to ω-saturated ones. Consider m pairwise distinct ultrafilters $b_1, \dots, b_m$ such that $b_i \in \operatorname{sp} \mathcal{A}_{l_i}$ and there exists an ultrafilter $c_i \in \overline{Z} \setminus Z$ which is $\mathcal{A}_{l_i}$-similar to b_i; besides, let

$$\bigcup_{k < l_m} \ker \mathcal{A}'_k \cap \{b_1, \dots, b_m\} = \varnothing.$$

Consider a countable sequence of algebras $\widehat{\mathcal{A}}_1, \dots, \widehat{\mathcal{A}}_k, \dots$. If the algebra $\mathcal{A}'_k$ is considered, then $\widehat{\mathcal{A}}_k = \mathcal{A}'_k$. For each l_i, $i \in [1, m]$, $\ker \widehat{\mathcal{A}}_{l_i} = \{b_i\}$, and, obviously, b_i is an $\widehat{\mathcal{A}}_{l_i}$-special ultrafilter. Let us construct sets of ultrafilters S_k, T_k corresponding to the algebras $\widehat{\mathcal{A}}_1, \dots, \widehat{\mathcal{A}}_k$ by induction. The sets S_1, T_1 are constructed in an obvious way. Suppose that we have constructed S_k, T_k. If $\widehat{\mathcal{A}}_{k+1} = \mathcal{A}'_{k+1}$, then let us argue as in the proof of Theorem 2.1, and construct S_{k+1}, T_{k+1}. If $\ker \widehat{\mathcal{A}}_{k+1} = \{b_i\}$, then $b_i \notin S_k \cup T_k$ and we put $S_{k+1} = S_k \cup \{b_i\}$ (i.e., $s_{k+1} = b_i$), $T_{k+1} = T_k$. It is important that, by Remark 4.17, each bush of (S_k, T_k) contains at most one $\widehat{\mathcal{A}}_i$-special ultrafilter s_i, $i \le k$. Put

$$\tau_k = \max\bigl\{\#(K) \mid K \text{ is a bush of } (S_k, T_k)\bigr\}.$$

There are two possible subcases:

(1) $\lim \tau_k < \infty$;
(2) $\lim \tau_k = \infty$.

Subcase (1). As in 4.18, let us construct sets of ultrafilters S and T. Let us construct a set $S^0 \subset S \cup T$. The set $S \cup T$ splits into bushes. If K is such a bush and $b_i \in K$ and $c_i \in \overline{S}$, then $S_0 \cap K = K \cap T$. In all other cases, $S^0 \cap K = K \cap S$. Let $Q \subset X$ and $\overline{Q} \cap Z = S^0$. Then $Q \notin \bigcup \mathcal{A}'_k \cup \bigcup_{i=1}^m \mathcal{A}_{l_i}$.

Subcase (2). Recall that for $k > l_m$ we construct sets S_k, T_k as in the proof of Theorem 2.1. By Lemma 4.30 (see also Remark 4.31), there exist $k_0 > l_m$ and $p > 2m$ such that if we have constructed S_k, T_k ($k \ge k_0$), then there exist ultrafilters

$$q_1, \dots, q_p \in \ker \widehat{\mathcal{A}}_{k+1},$$

each of which can be used for constructing the sets S_{k+1}, T_{k+1}. This means that for each $j \in [1, p]$ one of the two following conditions holds:

1) $q_j \notin S_k \cup T_k$;
2) q_j is the root of a bush of the second or third kind K_j of (S_k, T_k), and K_j does not contain an $\widehat{\mathcal{A}}_i$-special ultrafilter s_i, $i \le k$.

Let us take pairwise distinct ultrafilters $\widehat{b}_1, \dots, \widehat{b}_m \subset Z$ such that

1°. $\widehat{b}_i \in \operatorname{sp} \mathcal{A}_{l_i}$, and $\widehat{b}_i$ has an $\mathcal{A}_{l_i}$-similar ultrafilter $\widehat{c}_i \in \overline{Z} \setminus Z$;
2°. $\{\widehat{b}_1, \dots, \widehat{b}_m\} \cap \bigcup_{n \le k_0} \ker \widehat{\mathcal{A}}_n = \varnothing$.

Assume that we have constructed S_k, T_k ($k > k_0$) and

$$\{\widehat{b}_1, \dots, \widehat{b}_m\} \cap (S_k \cup T_k) = \varnothing.$$

Let us show that we can construct sets S_{k+1}, T_{k+1}, following the proof of Theorem 2.1, and

$$\{\widehat{b}_1, \dots, \widehat{b}_m\} \cap (S_{k+1} \cup T_{k+1}) = \varnothing.$$

Indeed, let us take the corresponding ultrafilters

$$q_1, \dots, q_p \in \ker \widehat{\mathcal{A}}_{k+1}.$$

Among them, let us take $m+1$ ultrafilters that do not belong to $\{\widehat{b}_1, \dots, \widehat{b}_m\}$. This is possible since $p > 2m$. Assume that, for example,

$$\{\widehat{b}_1, \dots, \widehat{b}_m\} \cap \{q_1, \dots, q_{m+1}\} = \varnothing.$$

If each ultrafilter q_j ($j \le m+1$) has an $\widehat{\mathcal{A}}_{k+1}$-similar ultrafilter in $\{\widehat{b}_1, \dots, \widehat{b}_m\}$, then there exists a pair of $\widehat{\mathcal{A}}_{k+1}$-similar ultrafilters q_{j_1}, q_{j_2}, $j_1, j_2 \in [1, m+1]$.[6] Then put $q = q_{j_1}$, $r = q_{j_2}$, and let us argue as in the proof of Theorem 2.1. Assume that

[6] See 3.8.

q_{j_0} ($j_0 \le m+1$) does not have an $\widehat{\mathcal{A}}_{k+1}$-similar ultrafilter in $\{\widehat{b}_1, \dots, \widehat{b}_m\}$. Then put $q = q_{j_0}$, and let us argue as in the proof of Theorem 2.1. Clearly, in all cases

$$\{\widehat{b}_1, \dots, \widehat{b}_m\} \cap (S_{k+1} \cup T_{k+1}) = \varnothing.$$

As in 4.18, let us construct sets S and T. Obviously,

$$\{\widehat{b}_1, \dots, \widehat{b}_m\} \cap (S \cup T) = \varnothing.$$

Put

$$S^{\diamond} = S \cup \{\widehat{b}_i \mid \widehat{c}_i \notin \overline{S}\}.$$

Let $Q \subset X$ and $\overline{Q} \cap Z = S^{\diamond}$. Then $Q \notin \bigcup \mathcal{A}'_k \cup \bigcup_{i=1}^{m} \mathcal{A}_{l_i}$. □

7.19. REMARK. The analysis of the proof of Theorem 2.8 easily implies the statement of footnote 2.4.

7.20. Here it is appropriate to return to systems of distinct representatives (see 3.26–3.29 and 5.15). There exists an infinite analog of Theorem 3.27 (see [**Ev-Wh**], [**Hal1**], [**Hal2**], [**Halm-V**]).

THEOREM. *Let $\{\Theta_\lambda\}_{\lambda \in \Lambda}$ be a family of finite sets. Then this family has a system of distinct representatives if and only if the following condition holds*:

$$\#\left(\bigcup_{\lambda \in M} \Theta_\lambda\right) \ge \#(M)$$

for any finite subset $M \subset \Lambda$.

7.21. In [**Halm-V**] it is shown that Theorem 7.20 follows from Theorem 3.27 by a general argument (this general argument is the Tychonoff theorem (proved in [**Ty**]), which states that the product of compact sets is a compact set). For our purposes one can consider the situation in which $\#(\Lambda) \le \aleph_0$. For the case in which $\#(\Lambda) = \aleph_0$, Theorem 7.20 follows from Theorem 3.27 and from an argument given in the proof of Lemma 7.12. Let us describe a result, the idea of the proof of which is taken from [**Halm-V**], in which a family of algebras of arbitrary cardinality is considered:

STATEMENT. *Let $\{\mathcal{A}_\lambda\}_{\lambda \in \Lambda}$ be a family of ω-saturated algebras; if $M \subset X$ and $\{x\} \in \mathcal{A}_\lambda$ for all $x \in M$, then $M \in \mathcal{A}_\lambda$. Assume that $\bigcup_{\lambda \in \Lambda'} \mathcal{A}_\lambda \ne \mathfrak{P}(X)$ for each nonempty finite $\Lambda' \subset \Lambda$. Then $\bigcup_{\lambda \in \Lambda} \mathcal{A}_\lambda \ne \mathfrak{P}(X)$.*

PROOF. Clearly, $\ker \mathcal{A}_\lambda \subset X$. By the Tychonoff theorem, we can consider the compact set K which is the product of all the compact sets $\ker \mathcal{A}_\lambda$. (Each element $s \in K$ is of the form $\{s_\lambda\}_{\lambda \in \Lambda}$ where $s_\lambda \in \ker \mathcal{A}_\lambda$, and, of course, it is possible that $s_\lambda = s'_\lambda$ though $\lambda \ne \lambda'$.) For each nonempty finite $\Lambda' \subset \Lambda$, consider the set S of all elements $s = \{s_\lambda\}_{\lambda \in \Lambda} \subset K$ such that if $\lambda \subset \Lambda'$ and s_λ is not an $\mathcal{A}_\lambda$ special ultrafilter, then there exists an ultrafilter t_λ which is $\mathcal{A}_\lambda$-similar to s_λ, and $s_{\lambda'} \ne t_\lambda$ for all $\lambda' \in \Lambda'$. Obviously, S is a closed subset of K, and the intersection of a finite number of sets of the type of S is nonempty. Hence there exists an element $s^0 = \{s^0_\lambda\}_{\lambda \in \Lambda} \in K$ that lies in the intersection of all sets of kind S. Let us take the subset $Q = \bigcup_{\lambda \in \Lambda} \{s^0_\lambda\}$ of X. Clearly, $Q \notin \mathcal{A}_\lambda$ for all $\lambda \in \Lambda$. □

7.22. THEOREM. *Suppose that $\varphi\colon \mathbb{N}^+ \to \mathbb{N}^+$, $\varphi(n) \geq n$ for all $n \in \mathbb{N}^+$, and $\varphi(n) - n \to \infty$. Let $\Theta_1, \ldots, \Theta_k, \ldots$ be a countable family of sets, and*

$$\#\left(\bigcup_{k\in M} \Theta_k\right) \geq \varphi(\#(M))$$

for any nonempty finite subset $M \subset \mathbb{N}^+$. Then this family has a system of distinct representatives.

PROOF. Put

$$\mathbb{N}_1 = \{k \mid \#(\Theta_k) < \aleph_0\}, \quad \mathbb{N}_2 = \{k \mid \#(\Theta_k) \geq \aleph_0\}.$$

Let us assume, without loss of generality, that $\mathbb{N}_1, \mathbb{N}_2$ are infinite sets. Let

$$\mathbb{N}_2 = \{l_1, \ldots, l_m, \ldots\}.$$

For any nonempty finite set $M \subset \mathbb{N}_1$, put

$$\mu_1(M) = \#\left(\bigcup_{k\in M} \Theta_k\right) - \#(M).$$

Assume that $\mu_1(M) > 0$ for all M. Then put $M_1 = A_1 = \varnothing$. Otherwise let us take the maximal set M_1 such that $\mu_1(M_1) = 0$. The family of sets $\{\Theta_k\}_{k\in M_1}$ has a system of distinct representatives, namely, $A_1 = \bigcup_{k\in M_1} \Theta_k$. In any case, let us choose a $b_1 \in \Theta_{l_1}$ such that $b_1 \notin A_1$, and for each $k \in \mathbb{N}_1 \setminus M_1$ put

$$\Theta_k^{(1)} = \Theta_k \setminus (A_1 \cup \{b_1\}).$$

It is clear that the family of sets $\{\Theta_k^{(1)}\}_{k\in\mathbb{N}_1\setminus M_1}$ satisfies the conditions of Theorem 7.20.

For any nonempty finite set $M \subset \mathbb{N}_1 \setminus M_1$, put

$$\mu_2(M) = \#\left(\bigcup_{k\in M} \Theta_k^{(1)}\right) - \#(M).$$

Assume that $\mu_2(M) > 0$ for all M. Then put $M_2 = A_2 = \varnothing$. Otherwise, let us take the maximal set M_2 such that $\mu_2(M_2) = 0$. The family of sets $\{\Theta_k^{(1)}\}_{k\in M_2}$ has a system of distinct representatives

$$A_2 = \bigcup_{k\in M_2} \Theta_k^{(1)}.$$

In any case, let us choose a $b_2 \in \Theta_{l_2}$ such that

(i) $b_2 \neq b_1$,
(ii) $b_2 \notin A_1 \cup A_2$;

and for each $k \in \mathbb{N}_1 \setminus (M_1 \cup M_2)$ put

$$\Theta_k^{(2)} = \Theta_k^{(1)} \setminus (A_2 \cup \{b_2\}).$$

Clearly, the family of sets $\{\Theta_k^{(2)}\}_{k\in\mathbb{N}_1\setminus(M_1\cup M_2)}$ satisfies the conditions of Theorem 7.20.

Continuing our construction in the natural way, we obtain:

1) a countable sequence of pairwise disjoint sets

$$M_1, \ldots, M_n, \ldots \subset \mathbb{N}_1;$$

2) for $M_n \neq \varnothing$, a system of distinct representatives A_n for the family of sets $\{\Theta_k\}_{k \in M_n}$;
3) for each m, an element $b_m \in \Theta_{l_m}$;
4) the sets $A_1, \dots, A_n, \dots, \{b_1\}, \dots, \{b_m\}, \dots$ are pairwise disjoint;
5) if $M_0 = \mathbb{N}_1 \setminus \bigcup_{n=1}^{\infty} M_n \neq \varnothing$, then the family of sets

$$\Theta_k^{(0)} = \Theta_k \setminus \left(\bigcup_{n=1}^{\infty} A_n \cup \bigcup_{m=1}^{\infty} \{b_m\} \right), \quad \text{where} \quad k \in M_0,$$

satisfies the conditions of Theorem 7.20.

If $M_0 = \varnothing$, then put $A_0 = \varnothing$. If $M_0 \neq \varnothing$, then, by Theorem 7.20, let us construct a system of distinct representatives A_0 for the family of sets $\{\Theta_k^{(0)}\}_{k \in M_0}$. We have constructed a system of distinct representatives

$$\bigcup_{n=0}^{\infty} A_n \cup \bigcup_{m=1}^{\infty} \{b_m\}$$

for the family of sets $\{\Theta_k\}_{k \in \mathbb{N}^+}$. □

7.23. Let us show that the conditions of Theorem 7.22 cannot be weakened. Suppose that $0 \leq m_0 < m_1 < \cdots < m_k < \cdots$ is an infinite sequence of integers. For each $n \in \mathbb{N}^+$ put

$$\varphi(n) = \min\{m_k \mid n \leq m_k - m_0\}.$$

Obviously, we have $\varphi(n) \geq n$ and

$$\varliminf_{n \to \infty} (\varphi(n) - n) = m_0, \quad \varlimsup_{n \to \infty} (\varphi(n) - n) = \varlimsup_{k \to \infty} (m_{k+1} - m_k + m_0 - 1).$$

Let $\Theta^1, \dots, \Theta^{m_0+1}$ be arbitrary infinite subsets of $\mathbb{N}^+$. For each $n \in \mathbb{N}^+$ put $\Theta_n = \{1, \dots, \varphi(n)\}$. Consider the family of sets

$$\Sigma = \{\Theta^1, \dots, \Theta^{m_0+1}, \Theta_1, \dots, \Theta_n, \dots\}.$$

If $\Sigma' \subset \Sigma$ is nonempty and finite, then

$$\#\left(\bigcup \Sigma'\right) \geq \varphi(\#(\Sigma')).$$

But by Theorem 7.20 Σ does not have a family of distinct representatives. □

7.24. Due to the above theorems the following two results hold.

THEOREM. *Let $\mathcal{A}_1, \dots, \mathcal{A}_k, \dots$ be a countable sequence of ω-saturated σ-algebras.*[7] *This sequence has a system of distinct representatives if and only if for any nonempty finite $M \subset \mathbb{N}^+$ there exist pairwise disjoint sets*

$$V_M^1, \dots, V_M^{\#(M)} \in \bigcup_{j \in M} (\mathfrak{P}(X) \setminus \mathcal{A}_j).$$

[7] Since we deal with ω-saturated algebras, there is no difference between σ-algebras and almost σ-algebras.

7.25. THEOREM. *Let* $\varphi\colon \mathbb{N}^+ \to \mathbb{N}^+$, $\varphi(n) \geq n$ *for all* $n \in \mathbb{N}^+$, *and* $\varphi(n) - n \to \infty$. *Let* $\mathcal{A}_1, \ldots, \mathcal{A}_k, \ldots$ *be a countable sequence of almost* σ*-algebras, and for any nonempty finite* $M \subset \mathbb{N}^+$ *let there exist pairwise disjoint sets*

$$V_M^1, \ldots, V_M^{\varphi(\#(M))} \in \bigcup_{j \in M} (\mathfrak{P}(X) \setminus \mathcal{A}_j).$$

Then this sequence of algebras has a family of distinct representatives.

CHAPTER 8

Theorems on Almost σ-Algebras. Proof of Theorem 2.9

8.1. The proof of Theorem 2.9 is preceded by five lemmas. The first of them is obvious.

LEMMA. *Let s be an irregular ultrafilter, and let $U_1, \dots, U_k, \dots$ be a countable sequence of sets, each belonging to s. Then $\bigcap U_k \in s$.*

8.2. LEMMA. *Let s be an irregular ultrafilter, $U \in s$, and let $\mathcal{A}_1, \dots, \mathcal{A}_k, \dots$ be a countable sequence of algebras. Then there exists a $V \in s \cap U$ such that for each algebra $\mathcal{A}_k$ either $V \notin \mathcal{A}_k$ or $\mathfrak{P}(V) \subset \mathcal{A}_k$.*

PROOF. Let $\mathbb{N}^*$ be a maximal subset of $\mathbb{N}^+$ such that for each $k \in \mathbb{N}^*$ there exists a set $U_k^* \in s \cap U$ and either $\mathfrak{P}(U_k^*) \subset \mathcal{A}_k$ or $W \notin \mathcal{A}_k$ for each $W \in s \cap U_k^*$. If $\mathbb{N}^* = \varnothing$, then we set $U^* = U$. If $\mathbb{N}^* \neq \varnothing$, then, by Lemma 8.1,

$$s \ni U^* = \bigcap_{k \in \mathbb{N}^*} U_k^*.$$

If $\mathbb{N}^* = \mathbb{N}^+$, put $V = U^*$. Otherwise there exists a family $\{U_k, V_k\}_{k \notin \mathbb{N}^*}$ of pairwise disjoint subsets of U^* such that $s \not\ni \bigcup V_k$ and for each pair U_k, V_k there exist $\mathcal{A}_k$-similar ultrafilters s_k, t_k ($U_k \in s_k$, $V_k \in t_k$). Put $V = U^* \setminus \bigcup V_k$. □

8.3. LEMMA. *Let*

$$\begin{pmatrix} U_1^1 & & \\ \dots & \dots & \dots \\ U_1^k & \dots & U_k^k \\ \dots & \dots & \dots \end{pmatrix}$$

be a matrix of pairwise disjoint sets not belonging to an algebra $\mathcal{A}$. Then there exists a sequence $U_{\beta_1}^1, \dots, U_{\beta_k}^k, \dots$ such that $\mathcal{A} \not\ni \bigcup_{k \in \mathbb{N}^*} U_{\beta_k}^k$ for any nonempty subset $\mathbb{N}^* \subset \mathbb{N}^+$.

PROOF. Put $U_{\beta_1}^1 = U_1^1$ (we have no other choice). Now let us argue as in the proof of Theorem 5.2. Take (if possible) a finite or infinite sequence of sets

$$U_{\beta_1^1}^{\alpha_1^1}, \dots, U_{\beta_m^1}^{\alpha_m^1}, \dots \quad (1 < \alpha_1^1 < \dots < \alpha_m^1 < \cdots)$$

such that

$$\mathcal{A} \ni U_{\beta_1}^1 \cup \bigcup_m U_{\beta_m^1}^{\alpha_m^1}.$$

Let us call all the sets $U_{\beta_1^1}^{\alpha_1^1}, \dots, U_{\beta_m^1}^{\alpha_m^1}, \dots$ *marked.* From the two sets U_1^2, U_2^2 let us choose one, say U_2^2, which is not marked. Put $U_{\beta_2}^2 = U_2^2$. Take (if possible) a finite or infinite sequence of sets

$$U_{\beta_1^2}^{\alpha_1^2}, \dots, U_{\beta_m^2}^{\alpha_m^2}, \dots \quad (2 < \alpha_1^2 < \dots < \alpha_m^2 < \dots)$$

such that

$$\mathcal{A} \ni U_{\beta_2}^2 \cup \bigcup_m U_{\beta_m^2}^{\alpha_m^2}.$$

Let us call all the sets $U_{\beta_1^2}^{\alpha_1^2}, \dots, U_{\beta_m^2}^{\alpha_m^2}, \dots$ marked. From the three sets U_1^3, U_2^3, U_3^3 let us choose one which is not marked, etc. In this way we construct the required sequence $U_{\beta_1}^1, \dots, U_{\beta_k}^k, \dots$. Let $\mathbb{N}^* \subset \mathbb{N}^+$, $\#(\mathbb{N}^*) > 1$, and $k_* = \min \mathbb{N}^*$. Clearly,

$$\mathcal{A} \not\ni U_{\beta_{k_*}}^{k_*} \cup \bigcup_{k \in \mathbb{N}^* \setminus \{k_*\}} U_{\rho_k}^k,$$

where $U_{\rho_k}^k$ are nonmarked sets. In particular,

$$\mathcal{A} \not\ni \bigcup_{k \in \mathbb{N}^*} U_{\beta_k}^k. \qquad \square$$

8.4. The following lemma is a refinement of Theorem 5.2.

LEMMA. *Let $\mathcal{A}_1, \dots, \mathcal{A}_k, \dots$ be a countable sequence of algebras, and let there exist a matrix*

$$\begin{pmatrix} U_1^1 & & \\ U_1^2 & U_2^2 & \\ \dots & \dots & \dots \\ U_1^k & \dots & U_k^k \\ \dots & \dots & \dots \end{pmatrix}$$

of pairwise disjoint sets such that $U_i^k \notin \mathcal{A}_k$. Let $W \subset X \setminus \bigcup_{k,i} U_i^k$. Then there exists a family of sets $\mathcal{U} \ni W$ $(1 \le \#(\mathcal{U}) \le \aleph_0)$; if $\#(\mathcal{U}) > 1$, then $\mathcal{U} \setminus \{W\} = \{U_{\beta_1}^{\alpha_1}, \dots, U_{\beta_i}^{\alpha_i}, \dots\}$ and $\alpha_1 < \dots < \alpha_i < \cdots$; for each $k \in \mathbb{N}^+$ there exists $V_k \in \mathcal{U}$ $(V_{\alpha_i} = U_{\beta_i}^{\alpha_i})$ such that if $V_k \subset Q \subset \bigcup \mathcal{U}$, then $Q \notin \mathcal{A}_k$.

PROOF. If U_i^k does not belong to an $\mathcal{A}_k$-special ultrafilter, then let us consider $\mathcal{A}_k$-similar ultrafilters s_i^k, t_i^k such that $s_i^k \ni U_i^k$, $t_i^k \not\ni U_i^k$.

Let us begin by considering the first row of our matrix, i.e., the set U_1^1. If U_1^1 belongs to an $\mathcal{A}_1$-special ultrafilter, then let us pass to the consideration of the second row of our matrix. Otherwise there are three possibilities:

(1) $W \in t_1^1$;
(2) there exists a sequence $U_{l_2^1}^2, U_{l_3^1}^3, \dots, U_{l_k^1}^k, \dots$ such that $t_1^1 \ni \bigcup_{k>1} U_{l_k^1}^k$;
(3) neither (1) nor (2) holds.

In case (1), put $V_1 = W$, and call the set U_1^1 marked. In case (2), call all the sets $U_{l_k^1}^k$ marked.

Consider the sets U_1^2, U_2^2. One of them, say U_2^2, is not marked. Let us call the set U_1^2 marked. (Possibly, we call it marked the second time.) If U_2^2 belongs to an $\mathcal{A}_2$-special ultrafilter, then we pass to the consideration of the third row of our matrix. Otherwise there are four possibilities:

(1) $W \in t_2^2$;
(2) there exists a sequence $U_{l_3^2}^3, U_{l_4^2}^4, \dots, U_{l_k^2}^k, \dots$ such that $t_2^2 \ni \bigcup_{k>2} U_{l_k^2}^k$;
(3) $U_1^1 \in t_2^2$;
(4) neither (1) nor (2) nor (3) holds.

In case (1), put $V_2 = W$, and call the set U_2^2 marked. In case (2), call all the sets $U_{l_k^2}^k$ marked. In case (3), if the set U_1^1 is not marked, put $V_2 = U_1^1$ and call the set U_2^2 marked.

Continuing our arguments, consider the sets $U_1^r, \dots, U_r^r$. Among these sets there is one nonmarked, say U_m^r. Let us call the remaining sets in the r-th row of our matrix marked. (Possibly some of them are called marked for the second time.) If U_m^r belongs to an $\mathcal{A}_r$-special ultrafilter, then we pass to the $(r+1)$-th row of the matrix. Otherwise there are four possibilities:

(1) $W \in t_m^r$;
(2) there exists a sequence $U_{l_{r+1}^r}^{r+1}, U_{l_{r+2}^r}^{r+2}, \dots, U_{l_k^r}^k, \dots$ such that $t_m^r \ni \bigcup_{k>r} U_{l_k^r}^k$;
(3) $U_p^n \in t_m^r$ and $n < r$;
(4) neither (1) nor (2) nor (3) holds.

In case (1), put $V_r = W$, and call the set U_m^r marked. In case (2), call all the sets $U_{l_k^r}^k$ marked. In case (3), if the set U_p^n is not marked, put $V_r = U_p^n$, and call the set U_m^r marked.

Consider all the rows of our matrix. The family $\mathcal{U}$ consists of W and all nonmarked sets U_i^k. If $U_i^k \in \mathcal{U}$, then $V_k = U_i^k$. □

8.5. LEMMA. *Let $\mathcal{A}_1, \dots, \mathcal{A}_k, \dots$ be a countable sequence of almost σ-algebras, and for each k let there exist a set of irregular points $H_k \subset \ker \mathcal{A}_k$ such that $\#(H_k) > \frac{4}{3}(k-1)$. Then there exists an at most countable set $H \subset \bigcup_k H_k$ such that for any k at least one of the following three conditions holds*:

(1) *H contains an $\mathcal{A}_k$-special ultrafilter s_k;*
(2) *there exist $\mathcal{A}_k$-similar ultrafilters s_k and t_k such that $s_k \in \overline{H}$, $t_k \notin \overline{H}$;*
(3) *H contains an ultrafilter s_k such that if $U \in s_k$, then there exist $\mathcal{A}_k$-similar ultrafilters $\widehat{s}, \widehat{t} \ni U$.*

PROOF. Let us assume that each set H_k is finite, and put $Z = \bigcup H_k$. Assume that for some k there exists a finite set $Z_k^* \subset Z$ satisfying the following conditions:

(a) $\#(Z_k^*) > \frac{4}{3}(k-1)$;
(b) each $z \in Z_k^*$ satisfies one of the following three requirements:
 (b$_1$) z is an $\mathcal{A}_k$-special ultrafilter;
 (b$_2$) z has an $\mathcal{A}_k$-similar ultrafilter in Z_k^*;
 (b$_3$) z has an $\mathcal{A}_k$-similar ultrafilter in $\beta X \setminus \overline{Z}$ provided z satisfies neither condition (b$_1$) nor condition (b$_2$).

Then let us define an ω-saturated σ-algebra $\mathcal{A}_k^\circ$: an ultrafilter z is $\mathcal{A}_k^\circ$-special if and only if $z \in Z_k^*$ and z satisfies either condition (b$_1$) or condition (b$_3$); ultrafilters a, b are $\mathcal{A}_k^\circ$-similar if and only if $a, b \in Z_k^*$ and they are $\mathcal{A}_k$-similar.[1] If for some k one cannot construct the algebra $\mathcal{A}_k^\circ$, then let us say that the algebra $\mathcal{A}_k$ is Z-*irreducible.* If the algebra $\mathcal{A}_k$ is Z-irreducible, then there exists an infinite set $Z_k \subset Z$ such that one of the following conditions holds:

(i) $Z_k \subset \ker \mathcal{A}_k$,
(ii) $Z_k \subset \overline{\ker \mathcal{A}_k} \setminus \ker \mathcal{A}_k$.

Moreover, in case (i), each ultrafilter $z \in Z_k$ has an $\mathcal{A}_k$-similar ultrafilter in $\overline{Z} \setminus Z$, whereas in case (ii), if $U \in z \in Z_k$, then there exist $\mathcal{A}_k$-similar ultrafilters $\widehat{s}, \widehat{t} \ni U$. In case (ii), let us take a finite set $\widehat{Z}_k \subset Z_k$ such that $\#(\widehat{Z}_k) > \frac{4}{3}(k-1)$. Construct

[1] Clearly, $\ker \mathcal{A}_k^\circ = Z_k^*$.

an ω-saturated algebra $\mathcal{B}_k^\circ$ by requiring that $\ker \mathcal{B}_k^\circ = \widehat{Z}_k$ and all ultrafilters in $\widehat{Z}_k$ are $\mathcal{B}_k^\circ$-special. For all algebras $\mathcal{A}_k^\circ$, $\mathcal{B}_k^\circ$ and all Z-irreducible algebras $\mathcal{A}_k$ for which we do not consider the corresponding algebras $\mathcal{B}_k^\circ$, we construct the set H by applying the arguments used in the proof of Theorems 2.7 and 2.8. $\square$

8.6. PROOF OF THEOREM 2.9.[2] **1)** Let us assign (if possible) to each set U_i^k an irregular point

$$s_i^k \in \overline{U_i^k} \cap \ker \mathcal{A}_k.$$

Let us call U_i^k a *set of the first kind* if for all sets $U_1^k, \dots, U_{m_k}^k$ there exist ultrafilters $s_1^k, \dots, s_{m_k}^k$. Let us call U_i^k a *set of the second kind* if it is not of the first kind but there exists an ultrafilter s_i^k. The remaining sets U_i^k are *sets of the third kind.* If U_i^k is a set of the third kind, then all points of $\overline{U_i^k} \cap \ker \mathcal{A}_k$ are regular and $\#(\overline{U_i^k} \cap \ker \mathcal{A}_k) \geq 2^{\mathfrak{c}}$ (see 5.14). Let us introduce the following notation:

$$\begin{aligned} Z^* &= \bigcup \{s_i^k\}, \\ Z^f &= \{s_i^k \mid s_i^k \text{ is assigned to a set of the first kind}\}, \\ \mathbb{N}^\circ &= \{k \mid \text{the sets } U_i^k \text{ are of the first kind}\}. \end{aligned}$$

Using Lemma 8.5, for the family $\{\mathcal{A}_k\}_{k \in \mathbb{N}^\circ}$ let us construct the set

$$H \subset Z^f \quad (H_k = \{s_1^k, \dots, s_{m_k}^k\}).$$

Without loss of generality, one can assume that for each $k \in \mathbb{N}^\circ$ either there exists an $\mathcal{A}_k$-special ultrafilter $s_k \in H$ or there exist $\mathcal{A}_k$-similar ultrafilters s_k, t_k ($s_k \in \overline{H}$). Put $T = \bigcup \{t_k\}$. We know that $\overline{H} \cap \overline{T} = \varnothing$. For each $z \in Z^*$, consider $U_z \in z$, and for each set U_i^k of the third kind consider a set $U_{k,i} \notin \mathcal{A}_k$ such that $U_{k,i} \subset U_i^k$. By the arguments from Chapter 5, one can assume that all sets $U_z, U_{k,i}$ are pairwise disjoint. Moreover, one can assume that the following conditions hold:

(1) $T \cap \overline{\bigcup_{z \in H} U_z} = \varnothing$;
(2) $U_{s_i^k} \subset U_i^k$;[3]
(3) for each k either $U_z \notin \mathcal{A}_k$, or $\mathfrak{P}(U_z) \subset \mathcal{A}_k$.[4]

We can already define the required sets $\widehat{U}_i^k$ for all sets of the first and the second kind by setting $\widehat{U}_i^k = U_{s_i^k}$.

2) Let $k \in \mathbb{N}^\circ$, $t_k \in T$, and $t_k \ni U_{m,i}$. Let us assume that this is possible if and only if $t_k \in \overline{\overline{U}_{m,i} \cap \ker \mathcal{A}_m}$ and $L_{m,i} = \overline{U}_{m,i} \cap \ker \mathcal{A}_m$ is an externally separable set. Let us show that there exists a set $M_{m,i}^k \in t_k \cap U_{m,i}$ such that $\overline{M_{m,i}^k} \cap \ker \mathcal{A}_m \subset \overline{\ker \mathcal{A}_k}$. Indeed, take a countable set $\Upsilon \subset \beta X$ such that $\overline{\Upsilon} \supset L_{m,i}$. Put

$$\Upsilon' = \{\nu \in \Upsilon \mid \text{there exists } O_\nu \in \nu \text{ such that } \mathfrak{P}(O_\nu) \subset \mathcal{A}_k\}.$$

---

[2]We will construct sets $\widehat{U}_i^k$ and Q in such a way that for each k there will exist $\widehat{U}_{\psi_k}^{\xi_k} \subset Q$ and $\widehat{U}_{\psi_k}^{\xi_k} \notin \mathcal{A}_k$ (see 2.34).

[3]Of course, it is possible that $z = s_i^k = s_j^l$ and $k \neq l$. Condition (2) can be assumed satisfied because of Lemma 8.1.

[4]Condition (3) can be satisfied by Lemma 8.2. At the end of the proof of our theorem it will become clear that condition (3) in this form is used only for $z \in H$. And if $s_i^k \in Z^* \setminus H$, then it suffices to know that $U_{s_i^k} \notin \mathcal{A}_k$.

Since $\mathcal{A}_k$ is an almost σ-algebra, $\mathfrak{P}\left(\bigcup_{\nu\in\Upsilon'} O_\nu\right) \subset \mathcal{A}_k$. Therefore $t_k \not\ni \bigcup_{\nu\in\Upsilon'} O_\nu$. It remains to choose an $M^k_{m,i} \in t_k \cap U_{m,i}$ such that

$$\overline{\Upsilon'} \cap \overline{M^k_{m,i}} = \varnothing.$$

Put

$$T_{m,i} = \left\{t_k \in T \mid t_k \ni U_{m,i}\right\}.$$

Consider the measures

$$\mu_{\mathcal{A}_m}^{U_{m,i}}, \{\mu_{\mathcal{A}_m}^{M^k_{m,i}}\}_{t_k\in T_{m,i}}$$

(see 5.13). It is clear that each $\mu_{\mathcal{A}_m}^{M^k_{m,i}}$-nonmeasurable set contains a subset which belongs neither to $\mathcal{A}_m$ nor to $\mathcal{A}_k$. Arguing as in 5.13, let us construct a family $\{Q^k_{m,i}\}_k$ of pairwise disjoint subsets of $U_{m,i}$ such that either $k = 0$, or $t_k \in T_{m,i}$; the set $Q^0_{m,i}$ is $\mu_{\mathcal{A}_m}^{U_{m,i}}$-nonmeasurable, and if $t_k \in T_{m,i}$, then the set $Q^k_{m,i}$ is $\mu_{\mathcal{A}_m}^{M^k_{m,i}}$-nonmeasurable. The sets $Q^k_{m,i}$ are needed to construct the matrix $\mathfrak{Q}$ introduced below.

3) For each algebra $\mathcal{A}_m$, $m \notin \mathbb{N}^\circ$, consider some set U_{m,i_m}. Let us define the additional required sets $\widehat{U}^k_i$ as follows: if $U_{m,j} \neq U_{m,i_m}$, put $\widehat{U}^m_j = U_{m,j}$. Denote by T° the union of all sets T_{m,i_m}. By Lemma 5.3, we can assume that the inclusions $t \in T$ and $t \ni \bigcup_m U_{m,i_m}$ imply that t contains some U_{m,i_m}, and hence $t \in T^\circ$. Denote by $\mathfrak{A}$ the family of all algebras $\mathcal{A}_m$, $m \notin \mathbb{N}^\circ$, and all algebras $\mathcal{A}_k$, $t_k \in T^\circ$. Let $\#(\mathfrak{A}) = \aleph_0$. Consider a matrix of pairwise disjoint sets

$$\mathfrak{Q} = \begin{pmatrix} Q^1_1 & & \\ \cdots & \cdots & \cdots \\ Q^p_1 & \cdots & Q^p_p \\ \cdots & \cdots & \cdots \end{pmatrix}.$$

There is a one-to-one correspondence between the rows of $\mathfrak{Q}$ and the algebras from $\mathfrak{A}$. The sets Q^p_j possess the following properties:

(a) if the p-th row of $\mathfrak{Q}$ corresponds to some algebra $\mathcal{A} \in \mathfrak{A}$, then $Q^p_j \notin \mathcal{A}$;
(b) if the p-th row of $\mathfrak{Q}$ corresponds to some algebra $\mathcal{A}_m$, $m \notin \mathbb{N}^\circ$, or to some algebra $\mathcal{A}_k$ such that $t_k \ni U_{m,i_m}$, then $Q^p_j \subset U_{m,i_m}$;
(c) if $\mathcal{V}$ is an arbitrary family of sets Q^p_j contained in U_{m,i_m}, then $\mathcal{A}_m \not\ni \bigcup \mathcal{V}$.[5]

Consider the set $W = \bigcup_{z\in H} U_z$, the matrix $\mathfrak{Q}$, and the family of algebras $\mathfrak{A}$. By Lemma 8.4 there exists a family of sets $\mathcal{U}$ consisting either of only W, or of W and a sequence (possibly finite) $Q^{\alpha_1}_{\beta_1}, \ldots, Q^{\alpha_i}_{\beta_i}, \ldots$ $(\alpha_1 < \cdots < \alpha_i < \cdots)$, such that for each algebra $\mathcal{A} \in \mathfrak{A}$ there exists a $V_{\mathcal{A}} \in \mathcal{U}$ such that $\mathcal{A} \not\ni V_{\mathcal{A}} \cup \bigcup \mathcal{U}'$ for any subfamily $\mathcal{U}' \subset \mathcal{U}$. Let us conclude the definition of the required sets $\widehat{U}^k_i$. Denote by $\widehat{U}^m_{i_m}$ the union of all sets from $\mathcal{U}$ which are contained in U_{m,i_m} (if such sets exist). If there are no such sets, then put $\widehat{U}^m_{i_m} = U_{m,i_m}$. Put $Q = \bigcup \mathcal{U}$. Clearly, $\widehat{U}^k_i \subset U^k_i$ and $\widehat{U}^k_i \notin \mathcal{A}_k$. Either $\widehat{U}^k_i = \widehat{U}^l_j$, or $\widehat{U}^k_i \cap \widehat{U}^l_j = \varnothing$. (If $\widehat{U}^k_i = \widehat{U}^l_j$ and $k \neq l$, then $\widehat{U}^k_i = \widehat{U}^l_j = U_z$.) Clearly, $Q \notin \bigcup \mathcal{A}_k$. Let us show that for each $k \in \mathbb{N}^+$ there exists a subset $\widehat{U}^{\xi_k}_{\psi_k} \subset Q$ such that $\widehat{U}^{\xi_k}_{\psi_k} \notin \mathcal{A}_k$. If $\mathcal{A}_k \in \mathfrak{A}$ and $V_{\mathcal{A}_k} \subset U_{m,i_m}$, put $\widehat{U}^{\xi_k}_{\psi_k} = \widehat{U}^m_{i_m}$. Let $\mathcal{A}_k \in \mathfrak{A}$ and $V_{\mathcal{A}_k} = W$. We know that $W \notin \mathcal{A}_k$ and $\mathcal{A}_k$ is an almost σ-algebra, while for each $U_z \subset W$ either $U_z \notin \mathcal{A}_k$, or $\mathfrak{P}(U_z) \subset \mathcal{A}_k$. Hence

[5] Property (c) can be satisfied by Lemma 8.3.

there exists a subset $U_{z_k} \subset W$ such that $U_{z_k} \notin \mathcal{A}_k$. Put $\widehat{U}^{\xi_k}_{\psi_k} = U_{z_k}$. Finally, let $\mathcal{A}_k \notin \mathfrak{A}$. Then again $W \notin \mathcal{A}_k$, and once more we find that $\widehat{U}^{\xi_k}_{\psi_k} = U_{z_k} \subset W$. □

8.7. REMARK. Taking into account the content of 8.1–8.6, we come to the conclusion that the statement of footnote 2.5 is true.

8.8. To formulate the next theorem, we first recall a definition:

DEFINITION. We say that a function $\varphi\colon \mathbb{N}^+ \to \mathbb{N}^+$ *grows faster than a linear function* if for each positive number α there exists $k_\alpha \in \mathbb{N}^+$ such that $\varphi(k) > \alpha \cdot k$ for all $k > k_\alpha$.

8.9. THEOREM. *Let $\mathcal{A}_1, \dots, \mathcal{A}_k, \dots$ be a countable sequence of almost σ-algebras, let $\varphi(k)$ be a function growing faster than a linear one, and suppose we have the inequality $\varphi(k) > \frac{4}{3}(k-1)$ for all k. Assume that for each k there exist pairwise disjoint sets $U^k_1, \dots, U^k_{\varphi(k)} \notin \mathcal{A}_k$. Then for each U^k_i one can choose a subset $\widehat{U}^k_i \subset U^k_i$ such that the following hold*:

(i) $\widehat{U}^k_i \notin \mathcal{A}_k$;
(ii) *either* $\widehat{U}^k_i = \widehat{U}^l_j$, *or* $\widehat{U}^k_i \cap \widehat{U}^l_j = \varnothing$;
(iii) *for each k there exists $\widehat{U}^{\xi_k}_{\psi_k}$ such that if $\widehat{U}^{\xi_k}_{\psi_k} \subset Q \subset \bigcup_j \widehat{U}^{\xi_j}_{\psi_j}$, then $Q \notin \mathcal{A}_k$.*[6]

PROOF. We will use the notions from the proof of Theorem 2.9. Let $\#(\mathbb{N}^\circ) = \aleph_0$. The assumptions of the theorem imply that there exist finite sets of ultrafilters S_*, T_*, an infinite matrix

$$\Theta = \begin{pmatrix} \tau^1_1 & \dots & \tau^1_{m_1} \\ \multicolumn{3}{c}{\dotfill} \\ \tau^n_1 & \dots & \tau^n_{m_n} \\ \multicolumn{3}{c}{\dotfill} \end{pmatrix}$$

($m_n \to \infty$), and $k_0 \in \mathbb{N}^+$ such that the following conditions can be satisfied:

(1) $S_* \cup \Theta = Z_* \subset Z^f$;
(2) $S_* \cap \Theta = \varnothing$;
(3) if $k < k_0$ and the sets U^k_i are of the first kind, then either there exists an $\mathcal{A}_k$-special ultrafilter $s_k \in S_*$, or $\mathcal{A}_k$-similar ultrafilters s_k, t_k are considered ($s_k \in S_*$, $t_k \in T_*$);
(4) if $k \geq k_0$, the sets U^k_i are of the first kind, and the natural number n is defined by the formula
$$n = n(k) = \#\bigl(\{l \mid l \in [k_0, k] \text{ and the sets } U^l_i \text{ are of the first kind}\}\bigr),$$
then we require that each ultrafilter τ^n_j coincides with s^k_i;
(5) $T_* \cap \overline{\bigcup_{z \in Z_*} U_z \cup \bigcup_{m \in \mathbb{N}^+ \setminus \mathbb{N}^\circ} U_{m,i_m}} = \varnothing$.

For each $m \in \mathbb{N}^+ \setminus \mathbb{N}^\circ$ consider the pairwise disjoint sets
$$O^m_1, \dots, O^m_m \subset U_{m,i_m}$$
such that $O^m_\iota \notin \mathcal{A}_m$. For each O^m_ι consider either an $\mathcal{A}_m$-special ultrafilter $q^m_\iota \ni O^m_\iota$, or $\mathcal{A}_m$-similar ultrafilters q^m_ι, r^m_ι ($q^m_\iota \ni O^m_\iota$, $r^m_\iota \not\ni O^m_\iota$). Since all points of $\overline{O^m_\iota} \cap \ker \mathcal{A}_m$ are regular, we have
$$\#(\overline{O^m_\iota} \cap \ker \mathcal{A}_m) \geq 2^{\mathfrak{c}}$$

[6]Possibly, $\widehat{U}^{\xi_k}_{\psi_k} = \widehat{U}^{\xi_l}_{\psi_l}$ for $k \neq l$.

(see 5.14). Hence one can assume that $r_\iota^m \not\ni U_z$. Besides, one can assume that if $r_\iota^m \ni \bigcup_{l,j} O_j^l$, then $r_\iota^m \ni O_j^l$. If the ultrafilter τ_j^n is not $\mathcal{A}_k$-special ($n = n(k)$), then there exists an ultrafilter η_j^n that is $\mathcal{A}_k$-similar to τ_j^n. Since for each k large enough there exist many sets U_i^k, one can assume that $\eta_j^n \not\ni \bigcup_{z\in Z_*} U_z$. Besides, one can assume that if $\eta_j^n \ni \bigcup_{m,\iota} O_\iota^m$, then $\eta_j^n \ni O_\iota^m$. Let us choose a countable sequence $\tau_{j_1}^1, \dots, \tau_{j_n}^n, \dots$ such that $r_\iota^m \not\ni \bigcup_n U_{\tau_{j_n}^n}$ for all r_ι^m. Put

$$\widetilde{Z} = S_* \cup \{\tau_{j_1}^1, \dots, \tau_{j_n}^n, \dots\}.$$

The analysis of the proof of Lemma 8.4 shows that one can find sets

$$O_{\beta_1}^{\alpha_1}, \dots, O_{\beta_i}^{\alpha_i}, \dots \quad (\alpha_1 < \dots < \alpha_i < \dots)$$

such that each $m \in \mathbb{N}^+ \setminus \mathbb{N}^\circ$ corresponds to some $O_{\beta_{i_m}}^{\alpha_{i_m}}$ so that the following holds: if

$$O_{\beta_{i_m}}^{\alpha_{i_m}} \subset Q^1 \subset \bigcup_j O_{\beta_j}^{\alpha_j}, \quad Q^2 \subset \bigcup_{z\in\widetilde{Z}} U_z,$$

then $Q^1 \cup Q^2 \notin \mathcal{A}_m$. Put $\widehat{U}_{\psi_m}^{\xi_m} = O_{\beta_{i_m}}^{\alpha_{i_m}}$. If $k < k_0$ and the sets U_i^k are of the first kind, then put $\widehat{U}_{\psi_k}^{\xi_k} = U_{s_k}$. If $\eta_{j_n}^n \ni \widehat{U}_{\psi_m}^{\xi_m}$ and $n = n(k)$, then put $\widehat{U}_{\psi_k}^{\xi_k} = \widehat{U}_{\psi_m}^{\xi_m}$. If the ultrafilter $\tau_{j_n}^n$ is $\mathcal{A}_k$-special ($n = n(k)$), or $\eta_{j_n}^n \not\ni \widehat{U}_{\psi_m}^{\xi_m}$, then put $\widehat{U}_{\psi_k}^{\xi_k} = U_{\tau_{j_n}^n}$. The sets $\widehat{U}_i^k$ are defined in the natural way. $\square$

8.10. In 2.25 the definition of the absolute density $p_a(\mathbb{N}^*)$ of a subset $\mathbb{N}^*$ of the set of natural numbers is given. The proof of the following Theorem 8.11 must be preceded by a more general definition.

DEFINITION. Let $L = \{l_1 < l_2 < \dots < l_k < \dots\}$ be a countable subset of the set of natural numbers, and let $L^* \subset L$. The number

$$\varliminf_{n\to\infty} \frac{\#(\{k \le n \mid l_k \in L^*\})}{n}$$

is called the *absolute density of L^* in L*, and denoted $p_a^L(L^*)$.

8.11. Now let us formulate and prove the following theorem which is, obviously, a generalization of Theorem 2.9.

THEOREM. *Let $\mathcal{A}_1, \dots, \mathcal{A}_k, \dots$ be a countable sequence of almost σ-algebras, for each k let there exist a finite sequence of pairwise disjoint sets $U_1^k, \dots, U_{m_k}^k \notin \mathcal{A}_k$, and let $m_k > \frac{4}{3}(k-1)$. Then for each U_i^k one can choose $\widehat{U}_i^k \subset U_i^k$, and for each k one can choose $\widehat{U}_{\psi_k}^{\xi_k}$ so that the following hold*:

(i) $\widehat{U}_i^k \notin \mathcal{A}_k$;
(ii) *either* $\widehat{U}_i^k = \widehat{U}_j^l$, *or* $\widehat{U}_i^k \cap \widehat{U}_j^l = \varnothing$;
(iii) *for any real number $\rho < 1$ and any finite $\mathbb{N}_* \subset \mathbb{N}^+$ one can find a subset $\mathbb{N}^* \subset \mathbb{N}^+$ such that $\mathbb{N}_* \subset \mathbb{N}^*$, $p_a(\mathbb{N}^*) > \rho$, and if $\widehat{U}_{\psi_k}^{\xi_k} \subset \widehat{Q} \subset \bigcup_{j\in\mathbb{N}^*} \widehat{U}_{\psi_j}^{\xi_j}$, then $\widehat{Q} \notin \mathcal{A}_k$*;
(iv) *there exists a set Q which is a union of sets of the type of $\widehat{U}_{\psi_l}^{\xi_l}$, such that $Q \notin \bigcup \mathcal{A}_k$.*

PROOF. 1) Let us make some changes in the proofs of Theorems 2.7 and 2.8. These changes concern Cases II and III. In Case II, put

$$S = \{s_1, \ldots, s_k, \ldots\},$$

where either s_k is an $\widetilde{\mathcal{A}}_k$-special ultrafilter, or s_k, t_k are $\widetilde{\mathcal{A}}_k$-similar ultrafilters and $t_k \in T$. By Lemma 4.25, we can assume that

$$\#\big(\{i \mid s_i = s_k\}\big) \leq 4 \quad \text{for all } s_k \in S.$$

Let us proceed to Case III. As in the proof of Theorem 2.8, let us construct sets of ultrafilters

$$S_1, T_1; S_2, T_2; \ldots; S_k, T_k; \ldots .$$

However, in constructing these sets we proceed as in the proof of Lemma 4.23 but not as in the proof of Theorem 2.1. Clearly, any bush of (S_k, T_k) contains not more than one $\widehat{\mathcal{A}}_i$-special ultrafilter s_i, $i \leq k$. Exactly as above, two subcases are considered. In Subcase (1), let us argue in exactly the same way as above. As for Subcase (2), the analysis of the proof of Lemma 4.23 implies that if we have a bush K of (S_k, T_k) and $\|K\|$ is a large number, then one can take ultrafilters $\widehat{b}_1, \ldots, \widehat{b}_m$, etc., as in the proof of Theorem 2.8 in Subcase (2).[7]

After we made changes in the proof of Theorems 2.7 and 2.8, let us discuss the construction of the set H in Lemma 8.5. Let us assume that if the algebra $\mathcal{A}_k$ is Z-irreducible, condition (ii) in the proof of Lemma 8.5 does not hold. We argue in exactly the same way as in the proof of Theorems 2.7 and 2.8 (taking into account the indicated changes), but instead of the algebras $\mathcal{A}'_k$ we consider the algebras $\mathcal{A}^\circ_k$, and instead of algebras which cannot be reduced to ω-saturated ones, we consider Z-irreducible algebras and the sets Z_k.

We also study three cases. In Case I, put $H = S_n \cup B^*$; in Case II, put $H = S \cup B'$; in Case III, put either $H = S^\circ$, or $H = S^\diamond$. Now we define the set of ultrafilters H^* in the modified proof of Lemma 8.5. In Case I, put $H^* = H$; in Case II, put $H^* = S \cup B^*$; in Case III, Subcase (1), put $H^* = H$, and in Subcase (2), put $H^* = S \cup \{\widehat{b}_1, \ldots, \widehat{b}_m\}$.

Let us proceed to the analysis of the proof of Theorem 2.9, taking into account the changes in the proofs of Theorems 2.7 and 2.8 and Lemma 8.5. Therefore we will assume (without loss of generality) that if s_i^k and t are $\mathcal{A}_k$-similar ultrafilters, U_i^k is a set of the first kind, and $t \in \overline{Z^f} \setminus Z^f$, then

$$\#(\ker \mathcal{A}_k \cap Z^f) = \aleph_0.$$

It is clear how to introduce the set of ultrafilters H^*. In the first part of the proof, let us require that

$$\{t_k \in T \mid s_k \in H\} \cap \overline{\bigcup_{z \in H^*} U_z} = \varnothing.$$

Now let us pass to the third part of the proof. Put

$$H' = \{s_k \in H \mid \mathcal{A}_k \notin \mathfrak{A}\}, \quad H'' = (H^* \setminus H) \cup H'.$$

For each set $Q_j^p \in \mathfrak{Q}$ that corresponds to the algebra $\mathcal{A}_\eta$ (this correspondence is mentioned in property (b)), consider either an $\mathcal{A}_\eta$-special ultrafilter q_j^p, or $\mathcal{A}_\eta$-similar ultrafilters q_j^p, r_j^p, $q_j^p \ni Q_j^p$, $r_j^p \not\ni Q_j^p$. Since

$$\#(\ker \mathcal{A}_\eta \cap \overline{Q_j^p}) \geq 2^{\mathfrak{c}}$$

[7]Instead of Lemmas 4.23 and 4.25, one can use simpler arguments from Remark 4.27.

(see 5.14), let us assume that either $r_j^p \not\ni \bigcup_{z\in H''} U_z$ or $r_j^p \in \overline{H'} \setminus H'$.[8] Also let us assume that if $r_j^p \in \overline{H'} \setminus H'$, then in the p-th row of $\mathfrak{Q}$ there is one more set Q_{p+1}^p (if $Q_{p+1}^p \cap Q_j^i \neq \varnothing$, then $i = p$, $j = p+1$) and

$$r_{p+1}^p \in \overline{H'} \setminus H'.$$

2) Using the proof of Lemma 8.4 with $W = \varnothing$, the matrix $\mathfrak{Q}$ without the sets Q_{p+1}^p, and the ultrafilters q_j^p, r_j^p, for the family of algebras $\mathfrak{A}$ we construct the corresponding family $\mathcal{U}^*$ of sets of the type of Q_j^p. (In Lemma 8.4 it is the family $\mathcal{U}$; but to avoid confusion we introduce the new notation.) In the construction of the sets $\widehat{U}_i^k$ let us make the following change. If U_{m,i_m} does not contain elements of $\mathcal{U}^*$, then put $\widehat{U}_{i_m}^m = U_{m,i_m}$. Otherwise put

$$\widehat{U}_{i_m}^m = \bigcup\{Q_j^p \in \mathcal{U}^* \mid Q_j^p \subset U_{m,i_m}\}.$$

If $\mathcal{A}_\eta \in \mathfrak{A}$, then one can find a $V_\eta \in \mathcal{U}^*$ such that if $V_\eta \subset \widehat{Q} \subset \bigcup\mathcal{U}^*$, then $\widehat{Q} \notin \mathcal{A}_\eta$. If $V_\eta \subset \widehat{U}_{i_m}^m$, then put $\widehat{U}_{\psi_\eta}^{\xi_\eta} = \widehat{U}_{i_m}^m$.

3) Let us define the rest of the sets $\widehat{U}_{\psi_k}^{\xi_k}$. If $s_k \in H$ and $t_k \not\ni U_{m,i_m}$, then put $\widehat{U}_{\psi_k}^{\xi_k} = U_{s_k}$.[9] Let $h \in H^* \setminus H$. Clearly, the ultrafilter h corresponds to a certain algebra $\mathcal{A}_k$, and h has an $\mathcal{A}_k$-similar ultrafilter $h' \in \overline{H} \setminus H$. Put $\widehat{U}_{\psi_k}^{\xi_k} = U_h$. It remains to consider the situation in which the construction of H comes from Subcase (1) of Case III considered in the proof of Theorem 2.8. We must consider a finite family of algebras $\mathcal{A}_k$ such that $s_k \in \overline{H} \setminus H$ and $t_k \in Z^f \setminus H$. For each such k consider $\mathcal{A}_k$-similar ultrafilters a_k, d_k such that $a_k \ni U_{h_k}$ ($h_k \in H$) and $d_k \notin H$. One can assume that

1) $d_k \not\ni U_z$ if $z \in H$; 2) $d_k \not\ni U_{m,i_m}$; 3) if $d_k \ni \bigcup_{z\in H'} U_z$, then $d_k \in \overline{H'} \setminus H'$; 4) if $s_k \in \overline{H \setminus H'}$, then $h_k \in H \setminus H'$; 5) all ultrafilters h_k are pairwise distinct; 6) $U_{h_k} \notin r_j^p$.

Put $\widehat{U}_{\psi_k}^{\xi_k} = U_{h_k}$.

4) Denote by H_* the set of all ultrafilters of the type of r_j^p, h', and d_k which are contained in $\overline{H'} \setminus H'$. Put

$$L = \{k \in \mathbb{N}^\circ \mid \mathcal{A}_k \notin \mathfrak{A}\}.$$

Assume, without loss of generality, that $\#(L) = \aleph_0$. Take an arbitrary real number $\rho < 1$. For any finite $L_* \subset L$ there exists a subset $L^* \subset L$ such that $L_* \subset L^*$, $p_a^L(L^*) > \rho$, and

$$\overline{\bigcup_{k\in L^*} U_{\psi_k}^{\xi_k}} \cap H_* = \varnothing.$$

If $\mathbb{N}^* = (\mathbb{N}^+ \setminus L) \cup L^*$, $k \in \mathbb{N}^*$, and

$$\widehat{U}_{\psi_k}^{\xi_k} \subset \widehat{Q} \subset \bigcup_{i\in\mathbb{N}^*} \widehat{U}_{\psi_i}^{\xi_i},$$

then $\widehat{Q} \notin \mathcal{A}_k$. Properties (i), (ii), and (iii) are proved.

---

[8]One can achieve this in Case II by choosing an appropriate sequence of ultrafilters B^* and, clearly, one can also achieve this in Case III since in this case $\#(H^* \setminus H) < \aleph_0$.

[9]Clearly, $s_k = s_{\psi_k}^{\xi_k}$.

5) It remains to construct the set Q. If the construction of H comes from Subcase (1) of Case III considered in the proof of Theorem 2.8, then put

$$Q = \bigcup_{\mathcal{A}_k \in \mathfrak{A}} \widehat{U}_{\psi_k}^{\xi_k} \cup \bigcup_{z \in H'} U_z \cup \bigcup \Big\{ U_{h_k} \mid d_k \not\ni \bigcup_{z \in H'} U_z \Big\}.$$

In all other cases put

$$Q = \bigcup_{\mathcal{A}_k \in \mathfrak{A}} \widehat{U}_{\psi_k}^{\xi_k} \cup \bigcup_{z \in H'} U_z \cup \bigcup \Big\{ U_h \mid h \in H^* \setminus H \text{ and } h' \not\ni \bigcup_{z \in H'} U_z \Big\}.$$

In conclusion, let us note that, similarly to Theorem 2.9, for each $k \in \mathbb{N}^+$ there exists a subset $\widehat{U}_{\gamma_k}^{\beta_k} \subset Q$ such that $\widehat{U}_{\gamma_k}^{\beta_k} \notin \mathcal{A}_k$. $\square$

8.12. Let us return to the study of the finite sequences of algebras mentioned in Remark 4.44. But first let us give a definition.

DEFINITION. A bush K is said to be *simple* if $\|K\| = 1$. A bush K is said to be *composite* if $\|K\| > 1$.

If K is a simple bush, then either K contains one $\mathcal{A}_i$-special ultrafilter s_i and K is a bush of the first kind, or K contains one pair of $\mathcal{A}_i$-similar ultrafilters s_i, t_i and K is a bush of the second kind.

8.13. Consider a finite sequence of algebras $\mathcal{A}_1, \dots, \mathcal{A}_n$. Let $\bigcup_{i=1}^n \mathcal{A}_i \neq \mathfrak{P}(X)$, i.e., there exist the corresponding sets of ultrafilters S_n, T_n. Put

$$n'_0 = \#\big(\{i \in [1,n] \mid s_i \text{ belongs to a composite bush of } (S_n, T_n)\}\big).$$

Assume that there are l'_0 composite bushes of (S_n, T_n). By the property from 4.11,

$$n'' = \#\big(\{q \mid q \text{ is contained in a composite bush of } (S_n, T_n)\}\big) \le n'_0 + l'_0.$$

Any composite bush contains some ultrafilters s_i, s_j with $i \neq j$. Hence

$$n'' \le n'_0 + \frac{n'_0}{2} = \frac{3}{2} n'_0.$$

Put

$$C' = \big\{q \mid q \text{ is the root of a simple bush of the second kind of } (S_n, T_n)\big\}.$$

Clearly,

$$\#((S_n \cup T_n) \setminus C') \le \frac{3}{2} n'_0 + (n - n'_0) \le \frac{3}{2} n.$$

8.14. LEMMA. *Let $\mathcal{A}_1, \dots, \mathcal{A}_n$ be a finite sequence of algebras, and for each $k \in [1,n]$ let there exist pairwise distinct ultrafilters $s_1^k, \dots, s_{m_k}^k \in \ker \mathcal{A}_k$ with $m_k > \frac{3}{2}(k-1)$. Then there exist corresponding sets of ultrafilters S_n, T_n, possessing the following additional property: if $s \in S_n$, then there exists $s_j = s = s_{i_j}^j$.*

PROOF. If $n = 1$, the lemma is true. Assume that it is true for n algebras, and let us prove that it is true for $n+1$ algebras. By the arguments from 8.13, there exists an ultrafilter $q = s_i^{n+1}$, and either $q \notin S_n \cup T_n$, or $q \in C'$; either q is an $\mathcal{A}_{n+1}$-special ultrafilter, or q has an $\mathcal{A}_{n+1}$-similar ultrafilter r. Let $q \notin S_n \cup T_n$; if q is an $\mathcal{A}_{n+1}$-special ultrafilter or $r \notin S_n$, then put $s_{n+1} = q$; in the second case put $t_{n+1} = r$. If $r \in S_n$, then always put $s_{n+1} = r$, $t_{n+1} = q$. It remains to consider the situation in which $q = t_j \in C'$ and either q is an $\mathcal{A}_{n+1}$-special ultrafilter, or $r \notin S_n$. Let us turn the bush $\{s_j, t_j\}$ over and put $s_{n+1} = q$; if r is considered, then put $t_{n+1} = r$. We have constructed sets S_{n+1}, T_{n+1} which possess the additional property of our lemma. □

8.15. REMARK. The sets of ultrafilters S_n, T_n from the previous lemma are constructed so that any bush K of (S_n, T_n) can contain only one $\mathcal{A}_i$-special ultrafilter s_i, $i \leq n$. If K contains such an ultrafilter and $t \in T_n \cap K$, then there exists $t_j = t = s_{i_j}^j$.

8.16. The following lemma is a refinement of Lemma 8.14. Its proof is similar to the arguments in Remark 4.27.

LEMMA. *Let $\mathcal{A}_1, \dots, \mathcal{A}_n, \mathcal{A}_{n+1}$, $n \in \mathbb{N}^+$, be a sequence of algebras, and for each $k \in [1, n+1]$ let there exist pairwise distinct ultrafilters $s_1^k, \dots, s_{m_k}^k \in \ker \mathcal{A}_k$ with $m_k > \frac{3}{2}(k-1)$. Then, considering the algebras $\mathcal{A}_1, \dots, \mathcal{A}_n$ and taking the corresponding sets of ultrafilters S_n, T_n which satisfy the additional property of Lemma 8.14, one can construct the corresponding sets of ultrafilters S_{n+1}, T_{n+1} so that the following hold:*

(i) *if $s \in S_{n+1}$, then there exists $s_j = s = s_{i_j}^j$ $(j \leq n+1)$;*
(ii) *if $\iota_n(s) \geq 4$, then $s \in S_{n+1}$ and $\iota_{n+1}(s) = \iota_n(s)$;*
(iii) *if $\iota_n(s) < 4$ and $s \in S_{n+1}$, then $\iota_{n+1}(s) \leq \iota_n(s) + 1$;*
(iv) *if $s \in S_{n+1} \setminus S_n$, then $\iota_{n+1}(s) \leq 2$.*

PROOF. We will use the notation and formulas from 8.13. Let

$$\{K_1, \dots, K_\alpha\} = \big\{K \text{ is a bush of } (S_n, T_n) \mid \text{there exists an } s \in K \text{ such that } \iota_n(s) \geq 4\big\}.$$

We have

$$n'' - \|K_1\| - \dots - \|K_\alpha\| - \alpha \leq \frac{3}{2}(n'_0 - \|K_1\| - \dots - \|K_\alpha\|).$$

Hence

$$\begin{aligned} n'' &\leq \frac{3}{2}(n'_0 - \|K_1\| - \dots - \|K_\alpha\|) + \|K_1\| + \dots + \|K_\alpha\| + \alpha \\ &= \frac{3}{2}n'_0 - \frac{\|K_1\| + \dots + \|K_\alpha\|}{2} + \alpha. \end{aligned}$$

Put

$$\gamma = \#\big(\{s \in S_n \mid \iota_n(s) \geq 4\}\big).$$

Obviously,

$$n'' \leq \frac{3}{2}n'_0 - 2\gamma + \alpha \leq \frac{3}{2}n'_0 - \gamma.$$

Hence

$$\#((S_n \cup T_n) \setminus C') \leq \frac{3}{2}n'_0 - \gamma + (n - n'_0).$$

In the construction of S_{n+1}, T_{n+1}, we will use one of the ultrafilters from the set

$$\{s_1^{n+1}, \dots, s_{m_{n+1}}^{n+1}\} \setminus ((S_n \cup T_n) \setminus C').$$

Clearly,

$$\nu = \#(\{s_1^{n+1}, \dots, s_{m_{n+1}}^{n+1}\} \setminus ((S_n \cup T_n) \setminus C')) > \frac{3}{2}n - \frac{3}{2}n_0' + \gamma - (n - n_0') \geq \gamma,$$

and each of these ν ultrafilters either belongs to C', or does not belong to $S_n \cup T_n$. Now, arguing as in the proof of Lemma 8.14, with the right to assume that $\iota_n(r) < 4$ for $r \in S_n$, we obtain the required statement. □

8.17. The first part of the following lemma is an obvious corollary of Lemma 8.16; the second part of the lemma is given for the sake of completeness.

LEMMA. (1) *Let $\mathcal{A}_1, \dots, \mathcal{A}_n$ be a finite sequence of algebras, and for each $k \in [1, n]$ let there exist pairwise distinct ultrafilters $s_1^k, \dots, s_{m_k}^k \in \ker \mathcal{A}_k$ with $m_k > \frac{3}{2}(k-1)$. Then there exist corresponding sets of ultrafilters S_n, T_n such that the following also hold*:

(i) *if $s \in S_n$, then there exists $s_j = s = s_{i_j}^j$;*
(ii) *$\iota_n(s) \leq 4$ for all $s \in S_n$.*

(2) *The upper bound 4 is the best possible in the following sense. One can construct algebras $\mathcal{A}_1, \dots, \mathcal{A}_n$ (in our construction $n = 4$) and take the corresponding ultrafilters s_i^k ($m_k > \frac{3}{2}(k-1)$) so that the corresponding sets of ultrafilters S_n, T_n for which condition (i) is satisfied, together with the condition that $\iota_n(s) \leq 3$ for all $s \in S_n$, do not exist.*

PROOF OF PART (2). Let us take five ultrafilters, $s, \tau_1, \tau_2, \tau_3, \tau_4$, and define the algebras $\mathcal{A}_1, \mathcal{A}_2, \mathcal{A}_3, \mathcal{A}_4$, each containing X, so that

$\ker \mathcal{A}_1 = \{s, \tau_1\}$;
$\ker \mathcal{A}_2 = \{s, \tau_2\}$;
$\ker \mathcal{A}_3 = \{s, \tau_1, \tau_2, \tau_3\}$, and $\{s, \tau_3\}, \{\tau_1, \tau_2\}$ are $\mathcal{A}_3$-similar sets;
$\ker \mathcal{A}_4 = \{s, \tau_1, \tau_2, \tau_3, \tau_4\}$, and $\{s, \tau_4\}, \{\tau_1, \tau_2, \tau_3\}$ are $\mathcal{A}_4$-similar sets.

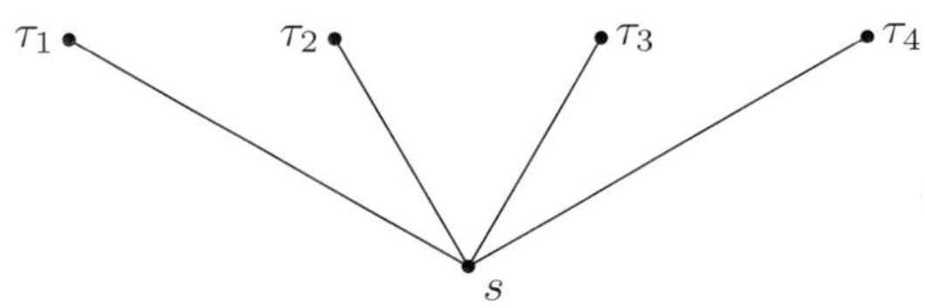

Let us list all the ultrafilters s_i^k:

$$s_1^1 = s;$$
$$s_1^2 = s,\ s_2^2 = \tau_2;$$
$$s_1^3 = s,\ s_2^3 = \tau_1,\ s_3^3 = \tau_2,\ s_4^3 = \tau_3;$$
$$s_1^4 = s,\ s_2^4 = \tau_1,\ s_3^4 = \tau_2,\ s_4^4 = \tau_3,\ s_5^4 = \tau_4.$$

There exists only one pair of corresponding sets of ultrafilters S_4 and T_4 for which condition (i) is satisfied: $S_4 = \{s\}$, $T_4 = \{\tau_1, \tau_2, \tau_3, \tau_4\}$. But $\iota_4(s) = 4$. □

8.18. Let us turn to Case II in the proof of Theorem 2.7. There we assumed that $\psi(n_{k_m}) - \psi(k_m) \geq m(m+1)$ and constructed the corresponding matrix B of pairwise distinct ultrafilters, and the m-th row of B consisted of m elements. For each $m \in \mathbb{N}^+$ set

$$\sigma(m) = \begin{cases} 1 & \text{if } m = 1, \\ \frac{3m^2+5m}{2} & \text{if } m > 1. \end{cases}$$

In this section we assume that $\psi(n_{k_m}) - \psi(k_m) \geq 2\sum_{j=1}^{m} \sigma(j)$ and the m-th row of B consists of $\sigma(m)$ elements. Let us make the following assumptions:

(1) for each ultrafilter $b_i^m \in B$ there exists $k(m,i)$ such that $b_i^m \in \ker \mathcal{A}'_{n_{k(m,i)}}$; we choose one such $k(m,i)$ for each b_i^m;
(2) the mapping $b_i^m \to k(m,i)$ is one-to-one;
(3) if $S = \{\widetilde{s}_1, \dots, \widetilde{s}_k, \dots\}$ and $s \in S$, then

$$\#(\{k \mid \widetilde{s}_k = s\}) \leq 4.\text{[10]}$$

As in the proof of Theorem 2.7, for each ultrafilter b_i^m there exists an $\mathcal{A}_{l_m}$-similar ultrafilter $c_i^m \in \overline{Z} \setminus Z$. Take a set

$$\mathfrak{C} \subset \overline{Z} \setminus Z \quad \text{such that } \#(\mathfrak{C}) = \aleph_0.$$

One can find ultrafilters b_j^2, $b_{j'}^2$ such that $\widetilde{s}_{k(1,1)}, \widetilde{s}_{k(2,j)}, \widetilde{s}_{k(2,j')}$ are pairwise distinct ultrafilters (because $\sigma(1) = 1$, $\sigma(2) = 11$, and assumption (3) is satisfied). Let $j = 1$, $j' = 2$. Let

$$\mathfrak{S} = \begin{pmatrix} \widetilde{s}_{k(1,1)} & & \\ \widetilde{s}_{k(2,1)} & \widetilde{s}_{k(2,2)} & \\ \cdots & \cdots & \cdots \\ \widetilde{s}_{k(m,1)} & \cdots & \widetilde{s}_{k(m,m)} \\ \cdots & \cdots & \cdots \end{pmatrix}$$

be a matrix of pairwise distinct ultrafilters. (Such a matrix or a similar one can be constructed since the $(m+1)$-th row of the matrix B consists of

$$\frac{3m(m+1)}{2} + 4(m+1)$$

elements.) For us it is important that we can take sequences of ultrafilters

$$S^* = \{\widetilde{s}_{k(1,p_1)}, \dots, \widetilde{s}_{k(m,p_m)}, \dots\},$$
$$B^* = \{b_{p_1}^1, \dots, b_{p_m}^m, \dots\}$$

such that

$$\overline{S^*} \cap \mathfrak{C} = \varnothing, \quad \overline{B^*} \cap \mathfrak{C} = \varnothing.$$

Now let us turn to the proof of Lemma 8.5. Let us construct the set H according to Case II in the proof of Theorem 2.7. The following simple arguments give the proof of the next theorem. Let $\mathcal{A}_{l_1}, \dots, \mathcal{A}_{l_m}, \dots$ be Z-irreducible algebras. As in the proof of Theorem 2.7, we construct a matrix of pairwise distinct ultrafilters with the m-th row consisting of the elements of the set Z_{l_m}, regardless of whether condition (i) or condition (ii) is satisfied. Now it is clear how to construct the set H.

Using the above arguments, the proof of Theorem 8.11, and the results of 8.12–8.17, it is not difficult to prove the following theorem.

[10] Recall that $B \cap S = \varnothing$.

THEOREM. *Let $\mathcal{A}_1,\dots,\mathcal{A}_k,\dots$ be a countable sequence of almost σ-algebras, and for each k let there exist a finite sequence of pairwise disjoint sets $U_1^k,\dots,U_{m_k}^k \notin \mathcal{A}_k$ with $m_k > \frac{3}{2}(k-1)$. Then for each U_i^k one can take $\widehat{U}_i^k \subset U_i^k$ and for each k one can take $\widehat{U}_{\psi_k}^{\xi_k}$ so that the following hold:*

(i) $\widehat{U}_i^k \notin \mathcal{A}_k$;
(ii) *either* $\widehat{U}_i^k = \widehat{U}_j^l$, *or* $\widehat{U}_i^k \cap \widehat{U}_j^l = \varnothing$;
(iii) *for any real number $\rho < 1$ and for any finite subset $\mathbb{N}_* \subset \mathbb{N}^+$ there exists $\mathbb{N}^* \subset \mathbb{N}^+$ such that $\mathbb{N}_* \subset \mathbb{N}^*$, $p_a(\mathbb{N}^*) > \rho$, $\xi_k \in \mathbb{N}^*$ for each $k \in \mathbb{N}^*$, and if $\widehat{U}_{\psi_k}^{\xi_k} \subset \widehat{Q} \subset \bigcup_{j\in\mathbb{N}^*} \widehat{U}_{\psi_j}^{\xi_j}$, then $\widehat{Q} \notin \mathcal{A}_k$;*
(iv) *there exists a set Q which is the union of sets of the type of $\widehat{U}_{\psi_l}^{\xi_l}$ and $Q \notin \bigcup \mathcal{A}_k$.*

8.19. In 2.34, a result on a countable sequence of σ-algebras was stated, including the estimate $\frac{3}{2}(k-1)$. But first one should obtain the corresponding result for a finite sequence of algebras. It is reasonable to place this latter result here, because we use the arguments from 8.12, 8.13 for its proof, which is similar to that of Lemma 4.23.

STATEMENT. *Let $\mathcal{A}_1,\dots,\mathcal{A}_n,\mathcal{A}_{n+1}$, $n \in \mathbb{N}^+$, be a sequence of algebras; for the algebras $\mathcal{A}_1,\dots,\mathcal{A}_n$ let the corresponding sets of ultrafilters S_n, T_n exist, and let $\#(\ker \mathcal{A}_{n+1}) > \frac{3}{2}n$. Then one can construct sets of ultrafilters S_{n+1} and T_{n+1} so that the following hold:*

1) *if $\iota_n(s) \geq 3$ and $s \in S_{n+1}$, then $\iota_{n+1}(s) = \iota_n(s)$;*
2) *if $\iota_n(s) < 3$ and $s \in S_{n+1}$, then $\iota_{n+1}(s) \leq \iota_n(s) + 1$;*
3) *if $s \in S_{n+1} \setminus S_n$, then $\iota_{n+1}(s) \leq 3$.*

PROOF. Let K be a bush of (S_n, T_n) possessing the following properties:

(i) there exists an ultrafilter $s \in K$ such that $\iota_n(s) \geq 3$;
(ii) $\#(K) = \|K\| + 1$;
(iii) there exists an ultrafilter $t \in K \cap T_n$ such that

$$\#(\{i \mid t_i = t\}) \geq 4.$$

Let us call such a bush *wide*. As in the proof of Lemma 4.23, define *sparse* and *thick* bushes and the number

$$\gamma(K) = \#(\{s \in K \cap S_n \mid \iota_n(s) \geq 3\}).$$

One easily checks that if K is a wide bush, then

$$\|K\| \geq 2 \cdot \gamma(K) + 4.$$

As in the proof of Lemma 4.23, consider the families of wide, sparse, and thick bushes and the numbers γ^j. We have (see the notation in 8.13):

$$n'' - \sum_{i=1}^{\alpha^1} \|K_i^1\| - \sum_{i=1}^{\alpha^2} \|K_i^2\| - \sum_{i=1}^{\alpha^3} \|K_i^3\| - \alpha^1 - \alpha^2$$
$$\leq \frac{3}{2}\Big(n_0' - \sum_{i=1}^{\alpha^1} \|K_i^1\| - \sum_{i=1}^{\alpha^2} \|K_i^2\| - \sum_{i=1}^{\alpha^3} \|K_i^3\|\Big).$$

Hence

$$\begin{aligned} n'' &\le \frac{3}{2}\Big(n_0' - \sum_{i=1}^{\alpha^1}\|K_i^1\| - \sum_{i=1}^{\alpha^2}\|K_i^2\| - \sum_{i=1}^{\alpha^3}\|K_i^3\|\Big) \\ &\quad + \sum_{i=1}^{\alpha^1}\|K_i^1\| + \sum_{i=1}^{\alpha^2}\|K_i^2\| + \sum_{i=1}^{\alpha^3}\|K_i^3\| + \alpha_1 + \alpha_2 \\ &= \frac{3}{2}n_0' - \frac{1}{2}\sum_{i=1}^{\alpha^1}\|K_i^1\| - \frac{1}{2}\sum_{i=1}^{\alpha^2}\|K_i^2\| - \frac{1}{2}\sum_{i=1}^{\alpha^3}\|K_i^3\| + \alpha_1 + \alpha_2 \\ &\le \frac{3}{2}n_0' - \gamma^1 - \gamma^3. \end{aligned}$$

Hence

$$\#((S_n \cup T_n) \setminus C') \le \frac{3}{2}n_0' - \gamma^1 - \gamma^3 + n - n_0'.$$

Now we proceed to the construction of S_{n+1} and T_{n+1}, which is exactly the same as that of S_{n+1} and T_{n+1} in the proof of Lemma 4.23. □

8.20. EXAMPLE. Let us construct algebras $\mathcal{A}_1, \ldots, \mathcal{A}_7$ and sets of ultrafilters S_6, T_6 (for algebras $\mathcal{A}_1, \ldots, \mathcal{A}_6$) such that $\iota_6(s) \le 2$ for all $s \in S_6$, $\#(\ker \mathcal{A}_7) = 10 > \frac{3}{2}(7-1)$ and there does not exist the corresponding sets of ultrafilters S_7, T_7 such that $\iota_7(s) \le 2$ for all $s \in S_7$.

CONSTRUCTION. Take pairwise distinct ultrafilters $s^1, s^2, s^3, t_1, t, t_5, t_6, q_1, q_2, q_3$ and put $s^1 = s_1 = s_2$, $s^2 = s_3 = s_4$, $s^3 = s_5 = s_6$, $t = t_2 = t_3 = t_4$; $\ker \mathcal{A}_i = \{s_i, t_i\}$, and s_i, t_i are $\mathcal{A}_i$-similar ultrafilters for all $i \in [1,6]$; $\{s^1, q_1\}$, $\{s^2, q_2\}$, $\{s^3, q_3\}$, $\{t_1, t, t_5, t_6\}$ are $\mathcal{A}_7$ similar sets.

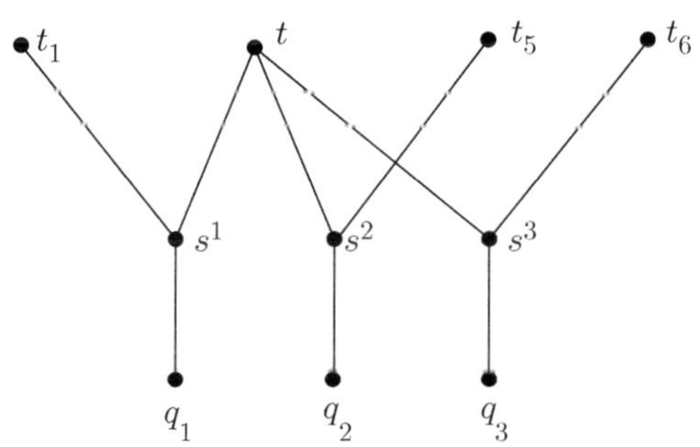

8.21. An obvious corollary of the previous statement is the following result.

STATEMENT. *Let $\mathcal{A}_1, \ldots, \mathcal{A}_n$ be a finite sequence of algebras, and let $\#(\ker \mathcal{A}_k) > \frac{3}{2}(k-1)$ for all $k \in [1,n]$. Then one can construct sets of ultrafilters S_n and T_n such that $\iota_n(s) \le 3$ for all $s \in S_n$.*

CHAPTER 9

Finite Sequences of Algebras (2). The Function $\mathfrak{g}(n)$

9.1. STATEMENT. $\mathfrak{g}(n+1) - \mathfrak{g}(n) \leq 2$ *for all* $n \in \mathbb{N}^+$.

PROOF. Let $\mathcal{A}_1, \dots, \mathcal{A}_n, \mathcal{A}_{n+1}$ be a sequence of algebras, and for each $k \in [1, n+1]$ let there exist $\mathfrak{g}(n) + 2$ pairwise disjoint sets not belonging to $\mathcal{A}_k$. We must prove that

$$\bigcup_{k=1}^{n+1} \mathcal{A}_k \neq \mathfrak{P}(X).$$

Let us assume, without loss of generality, that all the algebras are ω-saturated; and, using Theorem 3.24, let us assume that they all contain X. Let $q \in \ker \mathcal{A}_{n+1}$. For each $k \in [1, n]$ we construct an algebra $\mathcal{A}'_k > \mathcal{A}_k$ such that $q \notin \ker \mathcal{A}'_k$ and $\#(\ker \mathcal{A}'_k) \geq \mathfrak{g}(n)$. Consider three cases:

1) $q \notin \ker \mathcal{A}_k$; then $\mathcal{A}'_k = \mathcal{A}_k$.
2) $q \in M$, M is an $\mathcal{A}_k$-similar set, and $\#(M) = 2$; then $\ker \mathcal{A}'_k = \ker \mathcal{A}_k \setminus M$.
3) $q \in M$, M is an $\mathcal{A}_k$-similar set, and $\#(M) > 2$; then $\ker \mathcal{A}'_k = \ker \mathcal{A}_k \setminus \{q\}$.

By the definition of the function $\mathfrak{g}(n)$,

$$\bigcup_{k=1}^{n} \mathcal{A}'_k \neq \mathfrak{P}(X),$$

i.e., for the algebras $\mathcal{A}'_1, \dots, \mathcal{A}'_n$ we can construct the corresponding sets of ultrafilters S_n and T_n, which are also sets of ultrafilters corresponding to the algebras $\mathcal{A}_1, \dots, \mathcal{A}_n$. Since $q \not\in (S_n \cup T_n)$, one has

$$\bigcup_{k=1}^{n+1} \mathcal{A}_k \neq \mathfrak{P}(X). \qquad \square$$

9.2. DEFINITION. Let $\mathcal{A}, \mathcal{B}$ be algebras such that $\overline{\ker \mathcal{A}} \cap \overline{\ker \mathcal{B}} = \varnothing$. Then we denote

$$\mathcal{A} * \mathcal{B} = \mathcal{A} \cap \mathcal{B}.$$

Obviously, the system of all $(\mathcal{A} * \mathcal{B})$-similar sets is the system of all $\mathcal{A}$-similar sets and all $\mathcal{B}$-similar sets; and the system of all $(\mathcal{A} * \mathcal{B})$-special ultrafilters is the system of all $\mathcal{A}$-special ultrafilters and all $\mathcal{B}$-special ultrafilters. Evidently, we have

$$\ker(\mathcal{A} * \mathcal{B}) = \ker \mathcal{A} \cup \ker \mathcal{B}, \quad \#(\ker(\mathcal{A} * \mathcal{B})) = \#(\ker \mathcal{A}) + \#(\ker \mathcal{B}).$$

9.3. The proof of the following theorem is almost obvious.

THEOREM. *Suppose $\mathcal{A}_1, \ldots, \mathcal{A}_n$ and $\mathcal{B}_1, \ldots, \mathcal{B}_m$ are two finite sequences of algebras,*

$$\left(\bigcup_{k=1}^{n} \overline{\ker \mathcal{A}_k}\right) \cap \left(\bigcup_{k=1}^{m} \overline{\ker \mathcal{B}_k}\right) = \varnothing, \quad \bigcup_{k=1}^{n} \mathcal{A}_k = \bigcup_{k=1}^{m} \mathcal{B}_k = \mathfrak{P}(X),$$

and $\nu < n$, $\mu < m$ are nonnegative integers. Then

$$\bigcup\{\mathcal{A}_k \mid k \le \nu\} \cup \bigcup\{\mathcal{B}_k \mid k \le \mu\}$$
$$\cup \bigcup\{\mathcal{A}_k * \mathcal{B}_l \mid k \in [\nu+1, n],\ l \in [\mu+1, m]\} = \mathfrak{P}(X).$$

9.4. Assume that $\#(\ker \mathcal{A}_k) \ge d$ (resp. $= d$) for all $k \in [1, n]$, and $\#(\ker \mathcal{B}_k) \ge e$ (resp. $= e$) for all $k \in [1, m]$. Then we have $\nu + \mu + (n - \nu)(m - \mu)$ algebras, and among them ν have a kernel of cardinality not less than d (resp. equal to d), μ have a kernel of cardinality not less than e (resp. equal to e), and $(n-\nu)(m-\mu)$ have a kernel of cardinality not less than $d+e$ (resp. equal to $d+e$). Also, the union of all these algebras is equal to $\mathfrak{P}(X)$.

9.5. By Theorem 9.3, the function $\mathfrak{g}(n)$ has the property mentioned in 2.12; this property is given here in a general form.

STATEMENT.

$$\mathfrak{g}\left(\prod_{i=1}^{l} n_i\right) > \sum_{i=1}^{l} \mathfrak{g}(n_i) - l \quad \textit{for all } n_1, \ldots, n_l \in \mathbb{N}^+.$$

9.6. Let us begin to estimate $\mathfrak{g}(n)$. Obviously, $\mathfrak{g}(1) = \mathfrak{g}(2) = 1$. By Example 1.13 and Theorem 2.2(1), we have $\mathfrak{g}(3) = 3$.

STATEMENT. $\mathfrak{g}(5) \ge 5$.

PROOF. If suffices to construct the algebras $\mathcal{A}_1, \ldots, \mathcal{A}_5$ so that

$$\#(\ker \mathcal{A}_k) = 4 \quad \text{for all} \quad k \le 5 \quad \text{and} \quad \bigcup_{k=1}^{5} \mathcal{A}_k = \mathfrak{P}(X).$$

Here is this construction. Let a_0, a_1, a_2, a_3, a_4 be pairwise distinct ultrafilters, where

$\{a_0, a_1\}$, $\{a_2, a_3\}$ are the $\mathcal{A}_1$-similar sets;
$\{a_0, a_2\}$, $\{a_1, a_4\}$ are the $\mathcal{A}_2$-similar sets;
$\{a_0, a_3\}$, $\{a_2, a_4\}$ are the $\mathcal{A}_3$-similar sets;
$\{a_0, a_4\}$, $\{a_1, a_3\}$ are the $\mathcal{A}_4$-similar sets;
$\{a_1, a_2\}$, $\{a_3, a_4\}$ are the $\mathcal{A}_5$-similar sets.

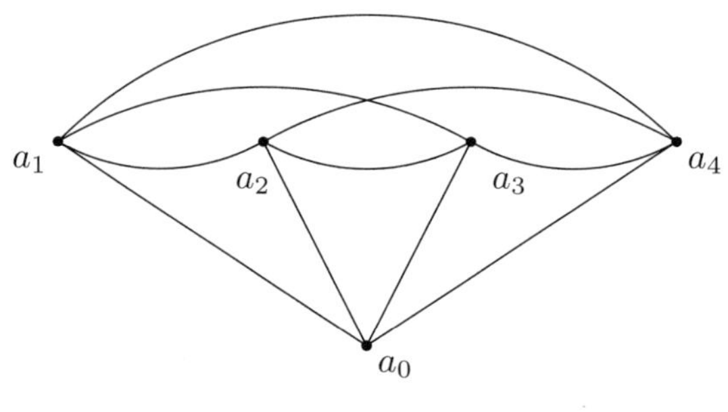

□

9.7. By Statements 9.5 and 9.6, the lower estimate for $\mathfrak{g}(n)$ from Statement 2.11 holds.

9.8. STATEMENT. $\mathfrak{g}(12) \geq 6$.

PROOF. Put

$$X = \{x_1, x_2, y_1, y_2, y_3, y_4\}.$$

Let us construct 12 algebras, each containing X, as follows:

the $\mathcal{A}_1$-similar sets are $\{x_1, y_1, y_2\}$, $\{x_2, y_4\}$;

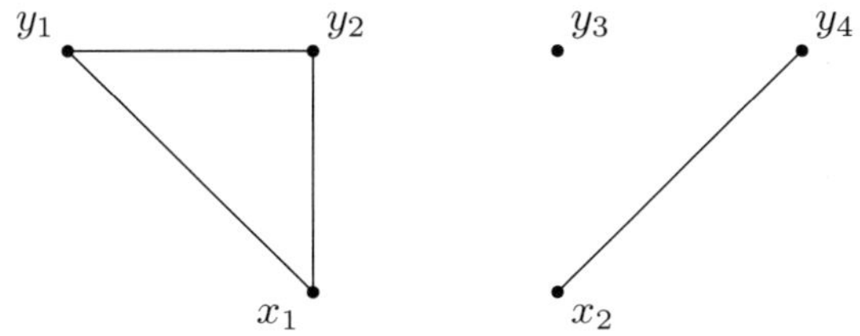

the $\mathcal{A}_2$-similar sets are $\{x_1, y_2, y_3\}$, $\{x_2, y_1\}$;

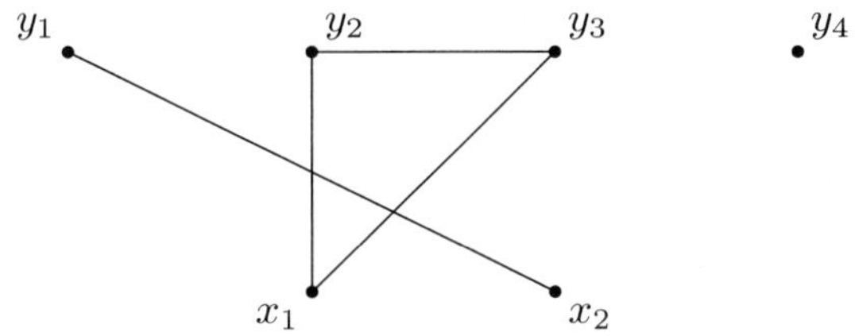

the $\mathcal{A}_3$-similar sets are $\{x_1, y_3, y_4\}$, $\{x_2, y_2\}$;

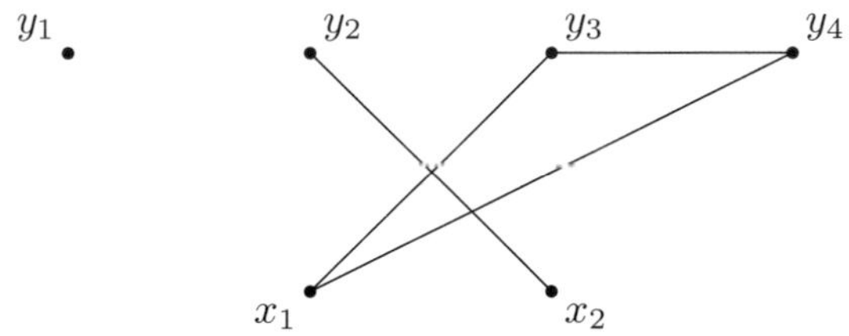

the $\mathcal{A}_4$-similar sets are $\{x_1, y_1, y_4\}$, $\{x_2, y_3\}$;

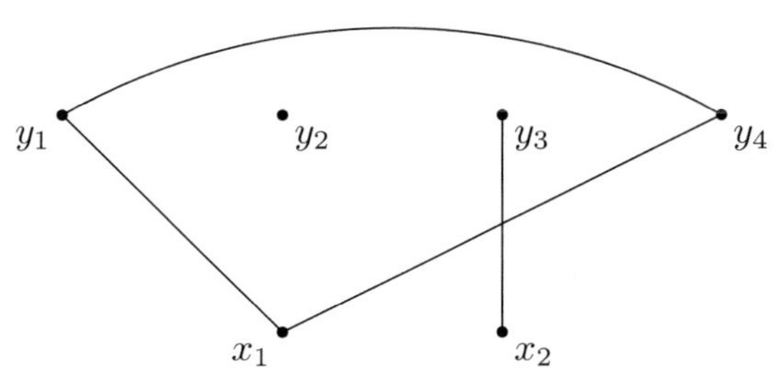

the $\mathcal{A}_5$-similar sets are $\{x_2, y_2, y_4\}$, $\{y_1, y_3\}$;

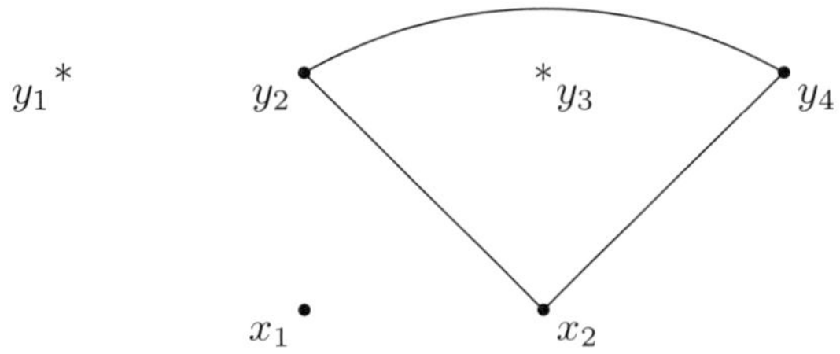

the $\mathcal{A}_6$-similar sets are $\{x_2, y_1, y_3\}$, $\{y_2, y_4\}$;

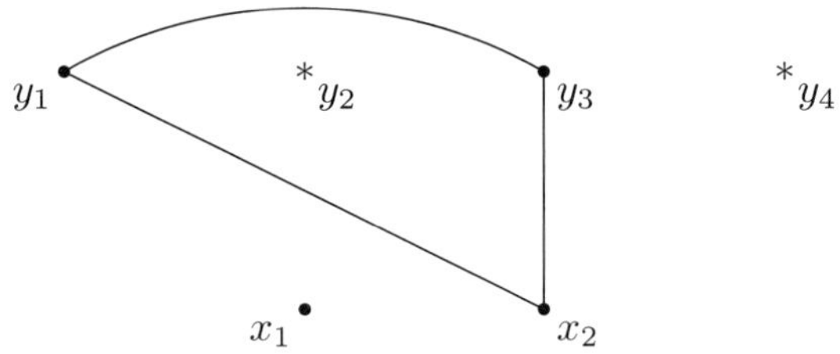

the $\mathcal{A}_7$-similar sets are $\{x_1, y_1\}$, $\{y_2, y_3, y_4\}$;

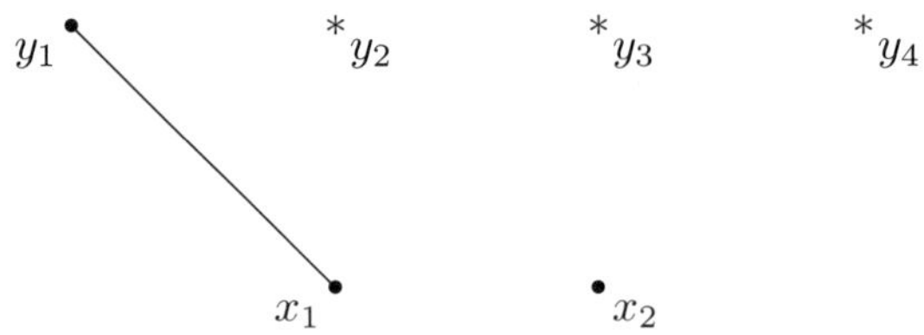

the $\mathcal{A}_8$-similar sets are $\{x_1, y_2\}$, $\{y_1, y_3, y_4\}$;

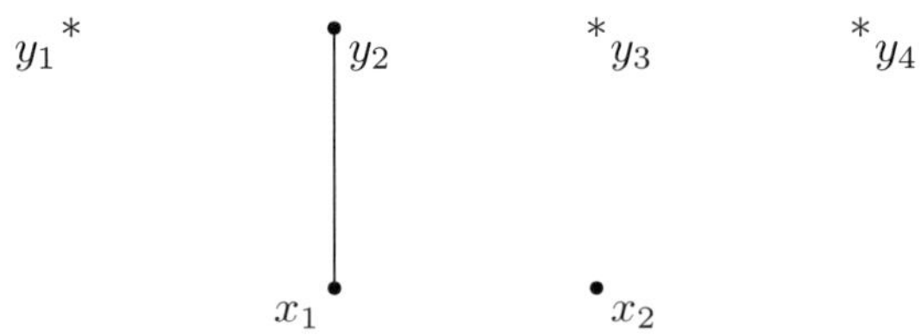

the $\mathcal{A}_9$-similar sets are $\{x_1, y_3\}$, $\{y_1, y_2, y_4\}$;

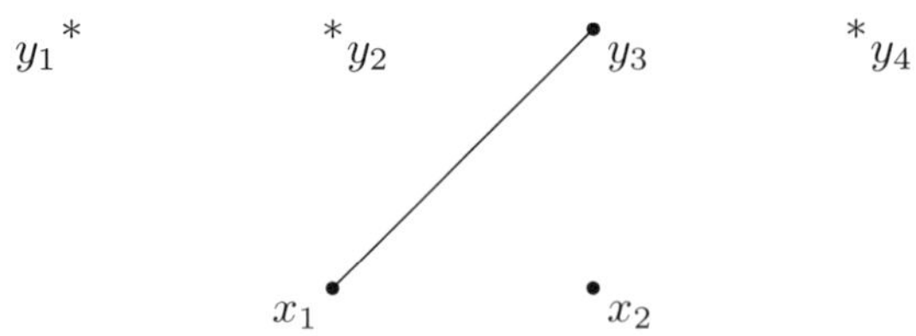

the $\mathcal{A}_{10}$-similar sets are $\{x_1, y_4\}$, $\{y_1, y_2, y_3\}$;

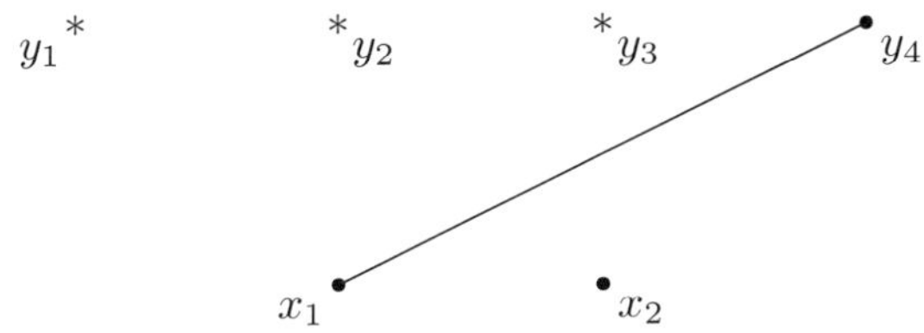

the $\mathcal{A}_{11}$-similar sets are $\{x_1, x_2\}$, $\{y_1, y_2\}$, $\{y_3, y_4\}$;

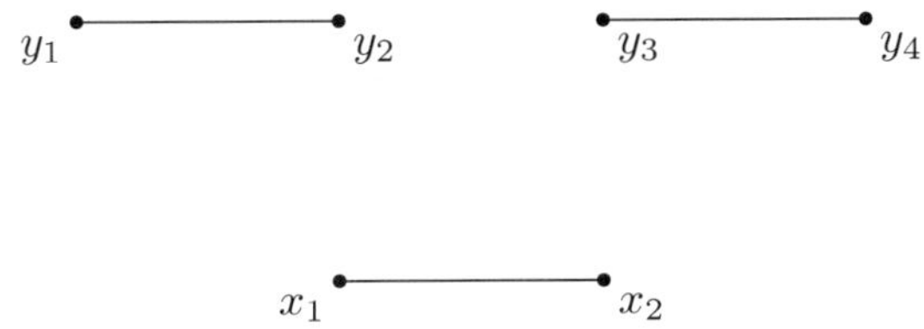

the $\mathcal{A}_{12}$-similar sets are $\{x_1, x_2\}$, $\{y_1, y_4\}$, $\{y_2, y_3\}$.

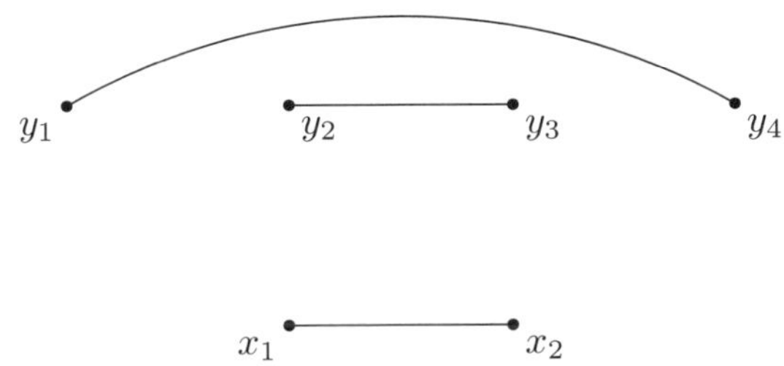

Since $\#(X) = 6$ and all the algebras contain X, we must check that if $M \subset X$ and $1 \leq \#(M) \leq 3$, then

$$M \in \bigcup_{k=1}^{12} \mathcal{A}_k.$$

A straightforward verification shows that it is true. Besides, we have $\#(\ker \mathcal{A}_k) \geq 5$ for all k. Hence $\mathfrak{g}(12) \geq 6$. □

9.9. STATEMENT. $\mathfrak{g}(13) \geq 7$.

PROOF. Put

$$X = \{x_1, x_2, y_1, y_2, y_3, y_4, y_5\}.$$

Let us construct 13 algebras, each of which contains X, as follows:

the $\mathcal{B}_1$-similar sets are $\{x_1, x_2\}$, $\{y_1, y_3\}$, $\{y_2, y_5\}$;

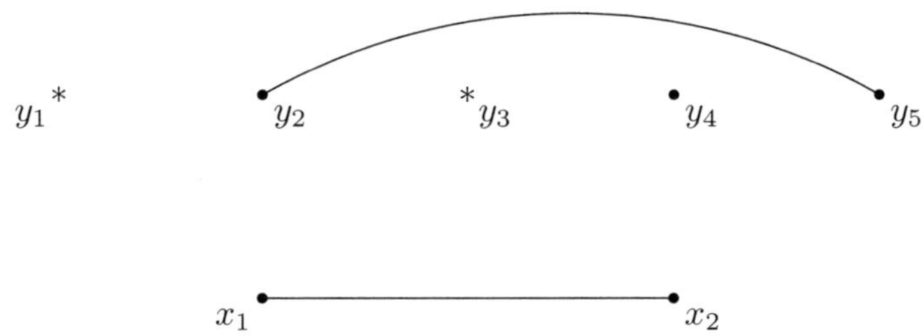

the $\mathcal{B}_2$-similar sets are $\{x_1, x_2\}$, $\{y_1, y_4\}$, $\{y_3, y_5\}$;

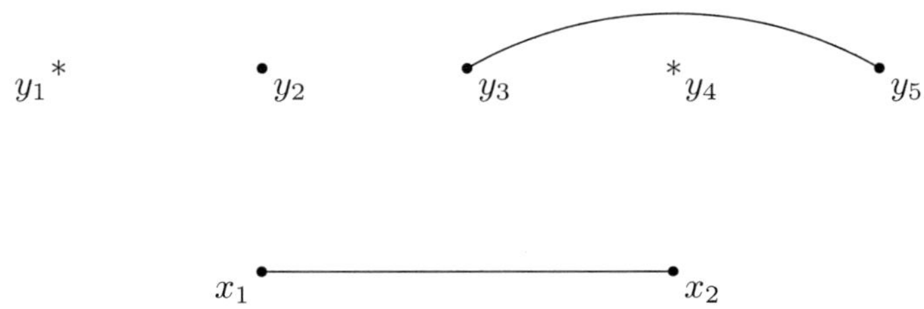

the $\mathcal{B}_3$-similar sets are $\{x_2, y_2, y_4\}$, $\{y_1, y_3, y_5\}$;

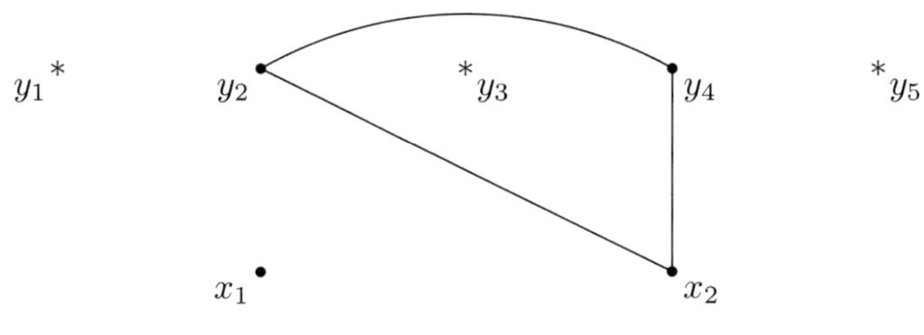

the $\mathcal{B}_4$-similar sets are $\{x_1, y_1\}$, $\{x_2, y_2\}$, $\{y_4, y_5\}$;

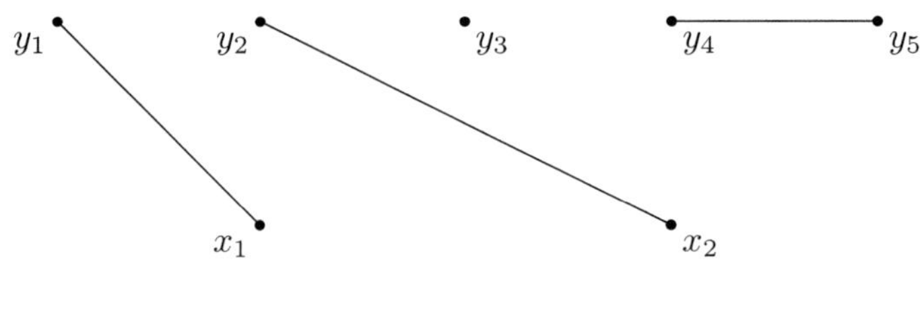

the $\mathcal{B}_5$-similar sets are $\{x_1, y_2\}$, $\{x_2, y_3\}$, $\{y_1, y_5\}$;

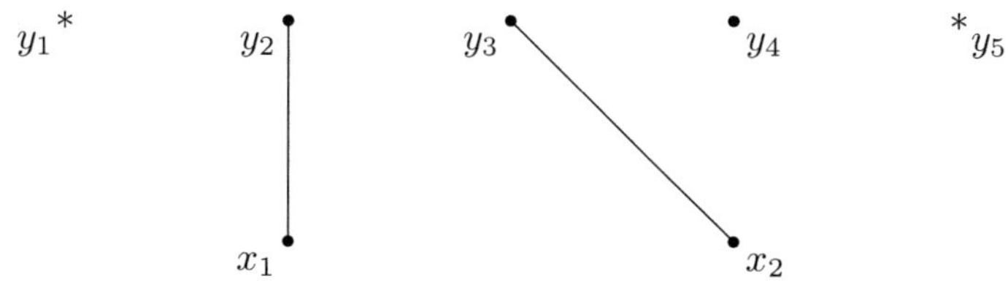

the $\mathcal{B}_6$-similar sets are $\{x_1, y_3\}$, $\{x_2, y_4\}$, $\{y_1, y_2\}$;

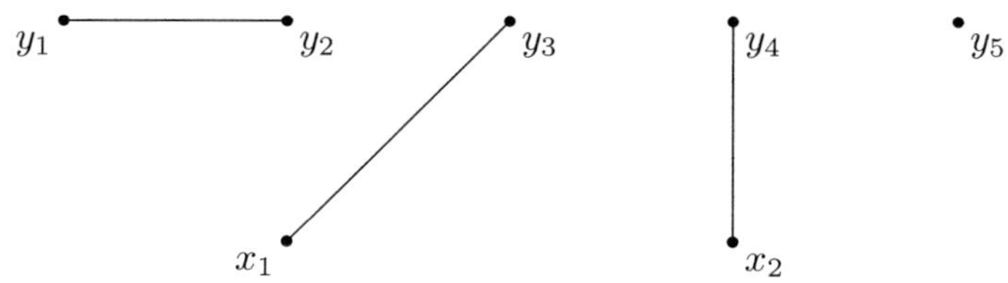

the $\mathcal{B}_7$-similar sets are $\{x_1, y_4\}$, $\{x_2, y_5\}$, $\{y_2, y_3\}$;

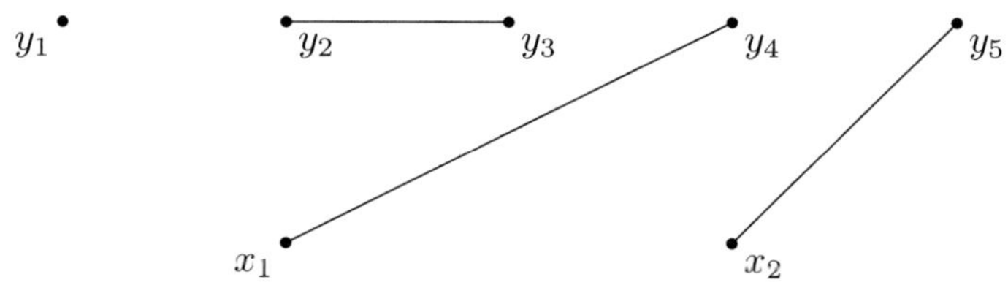

the $\mathcal{B}_8$-similar sets are $\{x_1, y_5\}$, $\{x_2, y_1\}$, $\{y_3, y_4\}$;

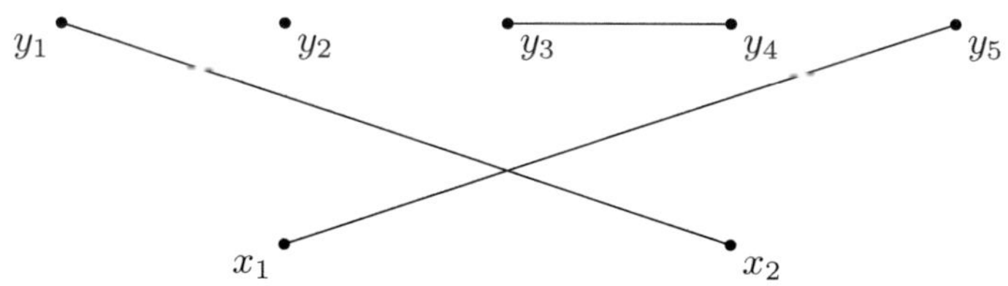

the $\mathcal{B}_9$-similar sets are $\{x_2, y_1\}$, $\{y_2, y_3\}$, $\{y_4, y_5\}$;

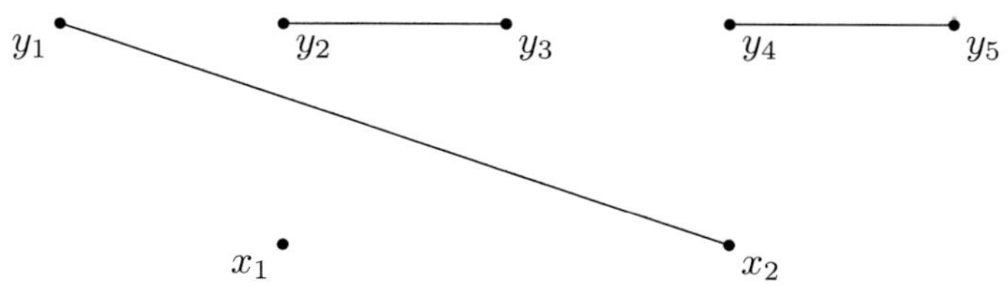

the $\mathcal{B}_{10}$-similar sets are $\{x_2, y_5\}$, $\{y_1, y_2\}$, $\{y_3, y_4\}$;

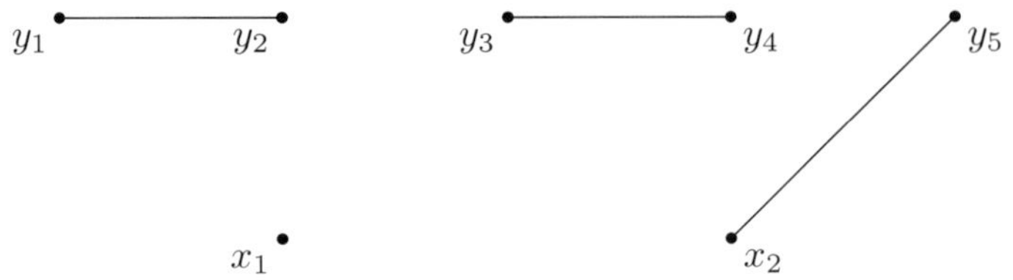

the $\mathcal{B}_{11}$-similar sets are $\{x_2, y_3\}$, $\{y_1, y_5\}$, $\{y_2, y_4\}$;

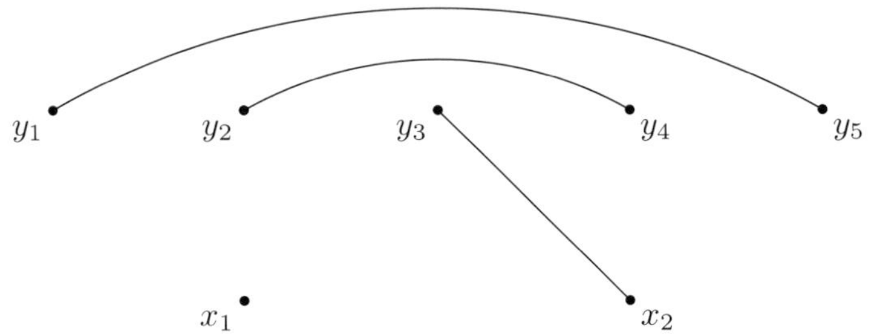

the $\mathcal{B}_{12}$-similar sets are $\{x_1, y_4\}$, $\{y_1, y_3\}$, $\{y_2, y_5\}$;

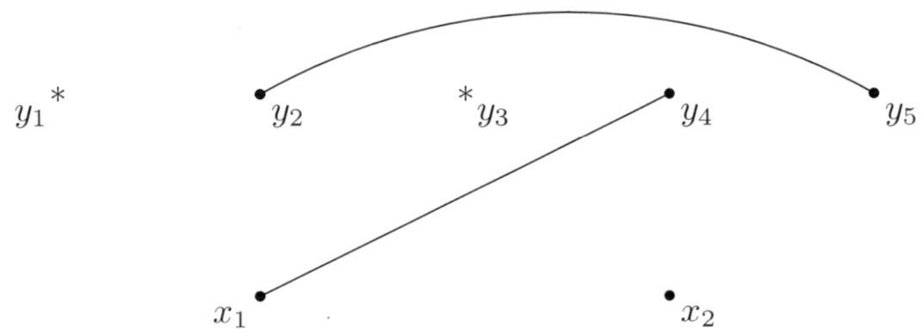

the $\mathcal{B}_{13}$-similar sets are $\{x_1, y_2\}$, $\{y_1, y_4\}$, $\{y_3, y_5\}$.

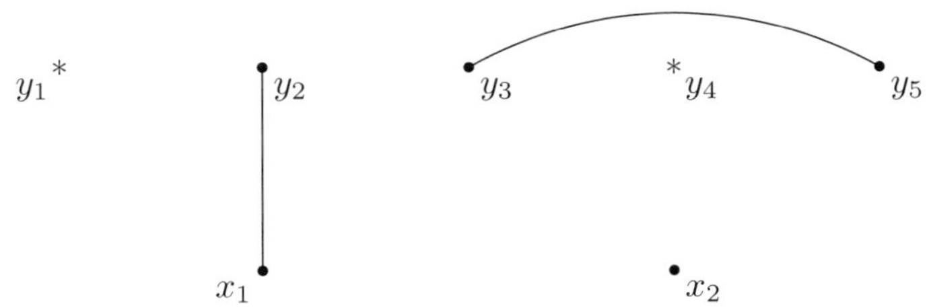

Since $\#(X) = 7$ and all algebras contain X, we must check that if $M \subset X$ and $1 \leq \#(M) \leq 3$, then

$$M \in \bigcup_{k=1}^{13} \mathcal{B}_k.$$

A straightforward verification shows that it is true. Besides, we have $\#(\ker \mathcal{B}_k) = 6$ for all k. Hence $\mathfrak{g}(13) \geq 7$. $\square$

9.10. DEFINITION. A sequence of pairs of natural numbers

$$A = \{a_1 = 1, b_1 = 1; a_2, b_2; \dots; a_n, b_n; \dots\}$$

is said to be *variational* if the following conditions are satisfied:

1) $\mathfrak{g}(a_n) \geq b_n$;
2) if $a_{i_1} \cdot a_{i_2} \cdots a_{i_k} \leq a_n$ and $a_{i_j} < a_n$, then $b_{i_1} + b_{i_2} + \cdots + b_{i_k} - k + 1 < b_n$.[1]

9.11. Having a variational sequence A, one can find a lower estimate for the function $\mathfrak{g}(n)$. Using Statement 9.5, let us construct a function $\mathfrak{g}_A(n) \leq \mathfrak{g}(n)$ as follows:

$$\mathfrak{g}_A(n) = \max\bigl\{b_{i_1} + b_{i_2} + \cdots + b_{i_k} - k + 1 \mid a_{i_1} \cdot a_{i_2} \cdots a_{i_k} \leq n\bigr\}.$$

Obviously, there exists a variational sequence A such that $\mathfrak{g}_A(n) = \mathfrak{g}(n)$ for all n. Let us summarize the above arguments. We have proved the following theorem:

THEOREM. $1, 1; 3, 3; 5, 5; 12, 6; 13, 7$ *is a variational sequence.*

9.12. First let us recall some of the arguments presented above. Consider a finite sequence of algebras $\mathcal{A}_1, \dots, \mathcal{A}_n$; for each $k \in [1, n]$ let there exist a finite number of pairwise disjoint sets $U_1^k, \dots, U_{m_k}^k \notin \mathcal{A}_k$. Let us divide X into a finite number of pairwise disjoint sets $Q_1, \dots, Q_\nu$, so that each set U_i^k is a union of some sets Q_j. It is natural to replace each set Q_j by a point x_j, putting $X = \{x_1, \dots, x_\nu\}$, and define our algebras on the new set X. Recall that

$$\ker \mathcal{A}_k = \bigl\{x_j \mid \{x_j\} \notin \mathcal{A}_k\bigr\}.$$

Put $\nu_k = \#(\ker \mathcal{A}_k)$. Clearly, $\nu_k \geq m_k$. If ν_k is even, then $\ker \mathcal{A}_k$ contains at most $\nu_k/2$ $\mathcal{A}_k$-similar sets. Hence

$$\#(\mathcal{A}_k) \leq 2^{\frac{\nu_k}{2}} \cdot 2^{\nu - \nu_k} = 2^{\nu - \frac{\nu_k}{2}} \leq 2^{\nu - \left[\frac{m_k + 1}{2}\right]}.$$

If ν_k is odd, then $\ker \mathcal{A}_k$ contains at most $\frac{\nu_k - 1}{2}$ $\mathcal{A}_k$-similar sets. Hence

$$\#(\mathcal{A}_k) \leq 2^{\frac{\nu_k - 1}{2}} \cdot 2^{\nu - \nu_k} = 2^{\nu - \frac{\nu_k + 1}{2}} \leq 2^{\nu - \left[\frac{m_k + 1}{2}\right]}.$$

THEOREM. *Let $\mathcal{A}_1, \dots, \mathcal{A}_n$ be a finite sequence of algebras, for each $k \in [1, n]$ let there exist m_k pairwise disjoint sets not belonging to $\mathcal{A}_k$, $m_k \in \mathbb{N}^+$, and let*

$$\sum_{k=1}^{n} 2^{-\left[\frac{m_k + 1}{2}\right]} \leq 1.$$

Then $\bigcup_{k=1}^{n} \mathcal{A}_k \neq \mathfrak{P}(X)$.[2]

PROOF. If $n = 1$, then the statement of the theorem is obvious. Let $n > 1$, let our algebras be defined on a finite set X, and let $\#(X) = \nu$. Taking into account the fact that $\varnothing \in \mathcal{A}_k$ and $\#(\mathcal{A}_k) \leq 2^{\nu - \left[\frac{m_k + 1}{2}\right]}$, we obtain:

$$\#\left(\bigcup_{k=1}^{n} \mathcal{A}_k\right) < \sum_{k=1}^{n} 2^{\nu - \left[\frac{m_k + 1}{2}\right]} = 2^\nu \sum_{k=1}^{n} 2^{-\left[\frac{m_k + 1}{2}\right]} \leq 2^\nu. \qquad \square$$

9.13. Let $m_k = 2k - 1$. Then all the assumptions of Theorem 9.12 hold. We come to a new proof of Theorem 3.2.

---

[1] Some of the numbers $a_{i_1}, \dots, a_{i_k}$ can coincide.

[2] Theorem 9.12 is Statement 2.14.

9.14. Let $m \in \mathbb{N}^+$, $n = 2^m$, and $m_k = 2m - 1$ for all $k \in [1, n]$. Then all the assumptions of Theorem 9.12 hold. This implies that the upper estimate for $\mathfrak{g}(n)$ in Statement 2.11 is true.

9.15. Let us present a general lemma.

LEMMA. *Let $\mathcal{A}_1, \ldots, \mathcal{A}_n$ be a finite sequence of algebras, and assume that for each $k \in [1, n-1]$ one considers either an $\mathcal{A}_k$-special ultrafilter s_k, or $\mathcal{A}_k$-similar ultrafilters s_k, t_k; $S^* = \{s_1, \ldots, s_{n-1}\}$, T^* is the set of all t_k, $S^* \cap T^* = \varnothing$, and* $\ker \mathcal{A}_n \subset T^*$. *Let*

$$t_1, t_2 \in \ker \mathcal{A}_1, \ker \mathcal{A}_2, \ker \mathcal{A}_n$$

be distinct ultrafilters; suppose $\{t_1, t_2\}$ is not an $\mathcal{A}_n$-similar set, and if $3 \leq k \leq n-1$, then $t_k \notin \{t_1, t_2\}$. Then

$$\bigcup_{k=1}^{n} \mathcal{A}_k \neq \mathfrak{P}(X).$$

PROOF. *Case* 1. *t_1, t_2 are $\mathcal{A}_1$-similar ultrafilters.* Put

$$S_n = S^* \cup \{t_1\}, \quad T_n = T^* \setminus \{t_1\}.$$

Case 2. *t_1, t_2 are $\mathcal{A}_2$-similar ultrafilters.* Put

$$S_n = S^* \cup \{t_2\}, \quad T_n = T^* \setminus \{t_2\}.$$

Case 3. *t_1 has an $\mathcal{A}_2$-similar ultrafilter $q \notin T^*$.* Put

$$S_n = S^* \cup \{q, t_2\}, \quad T_n = T^* \setminus \{t_2\}.$$

Case 4. *t_2 has an $\mathcal{A}_1$-similar ultrafilter $r \notin T^*$.* Put

$$S_n = S^* \cup \{r, t_1\}, \quad T_n = T^* \setminus \{t_1\}.$$

Case 5. *None of the previous cases holds.* Put

$$S_n = S^* \cup \{t_1, t_2\}, \quad T_n = T^* \setminus \{t_1, t_2\}.$$

□

9.16. To prove the more difficult result formulated in Section 2.12, we first prove the following somewhat easier result.

THEOREM. $\mathfrak{g}(2^m + 1) \leq 2m - 1$ *for all natural numbers* $m \geq 3$.

PROOF. Let us argue as we did in 9.12. So let $m \geq 3$ be an integer. Consider algebras

$$\mathcal{A}_1, \ldots, \mathcal{A}_{2^m+1}; \quad 2m - 1 \leq \#(\ker \mathcal{A}_k) \leq 2m$$ [3]

for all $k \in [1, 2^m + 1]$. By Theorem 3.24 we can assume that $X \in \mathcal{A}_k$ for all $k \in [1, 2^m + 1]$. We will also assume that each $\mathcal{A}_k$-similar set contains at most three elements. Let a, b be $\mathcal{A}_{2^m+1}$-similar ultrafilters. Consider an algebra $\mathcal{A}_k$ with $k \leq 2^m$ such that a, b are not $\mathcal{A}_k$-similar ultrafilters. The following cases are possible:

1) $\{a, b\} \cap \ker \mathcal{A}_k = \varnothing$;
2) $a \notin \ker \mathcal{A}_k$, and b belongs to an $\mathcal{A}_k$-similar set consisting of two elements;
3) $a \notin \ker \mathcal{A}_k$, and b belongs to an $\mathcal{A}_k$-similar set consisting of three elements;
4) a belongs to an $\mathcal{A}_k$-similar set consisting of two elements, and $b \notin \ker \mathcal{A}_k$;
5) a belongs to an $\mathcal{A}_k$-similar set consisting of two elements, and b belongs to an $\mathcal{A}_k$-similar set consisting of two elements;

[3]See 3.25.

6) a belongs to an $\mathcal{A}_k$-similar set consisting of two elements, and b belongs to an $\mathcal{A}_k$-similar set consisting of three elements;
7) a belongs to an $\mathcal{A}_k$-similar set consisting of three elements, and $b \notin \ker \mathcal{A}_k$;
8) a belongs to an $\mathcal{A}_k$-similar set consisting of three elements, and b belongs to an $\mathcal{A}_k$-similar set consisting of two elements;
9) a belongs to an $\mathcal{A}_k$-similar set consisting of three elements, and b belongs to an $\mathcal{A}_k$-similar set consisting of three elements.

It is easy to show that in all cases

$$\#(\{M \mid a \in M, b \notin M, M \in \mathcal{A}_k\}) \leq 2^{\nu-2-m}.$$

Hence

$$\#\left(\left\{M \mid a \in M, b \notin M, M \in \bigcup_{k=1}^{2^m} \mathcal{A}_k\right\}\right) < 2^{\nu-2}$$

if for some $k \in [1, 2^m]$ the ultrafilters a, b are $\mathcal{A}_k$-similar, or if for two distinct numbers $k_1, k_2 \in [1, 2^m]$ we have

$$a \notin \ker \mathcal{A}_{k_1}, \ker \mathcal{A}_{k_2}.$$

In these two cases

$$\bigcup_{k=1}^{2^m+1} \mathcal{A}_k \neq \mathfrak{P}(X),$$

since, obviously,

$$\#(\{M \subset X \mid a \in M, b \notin M\}) = 2^{\nu-2}.$$

It remains to consider only the following situation: if $k_0 \in [1, 2^m + 1]$ and $c \in \ker \mathcal{A}_{k_0}$, then there exists a set $L \subset [1, 2^m + 1] \setminus \{k_0\}$ such that $\#(L) = 2^m - 1$, and a set $L' \subset X \setminus \{c\}$ such that $\#(L') = 2^m$; also there exists a one-to-one correspondence between $L \cup \{k_0\}$ and L'; if $k \in L \cup \{k_0\}$ corresponds to $q_k \in L'$, then c, q_k are $\mathcal{A}_k$-similar ultrafilters. Let $k_0 = 1$, $L = [2, 2^m]$, and $m = 3$.

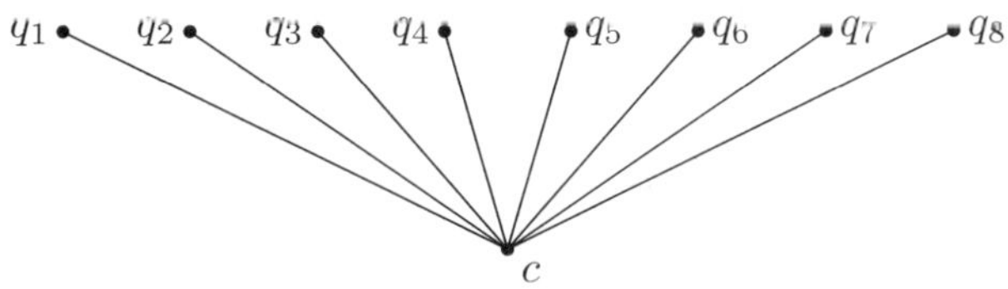

If $\ker \mathcal{A}_9 \setminus \{q_1, \dots, q_8\} \neq \varnothing$, then everything is simple. Therefore let us assume that

$$\ker \mathcal{A}_9 \subset \{q_1, \dots, q_8\}.$$

Since $\#(\ker \mathcal{A}_1) \leq 6$, we have

$$\#(\{q_2, \dots, q_8\} \setminus \ker \mathcal{A}_1) \geq 3.$$

Assume that $\ker \mathcal{A}_1 \cap \{q_6, q_7, q_8\} = \varnothing$. But then $\{q_6, q_7, q_8\} \subset \ker \mathcal{A}_k$ for all $k \in [6, 9]$. It remains to apply Lemma 9.15, obtaining

$$\bigcup_{k=1}^{9} \mathcal{A}_k \neq \mathfrak{P}(X).$$

Clearly, for $m > 3$ we also have

$$\bigcup_{k=1}^{2^m+1} \mathcal{A}_k \neq \mathfrak{P}(X). \qquad \square$$

9.17. From this section until the end of 9.49, we will be proving the statement from 2.12, namely the inequality

$$\mathfrak{g}(2^m + 2) \leq 2m - 1 \quad \text{for all } m \geq 3.$$

Of course, we assume that $\#(X) < \infty$. The notation $\#(X) = 2\mathfrak{n}$ means that $\mathfrak{n} \in \mathbb{N}^+$ and $\mathfrak{n}$ is a sufficiently large integer.[4] In the arguments below in this section (and also in Sections 9.18, 9.19, 9.20) we take

$$\#(X) = 2\mathfrak{n}.$$

In this section (and also in Sections 9.18, 9.19, 9.20) we consider algebras $\mathcal{A}^1, \mathcal{A}^2$ such that $X \in \mathcal{A}^1, \mathcal{A}^2$, and, as usual, $\mathcal{A}^1$- and $\mathcal{A}^2$-similar sets consist of two or three elements. Assume that each $\mathcal{A}^1$-similar set is an $\mathcal{A}^2$-similar set. Let

$$\mathcal{M}_1, \mathcal{M}_2 \subset X, \quad \mathcal{M}_1 \cap \mathcal{M}_2 = \varnothing.$$

It is assumed that if $x \in \mathcal{M}_1$ and $y \in \mathcal{M}_2$, then x, y are neither $\mathcal{A}^1$-similar nor $\mathcal{A}^2$-similar ultrafilters. Let

$$\widehat{M} = \ker \mathcal{A}^2 \setminus \ker \mathcal{A}^1,$$

where $\widehat{M}$ is an $\mathcal{A}^2$-similar set. In this section we assume that

$$\widehat{M} \cap (\mathcal{M}_1 \cup \mathcal{M}_2) = \varnothing.$$

Let L_1 be the union of all $\mathcal{A}^1$-similar sets which have nonempty intersection with $\mathcal{M}_1$. Define a set L:

1) either $L = \varnothing$; or
2) L is a union of some $\mathcal{A}^1$-similar sets and $L \cap (\mathcal{M}_1 \cup \mathcal{M}_2) = \varnothing$.

Put

$$l_* = \#(\mathcal{M}_1 \cup L_1 \cup L), \quad l^* = \#(\ker \mathcal{A}^1 \cup \mathcal{M}_1 \cup \mathcal{M}_2).$$

Clearly,

$$\#(\{M \mid M \cap (\ker \mathcal{A}^1 \cup \mathcal{M}_1 \cup \mathcal{M}_2) = \mathcal{M}_1 \cup L_1 \cup L, \#(M) = \mathfrak{n}\}) = \binom{2\mathfrak{n} - l^*}{\mathfrak{n} - l_*}.$$

Put

$$\delta = \#(\{M \mid \mathcal{M}_1 \cup L_1 \cup L \subset M \cap (\ker \mathcal{A}^2 \cup \mathcal{M}_1 \cup \mathcal{M}_2) \subset \mathcal{M}_1 \cup L_1 \cup L \cup \widehat{M}, \\ \#(M) = \mathfrak{n}, \text{ and } M \in \mathcal{A}^2\}).$$

If $\#(\widehat{M}) = 2$, then

$$\delta = \binom{2\mathfrak{n} - l^* - 2}{\mathfrak{n} - l_*} + \binom{2\mathfrak{n} - l^* - 2}{\mathfrak{n} - l_* - 2}.$$

If $\#(\widehat{M}) = 3$, then

$$\delta = \binom{2\mathfrak{n} - l^* - 3}{\mathfrak{n} - l_*} + \binom{2\mathfrak{n} - l^* - 3}{\mathfrak{n} - l_* - 3}.$$

---

[4]Below we will meet expressions which are true only for $\mathfrak{n}$ large enough; we will not mention this specifically.

The verification of the following inequalities does not cause difficulties since l_*, l^* are fixed numbers:

$$\binom{2\mathfrak{n} - l^*}{\mathfrak{n} - l_*} > 2\binom{2\mathfrak{n} - l^* - 2}{\mathfrak{n} - l_*} + 2\binom{2\mathfrak{n} - l^* - 2}{\mathfrak{n} - l_* - 2},$$
$$\binom{2\mathfrak{n} - l^*}{\mathfrak{n} - l_*} > 4\binom{2\mathfrak{n} - l^* - 3}{\mathfrak{n} - l_*} + 4\binom{2\mathfrak{n} - l^* - 3}{\mathfrak{n} - l_* - 3}.$$

Put

$$\gamma^i = \#(\{M \mid M \cap (\mathcal{M}_1 \cup \mathcal{M}_2) = \mathcal{M}_1, \#(M) = \mathfrak{n}, \text{ and } M \in \mathcal{A}^i\})$$

for $i = 1, 2$. In our situation,

$$\text{if } \#(\widehat{M}) = 2, \text{ then } \gamma^1 > 2\gamma^2;$$
$$\text{if } \#(\widehat{M}) = 3, \text{ then } \gamma^1 > 4\gamma^2.$$

9.18. Let us change the conditions from the previous section, assuming that $M \cap (\mathcal{M}_1 \cup \mathcal{M}_2) \neq \varnothing$ if M is an $\mathcal{A}^1$-similar set, and also that $\#(\widehat{M} \cap (\mathcal{M}_1 \cup \mathcal{M}_2)) = 1$. Here we put $l_* = \#(\mathcal{M}_1 \cup L_1)$ and define l^* in the same way as in the previous section. Clearly,

$$\gamma^1 = \binom{2\mathfrak{n} - l^*}{\mathfrak{n} - l_*}.$$

If $\widehat{M} \cap \mathcal{M}_1 \neq \varnothing$, then

$$\gamma^2 = \binom{2\mathfrak{n} - l^* - \#(\widehat{M}) + 1}{\mathfrak{n} - l_* - \#(\widehat{M}) + 1}.$$

If $\widehat{M} \cap \mathcal{M}_2 \neq \varnothing$, then

$$\gamma^2 = \binom{2\mathfrak{n} - l^* - \#(\widehat{M}) + 1}{\mathfrak{n} - l_*}.$$

Now it is easy to show that if $\#(\widehat{M}) = 2$, then $\gamma^1 > 2\gamma^2$ up to an infinitely small function with respect to γ^1.[5] If $\#(\widehat{M}) = 3$, then $\gamma^1 > 4\gamma^2$ up to an infinitely small function with respect to γ^1.

9.19. Let us return again to the conditions from 9.17. Let us make the following changes: $M \cap (\mathcal{M}_1 \cup \mathcal{M}_2) \neq \varnothing$ if M is an $\mathcal{A}^1$-similar set , $\#(\widehat{M}) = 3$, and $\#(\widehat{M} \cap (\mathcal{M}_1 \cup \mathcal{M}_2)) = 2$. The numbers l_*, l^* are defined as in the previous section, and

$$\gamma^1 = \binom{2\mathfrak{n} - l^*}{\mathfrak{n} - l_*}.$$

If $\widehat{M} \cap \mathcal{M}_1 \neq \varnothing$, then

$$\gamma^2 = \binom{2\mathfrak{n} - l^* - 1}{\mathfrak{n} - l_* - 1}.$$

If $\widehat{M} \cap \mathcal{M}_2 \neq \varnothing$, then

$$\gamma^2 = \binom{2\mathfrak{n} - l^* - 1}{\mathfrak{n} - l_*}.$$

As we have mentioned in the previous section, $\gamma^1 > 2\gamma^2$ up to an infinitely small function with respect to γ^1.

---

[5]This means the following: there exists a function $o(\mathfrak{n})$, $o(\mathfrak{n}) \to 0$ as $\mathfrak{n} \to \infty$, such that $(1 + o(\mathfrak{n})) \cdot \gamma^1(\mathfrak{n}) > 2 \cdot \gamma^2(\mathfrak{n})$.

9.20. Put

$$\gamma_0 = \#(\{M \subset X \mid M \cap (\mathcal{M}_1 \cup \mathcal{M}_2) = \mathcal{M}_1 \text{ and } \#(M) = \mathfrak{n}\}).$$

Assume that

$$2m - 1 \leq \#(\ker \mathcal{A}^1) \leq 2m,$$

where $m \in \mathbb{N}^+$. For each $\mathcal{A}^1$-similar set M, let

$$\#(M \cap (\mathcal{M}_1 \cup \mathcal{M}_2)) \leq 1.$$

Then $\gamma^1 < 2^{-m} \cdot \gamma_0$ up to an infinitely small function with respect to γ_0. Now let there exist an $\mathcal{A}^1$-similar set M_0 such that

$$\#(M_0 \cap (\mathcal{M}_1 \cup \mathcal{M}_2)) = 2,$$

and for all other $\mathcal{A}^1$-similar sets M (if they exist) let

$$\#(M \cap (\mathcal{M}_1 \cup \mathcal{M}_2)) \leq 1.$$

Then $\gamma^1 < 2^{1-m} \cdot \gamma_0$ up to an infinitely small function with respect to γ_0.

9.21. We consider the algebras $\mathcal{A}_1, \dots, \mathcal{A}_{2^m+2}$, where $m \geq 3$ is an integer; let

$$X \in \mathcal{A}_k, \quad 2m - 1 \leq \#(\ker \mathcal{A}_k) \leq 2m \quad \text{for all } k \in [1, 2^m + 2].$$

As usual, any $\mathcal{A}_k$-similar set consists of two or three elements. If there exists an $\mathcal{A}_k$-similar set consisting of three elements, then let us assume that

$$\#(\ker \mathcal{A}_k) = 2m - 1.$$

Our final goal is to prove that

$$\bigcup_{k=1}^{2^m+2} \mathcal{A}_k \neq \mathfrak{P}(X).$$

LEMMA. *Suppose a, b, c are pairwise distinct ultrafilters, a, b are $\mathcal{A}_1$- and $\mathcal{A}_2$-similar ultrafilters, and a, c are $\mathcal{A}_3$-similar ultrafilters. Assume that there does not exist $k \in [4, 2^m + 2]$ such that b, c are $\mathcal{A}_k$-similar ultrafilters. Then*

$$\bigcup_{k=1}^{2^m+2} \mathcal{A}_k \neq \mathfrak{P}(X).$$

PROOF. Let $\#(X) = 2\mathfrak{n}$. Let us assume that there does not exist $k \in [4, 2^m+2]$ such that either a, b or b, c are $\mathcal{A}_k$-similar ultrafilters. By 9.20, we have

$$\#\left(\left\{M \mid M \cap \{a,b,c\} = \{b,c\}, \#(M) = \mathfrak{n}, \text{ and } M \in \bigcup_{k=4}^{2^m+2} \mathcal{A}_k\right\}\right)$$
$$< 2^{-m} \cdot (2^m - 1) \cdot \#(\{M \subset X \mid M \cap \{a,b,c\} = \{b,c\} \text{ and } \#(M) = \mathfrak{n}\})$$

up to an infinitely small function with respect to the latter expression $\#(\dots)$. Hence there exists an $M \notin \mathcal{A}_k$ for all $k \in [1, 2^m + 2]$. □

Now we assume that there do not exist pairwise distinct numbers

$$k_1, k_2, k_3 \in [1, 2^m + 2]$$

and pairwise distinct ultrafilters a, b, c such that a, b are $\mathcal{A}_{k_1}$- and $\mathcal{A}_{k_2}$-similar ultrafilters, a, c are $\mathcal{A}_{k_3}$-similar ultrafilters, and there is no

$$k \in [1, 2^m + 2] \setminus \{k_1, k_2, k_3\}$$

such that b, c are $\mathcal{A}_k$-similar ultrafilters.

9.22. Let us interrupt our exposition, and turn to the following rather simple facts. Let $\mathcal{A}^1, \ldots, \mathcal{A}^n, \mathcal{A}^{n+1}$ be a sequence of algebras, $n \in \mathbb{N}^+$,

$$\bigcup_{k=1}^{n} \mathcal{A}^k \neq \mathfrak{P}(X), \quad \ker \mathcal{A}^{n+1} \setminus \bigcup_{k=1}^{n} \ker \mathcal{A}^k \neq \varnothing.$$

Then

$$\bigcup_{k=1}^{n+1} \mathcal{A}^k \neq \mathfrak{P}(X).$$

Indeed, for the algebras $\mathcal{A}^1, \ldots, \mathcal{A}^n$ the corresponding sets of ultrafilters S_n, T_n exist. There exists

$$q \in \ker \mathcal{A}^{n+1} \setminus (S_n \cup T_n),$$

and one easily constructs the corresponding sets of ultrafilters S_{n+1}, T_{n+1} for the algebras $\mathcal{A}^1, \ldots, \mathcal{A}^n, \mathcal{A}^{n+1}$.

Now let $\mathcal{A}^1, \ldots, \mathcal{A}^n, \mathcal{A}^{n+1}, \mathcal{A}^{n+2}$ be a sequence of algebras, $n \in \mathbb{N}^+$,

$$\bigcup_{k=1}^{n} \mathcal{A}^k \neq \mathfrak{P}(X),$$

and let there exist distinct ultrafilters

$$q_1, q_2 \notin \bigcup_{k=1}^{n} \ker \mathcal{A}^k$$

such that $q_i \in \ker \mathcal{A}^{n+i}$ for $i = 1, 2$. Then

$$\bigcup_{k=1}^{n+2} \mathcal{A}^k \neq \mathfrak{P}(X).$$

Indeed, for the algebras $\mathcal{A}^1, \ldots, \mathcal{A}^n$ the corresponding sets of ultrafilters S_n, T_n exist. Obviously, $q_1, q_2 \notin S_n \cup T_n$, and one easily constructs the corresponding sets of ultrafilters S_{n+2}, T_{n+2} for the algebras $\mathcal{A}^1, \ldots, \mathcal{A}^n, \mathcal{A}^{n+1}, \mathcal{A}^{n+2}$.

9.23. Let us return to our algebras $\mathcal{A}_1, \ldots, \mathcal{A}_{2^m+2}$. Consider the following situation: a, b, c are pairwise distinct ultrafilters, a, b are $\mathcal{A}_1$-similar ultrafilters, a, c are $\mathcal{A}_2$-similar ultrafilters, and there does not exist any $k \in [1, 2^m + 2]$ such that b, c are $\mathcal{A}_k$-similar ultrafilters. Assume that

$$\bigcup_{k=1}^{2^m+2} \mathcal{A}_k = \mathfrak{P}(X).$$

Put

$$\mathfrak{K} = \Bigg\{ q \in \bigcup_{k=1}^{2^m+2} \ker \mathcal{A}_k \setminus \{a, b, c\} \mid \text{there exists a } k \in [1, 2^m + 2]$$
$$\text{such that } a, q \text{ are } \mathcal{A}_k\text{-similar ultrafilters} \Bigg\}.$$

LEMMA. $\#\left(\bigcup_{k=1}^{2^m+2} \ker \mathcal{A}_k \setminus \mathfrak{K}\right) \leq 4.$

PROOF. Let $\#(X) = 2\mathfrak{n}$ and let

$$q \in \bigcup_{k=1}^{2^m+2} \ker \mathcal{A}_k \setminus \{a, b, c\} \quad \text{and} \quad \{k \mid q \in \ker \mathcal{A}_k\} \neq \{1, 2\}.$$

Let us show that $q \in \mathfrak{K}$. Suppose the contrary, i.e., a, q are not $\mathcal{A}_k$-similar ultrafilters for any $k \in [1, 2^m + 2]$. From the results of the previous section it follows that

$$\#(\{k \mid q \in \ker \mathcal{A}_k\}) \geq 2.$$

Therefore $q \in \ker \mathcal{A}_k$, where $k \geq 3$. Let $q \in \ker \mathcal{A}_3$. Suppose that q has an $\mathcal{A}_3$-similar ultrafilter in $\{b, c\}$. We can assume, without loss of generality, that q does not have an $\mathcal{A}_k$-similar ultrafilter in $\{b, c\}$ for $k \in [4, 2^m + 2]$.

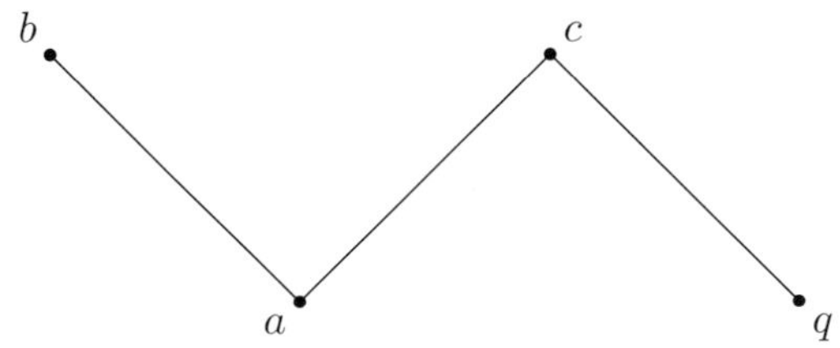

By the results of 9.20,

$$\#\left(\left\{M \mid M \cap \{a, b, c, q\} = \{b, c\}, \#(M) = \mathfrak{n}, \text{ and } M \in \bigcup_{k=4}^{2^m+2} \mathcal{A}_k\right\}\right)$$
$$< 2^{-m} \cdot (2^m - 1) \cdot \#(\{M \subset X \mid M \cap \{a, b, c, q\} = \{b, c\} \text{ and } \#(M) = \mathfrak{n}\})$$

up to an infinitely small function with respect to the latter expression $\#(\dots)$. We come to a contradiction:

$$\bigcup_{k=1}^{2^m+2} \mathcal{A}_k \neq \mathfrak{P}(X).$$

Let us now assume that q does not have an $\mathcal{A}_k$-similar ultrafilter in $\{b, c\}$ for any $k \in [3, 2^m + 2]$. Let q, q' be $\mathcal{A}_3$-similar ultrafilters. Assume first that q' does not have an $\mathcal{A}_k$-similar ultrafilter in $\{a, b, c\}$ for any $k \in [4, 2^m + 2]$. We can assume, without loss of generality, that q, q' are not $\mathcal{A}_k$-similar ultrafilters for any $k \in [4, 2^m + 2]$.

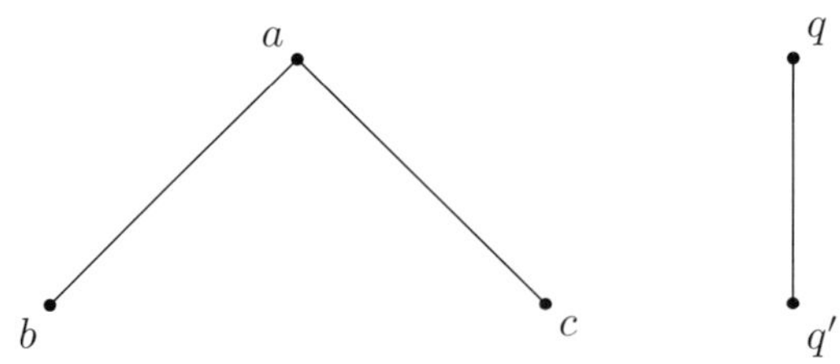

By the results of 9.20,

$$\#\left(\left\{M \mid M \cap \{a,b,c,q,q'\} = \{a,q\}, \#(M) = \mathfrak{n}, \text{ and } M \in \bigcup_{k=4}^{2^m+2} \mathcal{A}_k\right\}\right)$$
$$< 2^{-m} \cdot (2^m - 1) \cdot \#\big(\{M \subset X \mid M \cap \{a,b,c,q,q'\} = \{a,q\} \text{ and } \#(M) = \mathfrak{n}\}\big)$$

up to an infinitely small function with respect to the latter expression $\#(\dots)$. We come to a contradiction:

$$\bigcup_{k=1}^{2^m+2} \mathcal{A}_k \neq \mathfrak{P}(X).$$

Let us now assume that q' has $\mathcal{A}_k$-similar ultrafilters for some $k \in [4, 2^m + 2]$ in $\{a, b, c\}$.

Case 1. *a, q' are not $\mathcal{A}_k$-similar ultrafilters for any $k \in [4, 2^m + 2]$.* Put

$$P = \{k \in [4, 2^m + 2] \mid q' \text{ has an } \mathcal{A}_k\text{-similar ultrafilter in } \{b, c\}\}$$
$$\cup \{k \in [4, 2^m + 2] \mid q, q' \text{ are } \mathcal{A}_k\text{-similar ultrafilters}\}.$$

Comparing

$$\#\big(\{M \mid M \cap \{a,b,c,q,q'\} = \{a,q'\}, \#(M) = \mathfrak{n},$$
$$M \in \bigcup\{\mathcal{A}_k \mid k \in [4, 2^m + 2] \setminus P\}\}\big)$$

with

$$\#\big(\{M \subset X \mid M \cap \{a,b,c,q,q'\} = \{a,q'\},\ \#(M) = \mathfrak{n}\}\big),$$

we come to a contradiction:

$$\bigcup_{k=1}^{2^m+2} \mathcal{A}_k \neq \mathfrak{P}(X).$$

Case 2. *a, q' are $\mathcal{A}_4$-similar ultrafilters.* By the assumptions in 9.21, there does not exist $k \in [1, 2^m + 2] \setminus \{4\}$ such that a, q' are $\mathcal{A}_k$-similar ultrafilters (because we assume that a, q are not $\mathcal{A}_k$-similar ultrafilters for any $k \in [1, 2^m + 2]$). Similarly, there does not exist $k \in [1, 2^m + 2] \setminus \{3\}$ such that q, q' are $\mathcal{A}_k$-similar ultrafilters.

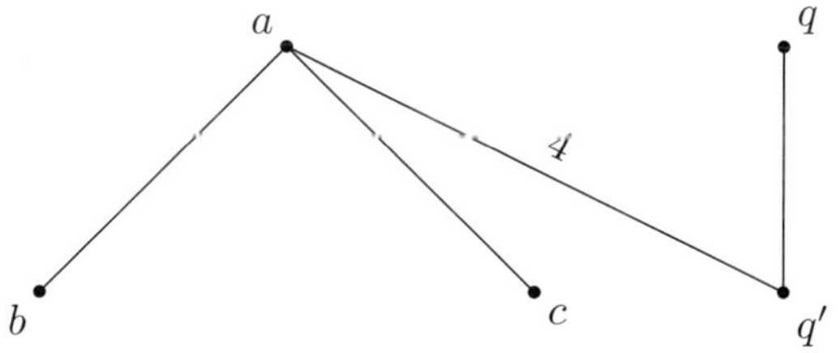

Put

$$P' = \{k \in [5, 2^m + 2] \mid q' \text{ has an } \mathcal{A}_k\text{-similar ultrafilter in } \{b, c\}\}.^6$$

If $P' = \varnothing$, then comparing

$$\#\left(\left\{M \mid M \cap \{a,b,c,q,q'\} = \{a,q\}, \#(M) = \mathfrak{n},\ M \in \bigcup_{k=5}^{2^m+2} \mathcal{A}_k\right\}\right)$$

[6] By the assumption from 9.21, q' cannot have an $\mathcal{A}_4$-similar ultrafilter in $\{b, c\}$ (because this implies that either a, b or a, c are $\mathcal{A}_4$-similar ultrafilters, whereas our main assumption is that b, c are not $\mathcal{A}_k$-similar ultrafilters for any $k \in [1, 2^m + 2]$).

with

$$\#(\{M \subset X \mid M \cap \{a,b,c,q,q'\} = \{a,q\},\ \#(M) = \mathfrak{n}\}),$$

we come to a contradiction:

$$\bigcup_{k=1}^{2^m+2} \mathcal{A}_k \neq \mathfrak{P}(X).$$

Let $P' \neq \varnothing$. Assume, without loss of generality, that $P' = \{5\}$. We know that there exists an $\mathcal{A}_4$-similar set M_0 such that $\#(M_0 \cap \{a,b,c,q,q'\}) = 2$. For other $\mathcal{A}_4$-similar sets M we have

$$\#(M \cap \{a,b,c,q,q'\}) \leq 1.$$

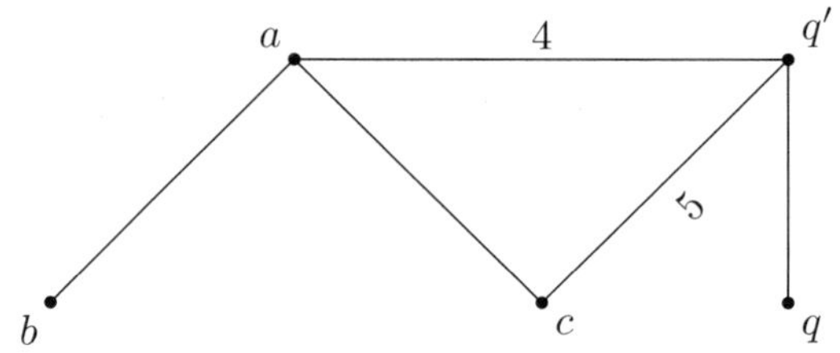

By the results of 9.20,

$$\#\left(\left\{M \mid M \cap \{a,b,c,q,q'\} = \{a,q'\}, \#(M) = \mathfrak{n}, \text{ and } M \in \mathcal{A}_4 \cup \bigcup_{k=6}^{2^m+2} \mathcal{A}_k\right\}\right)$$
$$< (2^{1-m} + 2^{-m} \cdot (2^m - 3))$$
$$\cdot \#(\{M \subset X \mid M \cap \{a,b,c,q,q'\} = \{a,q'\} \text{ and } \#(M) = \mathfrak{n}\})$$

up to an infinitely small function with respect to the latter expression $\#(\dots)$. We come to a contradiction:

$$\bigcup_{k=1}^{2^m+2} \mathcal{A}_k \neq \mathfrak{P}(X).$$

We have proved that $q \in \mathfrak{K}$. There exists at most one ultrafilter $r \in \bigcup_{k=1}^{2^m+2} \ker \mathcal{A}_k$ such that $\{k \mid r \in \ker \mathcal{A}_k\} = \{1,2\}$ (see 9.22). If r exists and $r \notin \{a,b,c\}$, then we have:

$$\#\left(\bigcup_{k=1}^{2^m+2} \ker \mathcal{A}_k \setminus \mathfrak{K}\right) \leq 4.$$

Otherwise

$$\mathfrak{K} = \bigcup_{k=1}^{2^m+2} \ker \mathcal{A}_k \setminus \{a,b,c\}. \qquad \square$$

9.24. Let $d, e \in \mathfrak{K}$ be distinct ultrafilters which are not $\mathcal{A}_k$-similar for any $k \in [1, 2^m+2]$. Let a, d be $\mathcal{A}_{k_1}$-similar ultrafilters, and a, e be $\mathcal{A}_{k_2}$-similar ultrafilters. The analysis of the proof of Lemma 9.23 implies that if $\{k_1, k_2\} \neq \{1,2\}$, then

$$\mathfrak{K} = \bigcup_{k=1}^{2^m+2} \ker \mathcal{A}_k \setminus \{a,b,c\}.$$

Further, let $d', e' \in \mathfrak{K}$, $\{d, e\} \cap \{d', e'\} = \varnothing$, and let there not exist a $k \in [1, 2^m + 2]$ such that d', e' are $\mathcal{A}_k$-similar ultrafilters. Then

$$\mathfrak{K} = \bigcup_{k=1}^{2^m+2} \ker \mathcal{A}_k \setminus \{a, b, c\}$$

because a $\mathcal{A}_1$-similar set contains at most three elements.

9.25. Let $q \in \mathfrak{K}$. If $k \in [1, 2]$ and a, q are $\mathcal{A}_k$-similar ultrafilters, then, obviously, q has an $\mathcal{A}_k$-similar ultrafilter in $\{b, c\}$. Let $k \in [3, 2^m + 2]$, and let a, q be $\mathcal{A}_k$-similar ultrafilters. By the arguments above, to avoid the contradiction

$$\bigcup_{k=1}^{2^m+2} \mathcal{A}_k \neq \mathfrak{P}(X),$$

we must assume that q has an $\mathcal{A}_{k'}$-similar ultrafilter in $\{b, c\}$.

9.26. Let us interrupt our exposition, and in this and the following sections consider two algebras $\mathcal{A}^1, \mathcal{A}^2$ such that $X \in \mathcal{A}^1, \mathcal{A}^2$. In addition, $\#(X) = 2\mathfrak{n}$.

Let K_1, K_2 be distinct $\mathcal{A}^1$-similar sets, each consisting of three elements. Let K_1', K_2', K_3' be $\mathcal{A}^2$-similar sets, each consisting of two elements. Besides, let

$$K_1 \cup K_2 = K_1' \cup K_2' \cup K_3', \quad \ker \mathcal{A}^1 = \ker \mathcal{A}^2,$$

and let each $\mathcal{A}^1$-similar set $M \subset \ker \mathcal{A}^1 \setminus (K_1 \cup K_2)$ (if such a set exists) be an $\mathcal{A}^2$-similar set. It is easy to show that

$$\#(\{M \mid \#(M) = \mathfrak{n} \text{ and } M \in \mathcal{A}^1\}) < \#(\{M \mid \#(M) = \mathfrak{n} \text{ and } M \in \mathcal{A}^2\}).$$

9.27. In this section we have other assumptions. Let K be an $\mathcal{A}^1$-similar set and $\#(K) = 3$. Let K', K'' be $\mathcal{A}^2$-similar sets and

$$\#(K') = \#(K'') = 2, \quad K \subset K' \cup K''.$$

Let $\ker \mathcal{A}^1 \setminus K = \ker \mathcal{A}^2 \setminus (K' \cup K'')$, and let each $\mathcal{A}^1$-similar set $M \subset \ker \mathcal{A}^1 \setminus K$ be an $\mathcal{A}^2$-similar set. It is easy to show that

$$\#(\{M \mid \#(M) = \mathfrak{n} \text{ and } M \in \mathcal{A}^1\}) < \#(\{M \mid \#(M) = \mathfrak{n} \text{ and } M \in \mathcal{A}^2\}).$$

Let $2m - 1 \leq \#(\ker \mathcal{A}^2) \leq 2m$, where $m \in \mathbb{N}^+$. The number

$$\#(\{M \mid \#(M) = \mathfrak{n} \text{ and } M \in \mathcal{A}^2\})$$

takes the maximal value when each $\mathcal{A}^2$-similar set consists of two elements, and then this value equals

$$\sum_{k=0}^{m} \binom{m}{k} \binom{2\mathfrak{n} - 2m}{\mathfrak{n} - 2k}.$$

9.28. We use the following general definition:

DEFINITION. Let $\{\mathcal{A}_\lambda\}_{\lambda \in \Lambda}$ be a family of algebras. A nonempty set

$$K \subset \bigcup_{\lambda \in \Lambda} \ker \mathcal{A}_\lambda$$

is called a *connected component* of this family if the following conditions are satisfied:

(i) if $a, b \in K$ and $a \neq b$, then there exists a λ such that a, b are $\mathcal{A}_\lambda$-similar ultrafilters;

(ii) if $q' \in \bigcup_{\lambda\in\Lambda} \ker \mathcal{A}_\lambda \setminus K$, then there does not exist λ' such that q' has an $\mathcal{A}_{\lambda'}$-similar ultrafilter in K.

Assume that for any pairwise distinct ultrafilters $a, b, c \in \bigcup_{\lambda\in\Lambda} \ker \mathcal{A}_\lambda$, if the following two conditions hold:

1) there exists a λ_1 such that a, b are $\mathcal{A}_{\lambda_1}$-similar ultrafilters,
2) there exists a λ_2 such that a, c are $\mathcal{A}_{\lambda_2}$-similar ultrafilters,

then there exists a λ_3 such that b, c are $\mathcal{A}_{\lambda_3}$-similar ultrafilters. In this case $\bigcup_{\lambda\in\Lambda} \ker \mathcal{A}_\lambda$ splits into connected components.

9.29. Let us return to our algebras $\mathcal{A}_1, \ldots, \mathcal{A}_{2^m+2}$.

LEMMA. *If $\bigcup_{k=1}^{2^m+2} \ker \mathcal{A}_k$ splits into connected components and their number is not less than three, then*

$$\bigcup_{k=1}^{2^m+2} \mathcal{A}_k \neq \mathfrak{P}(X).$$

PROOF. Let $\#(X) = 2\mathfrak{n}$. Let us take three connected components K_1, K_2, K_3. If, for example,

$$\#(\{k \mid \ker \mathcal{A}_k \cap K_1 \neq \varnothing\}) < 3,$$

then everything is simple. Therefore assume that a_i, b_i are $\mathcal{A}_i$-similar ultrafilters, $i = 1, 2, 3$, and $\{a_i, b_i\} \subset K_i$. By the results of 9.20,

$$\#\Bigl(\Bigl\{M \mid M \cap \{a_1, a_2, a_3, b_1, b_2, b_3\} = \{a_1, a_2, a_3\},\ \#(M) = \mathfrak{n},\ M \in \bigcup_{k=4}^{2^m+2} \mathcal{A}_k\Bigr\}\Bigr)$$
$$< 2^{-m}(2^m - 1) \cdot \#(\{M \subset X \mid M \cap \{a_1, a_2, a_3, b_1, b_2, b_3\} = \{a_1, a_2, a_3\} \text{ and } \#(M) = \mathfrak{n}\})$$

up to an infinitely small function with respect to the latter expression $\#(\ldots)$. □

9.30. Let us determine the maximal possible number kr_m of pairs of $\mathcal{A}_k$-similar ultrafilters for various $k \in [1, 2^m + 2]$:

$$\text{if } m \equiv 0 \pmod 3, \text{ then } kr_m = (2^m + 2)(2m - 2);$$
$$\text{if } m \equiv 1 \pmod 3, \text{ then } kr_m = (2^m + 2)(2m - 3);$$
$$\text{if } m \equiv 2 \pmod 3, \text{ then } kr_m = (2^m + 2)(2m - 1).$$

Let n_* be the minimal natural number such that $1 + \sqrt{1 + 4kr_m} \leq 2n_*$.

9.31. THEOREM. *$\mathfrak{g}(2^m + 2) \leq 2^m - 1$ for $m \in [4, 8]$.*

PROOF. We must prove that

$$\bigcup_{k=1}^{2^m+2} \mathcal{A}_k \neq \mathfrak{P}(X),$$

if $m \in [4, 8]$. A verification shows that

$$(2^m + 2) \cdot \sum_{k=0}^{m} \binom{m}{k} \cdot \binom{2n_* - 2m}{n_* - 2k} < \binom{2n_*}{n_*},$$

if $m \in [4, 8]$. Assume that $\#(X) = 2n_*$ and $m \in [4, 8]$. Since the results of 9.26 and 9.27 hold,

$$\bigcup_{k=1}^{2^m+2} \mathcal{A}_k \neq \mathfrak{P}(X).$$

Now let

$$X = \bigcup_{k=1}^{2^m+2} \ker \mathcal{A}_k \quad \text{and} \quad \#(X) \geq 2n_* + 1.$$[7]

Assume that

$$\bigcup_{k=1}^{2^m+2} \mathcal{A}_k = \mathfrak{P}(X).$$

Denote by ν^* the number of all pairwise distinct pairs of $\mathcal{A}_k$-similar ultrafilters for all $k \in [1, 2^m + 2]$.[8] As before, assume that there exist pairwise distinct ultrafilters a, b, c such that a, b are $\mathcal{A}_1$-similar ultrafilters, a, c are $\mathcal{A}_2$-similar ultrafilters, and b, c are not $\mathcal{A}_k$-similar ultrafilters for any $k \in [1, 2^m + 2]$. Let us use the arguments presented above. We know that $\#(X \setminus \mathfrak{K}) \leq 4$.[9] Suppose there exist at most ν pairwise disjoint pairs of distinct ultrafilters from $\mathfrak{K}$, and each of these pairs is not a pair of $\mathcal{A}_k$-similar ultrafilters for any $k \in [1, 2^m + 2]$. Let us fix ν such pairs. Let us start counting pairs of distinct ultrafilters which are $\mathcal{A}_k$-similar; the number k depends on the pair. Let $\mathfrak{K}'$ be the subset of all ultrafilters in $\mathfrak{K}$ which do not belong to any of the ν pairs. Each pair of distinct ultrafilters from $\{a\} \cup \mathfrak{K}'$ is a pair of $\mathcal{A}_k$-similar ultrafilters for some k. The number of such pairs is not less than

$$(n_* - \nu - 1)(2n_* - 2\nu - 3).$$

Each ultrafilter from $\mathfrak{K}'$ has some $\mathcal{A}_k$-similar ultrafilter in $\{b, c\}$ and in each of the ν pairs. We obtain at least

$$(\nu + 1)(2n_* - 2\nu - 3)$$

more pairs. Further, let us choose two distinct pairs from the collection of our ν pairs and the pair (b, c):

$$(x_1, x_2), \quad (y_1, y_2).$$

By the results of 9.25, x_i has some $\mathcal{A}_k$-similar ultrafilter in $\{y_1, y_2\}$. We obtain at least $\nu(\nu + 1)$ more pairs. Finally, if $q \in \{b, c\}$, or q belongs to one of the ν pairs, then there exists $k \in [1, 2^m + 2]$ such that a, q are $\mathcal{A}_k$-similar ultrafilters. We obtain $2(\nu + 1)$ more pairs. We do not count pairs which contain the ultrafilter

$$r \in X \setminus (\{a, b, c\} \cup \mathfrak{K}),$$

under the assumption that $\#(X \setminus \mathfrak{K}) = 4$. So we have

$$\begin{aligned}\nu^* &> (n_* - \nu - 1)(2n_* - 2\nu - 3) + (\nu + 1)(2n_* - 2\nu - 3) + \nu(\nu + 1) + 2(\nu + 1) \\ &\geq \frac{(x - 2\nu - 1)(x - 2\nu - 2)}{2} + (\nu + 1)(x - 2\nu - 2) + \nu(\nu + 1) + 2(\nu + 1),\end{aligned}$$

---

[7] Here one does not impose restrictions on m; i.e., m is an arbitrary natural number ≥ 3.

[8] If x, y are $\mathcal{A}_{k_1}$- and $\mathcal{A}_{k_2}$-similar ultrafilters and $k_1 \neq k_2$, then the pair (x, y) is counted once.

[9] Using the arguments from 9.24, one can easily show that $\mathfrak{K} = X \setminus \{a, b, c\}$. But for us it is enough to know that $\#(X \setminus \mathfrak{K}) \leq 4$.

where $x = \sqrt{1+4kr_m}$. Let us show that

$$\frac{(x-2\nu-1)(x-2\nu-2)}{2}+(\nu+1)(x-2\nu-2)+\nu(\nu+1)+2(\nu+1) > \frac{x^2-1}{4} = kr_m.$$

In other words, we must show that

$$4\nu^2+\nu(8-4x)+x^2-2x+5>0.$$

But this is clear, since if we consider ν as a variable, then the discriminant is

$$D=-32x-16.$$

We have obtained a contradiction: $\nu^* > kr_m$. Therefore the corresponding ultrafilters a, b, c cannot exist. We must conclude that the set

$$\bigcup_{k=1}^{2^m+2} \ker \mathcal{A}_k$$

splits into connected components. By Lemma 9.29, we can assume that the number of components is at most two. But then

$$\nu^* > n_*(n_*-1) \geq \frac{x+1}{2}\cdot\frac{x-1}{2} = kr_m.$$

Hence the assumption

$$\bigcup_{k=1}^{2^m+2} \mathcal{A}_k = \mathfrak{P}(X)$$

leads to a contradiction. □

9.32. REMARK. A verification shows that in contrast to the case when $m \in [4,8]$, we have

$$(2^m+2)\sum_{k=0}^{m}\binom{m}{k}\binom{2n_*-2m}{n_*-2k} > \binom{2n_*}{n_*},$$

if $m=3$ or 9.

9.33. THEOREM. $\mathfrak{g}(10) \leq 5$.

PROOF. As usual, we consider the algebras $\mathcal{A}_1, \ldots, \mathcal{A}_{10}$, etc. Let $\#(X)=12$. Then the results of 9.26 and 9.27 hold. Hence

$$\bigcup_{k=1}^{10} \mathcal{A}_k \neq \mathfrak{P}(X),$$

because

$$10\cdot\sum_{k=0}^{3}\binom{3}{k}\binom{6}{6-2k} < \binom{12}{6}.$$

Let us assume that

$$\omega = \#\left(\bigcup_{k=1}^{10} \ker \mathcal{A}_k\right) > 12.$$

We will follow the proof of Theorem 9.31 and assume that

$$\bigcup_{k=1}^{10} \mathcal{A}_k = \mathfrak{P}(X).$$

Let there exist corresponding ultrafilters a, b, c. We know that $\#(\mathfrak{K}) \geq \omega - 4$. To avoid an immediate contradiction, we must assume the existence of distinct ultrafilters d, e in $\mathfrak{K}$ such that d, e are not $\mathcal{A}_k$-similar ultrafilters for any $k \in [1, 10]$. Assume that for each pair of distinct ultrafilters from $\mathfrak{K} \setminus \{d, e\}$ there exists a $k \in [1, 10]$ such that these ultrafilters are $\mathcal{A}_k$-similar. We have

$$\nu^* > \frac{(\omega - 5)(\omega - 6)}{2} + 2(\omega - 6) + 2 + 4 > 40 = kr_3.$$

Due to the contradiction obtained and by the results of 9.24 we have $\mathfrak{K} = X \setminus \{a, b, c\}$. Now we can write

$$\nu^* \geq \frac{(\omega - 2\nu - 2)(\omega - 2\nu - 3)}{2} + (\nu + 1)(\omega - 2\nu - 3) + \nu(\nu + 1) + 2(\nu + 1).$$

Hence

$$2\nu^* - 80 \geq 2\nu^2 + \nu(6 - 2\omega) + \omega^2 - 3\omega - 76.$$

Viewing this as a quadratic expression in ν with the discriminant $-4\omega^2 + 644 < 0$, we conclude that $\nu^* > 40$. Due to the obtained contradiction,

$$\bigcup_{k=1}^{10} \ker \mathcal{A}_k$$

splits into at most two connected components. Let $\omega = 13$. Obviously,

$$\nu^* \geq \binom{6}{2} + \binom{7}{2} = 36.$$

Hence we must assume that there are not less than six algebras $\mathcal{A}_k$ such that

$$\#(\ker \mathcal{A}_k) = 5.$$

Put $\#(X) = 14$, and let us count:

$$\#\left(\left\{M \mid \#(M) = 7 \text{ and } M \in \bigcup_{k=1}^{10} \mathcal{A}_k\right\}\right) \leq 12\binom{9}{2} + 12\binom{9}{4} + 8\binom{8}{1} + 24\binom{8}{3}$$
$$< \binom{14}{7} = \#(\{M \subset X \mid \#(M) = 7\}).$$

So the assumption

$$\bigcup_{k=1}^{10} \mathcal{A}_k = \mathfrak{P}(X)$$

leads to a contradiction when $\omega = 13$. Let $\omega \geq 14$. We immediately get a contradiction since

$$\nu^* \geq \binom{7}{2} + \binom{7}{2} > 40. \qquad \square$$

9.34. Let us summarize the above arguments. We have proved that

$$\mathfrak{g}(2^m + 2) \leq 2m - 1$$

for all $m \in [3, 8]$. This allows us to conclude the description of the behavior of $\mathfrak{g}(n)$ on the interval $[1, 258]$ presented in 2.12.

9.35. Let $\#(X) = 2\mathfrak{n}$. Let a, b, c be pairwise distinct ultrafilters, and let $\mathcal{A}$ be an algebra such that neither a, b, nor a, c, nor b, c are $\mathcal{A}$-similar ultrafilters. Then let us say that $\mathcal{A}$ satisfies the *$\delta(a,b,c)$-conditions*, and put

$$\delta(a,b,c)(\mathcal{A}) = \#\big(\{M \mid M \cap \{a,b,c\} = \{b,c\},\ \#(M) = \mathfrak{n},\ \text{and } M \in \mathcal{A}\}\big).$$

Assume that $\#(\ker \mathcal{A}) = 11$ and $\ker \mathcal{A}$ splits into four $\mathcal{A}$-similar sets: three of them consist of three elements each, and one consists of two elements. We can choose two $\mathcal{A}$-similar sets M_1 and M_2 so that $\#(M_1) = \#(M_2) = 3$ and one of the following conditions holds:

(i) $M_i \cap \{b,c\} \neq \varnothing$ for $i = 1, 2$;
(ii) $M_1 \cap \{b,c\} \neq \varnothing$, $M_2 \cap \{a,b,c\} = \varnothing$;
(iii) $M_i \cap \{a,b,c\} = \varnothing$ for $i = 1, 2$.

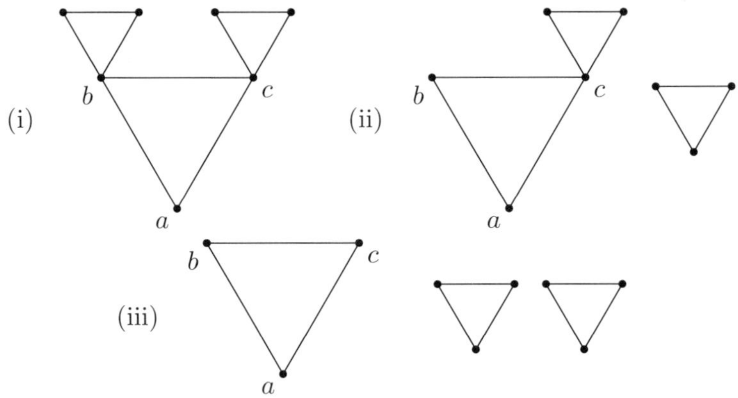

Let us construct an algebra $\mathcal{A}'$ which differs from $\mathcal{A}$ only in that instead of M_1 and M_2 there are three $\mathcal{A}'$-similar sets M_1', M_2', M_3', each consisting of two elements, and

$$M_1 \cup M_2 = M_1' \cup M_2' \cup M_3'.$$

The other two $\mathcal{A}$-similar sets are $\mathcal{A}'$-similar, and $\#(\ker \mathcal{A}') = 11$. The algebra $\mathcal{A}'$ can be constructed so that $\mathcal{A}'$ satisfies the $\delta(a,b,c)$-conditions and $\delta(a,b,c)(\mathcal{A}) < \delta(a,b,c)(\mathcal{A}')$.

9.36. Let $\#(X) = 2\mathfrak{n}$. We return to our algebras $\mathcal{A}_1, \ldots, \mathcal{A}_{2^m+2}$.

LEMMA. *If $\mathcal{A}_1$ satisfies the $\delta(a,b,c)$-conditions and $m = 6$, then*

$$64 \cdot \delta(a,b,c)(\mathcal{A}_1) < \#\big(\{M \subset X \mid M \cap \{a,b,c\} = \{b,c\} \text{ and } \#(M) = \mathfrak{n}\}\big) = \gamma.$$

PROOF. Consider an algebra $\mathcal{A}^3$, $\#(\ker \mathcal{A}^3) = 6$, and suppose that $\ker \mathcal{A}^3$ splits into three $\mathcal{A}^3$-similar sets, each consisting of two elements. Let $\mathcal{A}^3$ satisfy the $\delta(a,b,c)$-conditions. One easily shows that $\delta(a,b,c)(\mathcal{A}^3)$ takes its maximal value when each $\mathcal{A}^3$-similar set has a nonempty intersection with $\{a,b,c\}$ or when only two $\mathcal{A}^3$-similar sets have a nonempty intersection with $\{a, b, c\}$ and one of them contains a:

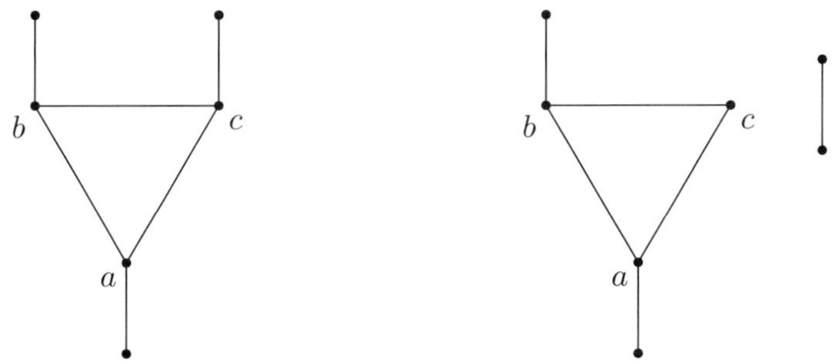

Then

$$\delta(a, b, c)(\mathcal{A}^3) = \binom{2n - 6}{n - 2}.$$

Therefore

$$8 \cdot \delta(a, b, c)(\mathcal{A}^3) < \gamma = \binom{2n - 3}{n - 1}.$$

Assume that $\#(\ker \mathcal{A}_1) = 12$, i.e., $\ker \mathcal{A}_1$ splits into six $\mathcal{A}_1$-similar sets, each consisting of two elements. By the arguments from 9.17, for $\mathcal{A}_1$ we can find an algebra $\mathcal{A}^3$ such that $\delta(a, b, c)(\mathcal{A}^3) > 8 \cdot \delta(a, b, c)(\mathcal{A}_1)$; therefore $64 \cdot \delta(a, b, c)(\mathcal{A}_1) < \gamma$.

Assume now that $\#(\ker \mathcal{A}_1) = 11$. By the results of 9.35, we can assume that $\ker \mathcal{A}_1$ consists of one $\mathcal{A}_1$-similar set M_0 constituted by three elements, and four more $\mathcal{A}_1$-similar sets, each consisting of two elements. Assume that $M_0 \cap \{a, b, c\} = \varnothing$. Let us construct an algebra $\mathcal{A}_1^*$: each $\mathcal{A}_1$-similar set consisting of two elements is an $\mathcal{A}_1^*$-similar set; there are two more $\mathcal{A}_1^*$-similar sets, M_1 and M_2;

$$\#(M_1) = \#(M_2) = 2, \quad M_0 \subset M_1 \cup M_2,$$
$$(M_1 \cup M_2) \cap \{a, b, c\} = \varnothing, \quad \#(\ker \mathcal{A}_1^*) = 12.$$

It is easy to check that $\delta(a, b, c)(\mathcal{A}_1) < \delta(a, b, c)(\mathcal{A}_1^*)$. But this implies that $64 \cdot \delta(a, b, c)(\mathcal{A}_1) < \gamma$. Therefore let us assume that

$$\#(M_0 \cap \{a, b, c\}) = 1.$$

Consider an algebra $\mathcal{A}_*^3$; $\#(\ker \mathcal{A}_*^3) = 5$, and there are two $\mathcal{A}_*^3$-similar sets, M_0^*, M_1^*; $\#(M_0^*) = 3$,

$$\#(M_0^* \cap \{a, b, c\}) = 1, \quad M_1^* \cap \{a, b, c\} = \varnothing.$$

It is easy to check that $8 \cdot \delta(a, b, c)(\mathcal{A}_*^3) < \gamma$. Let us now assume that among all $\mathcal{A}_1$-similar sets only one (which is necessarily M_0) has nonempty intersection with $\{a, b, c\}$. By the results of 9.17, we have $\delta(a, b, c)(\mathcal{A}_*^3) > 8 \cdot \delta(a, b, c)(\mathcal{A}_1)$, and hence $64 \cdot \delta(a, b, c)(\mathcal{A}_1) < \gamma$. Consider an algebra $\mathcal{A}^5$, $\#(\ker \mathcal{A}^5) = 9$. There are four $\mathcal{A}^5$-similar sets, $\widehat{M}_0, \widehat{M}_1, \widehat{M}_2, \widehat{M}_3$; $\#(\widehat{M}_0) = 3$ and $\#(\widehat{M}_0 \cap \{a, b, c\}) = 1$. There exists an $i_0 \in [1, 3]$ such that

$$\#(\widehat{M}_{i_0} \cap \{a, b, c\}) = 1;$$

if $i \in [1,3] \setminus \{i_0\}$, then $\widehat{M}_i \cap \{a,b,c\} = \varnothing$. A simple verification shows that $32 \cdot \delta(a,b,c)(\mathcal{A}^5) < \gamma$.[10] Now let us assume that there is only one $\mathcal{A}_1$-similar set (besides M_0) which has nonempty intersection with $\{a,b,c\}$. By the results of 9.17, $\delta(a,b,c)(\mathcal{A}^5) > 2 \cdot \delta(a,b,c)(\mathcal{A}_1)$; hence $64 \cdot \delta(a,b,c)(\mathcal{A}_1) < \gamma$.

It remains to consider the following situation: two $\mathcal{A}_1$-similar sets (besides M_0) have nonempty intersection with $\{a,b,c\}$. In this situation $\delta(a,b,c)(\mathcal{A}_1)$ takes its maximal value when $a \in M_0$, and this value equals

$$\binom{2\mathfrak{n}-11}{\mathfrak{n}-3} + \binom{2\mathfrak{n}-11}{\mathfrak{n}-4} + 2 \cdot \binom{2\mathfrak{n}-11}{\mathfrak{n}-5}.$$

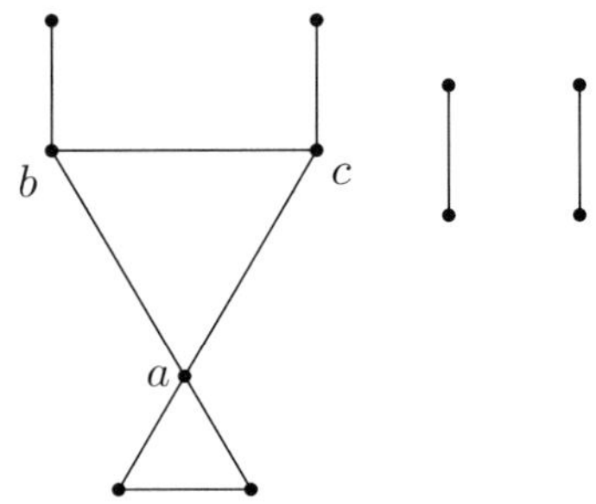

A straightforward verification shows that

$$64 \cdot \delta(a,b,c)(\mathcal{A}_1) < \gamma.$$[11]

□

9.37. REMARK. In principle, the validity of Lemma 9.36 can be checked by direct computations. The arguments presented in the proof of this lemma make these computations easier.

9.38. As was done at the beginning of 9.35, consider an algebra $\mathcal{A}^*$, and let us change the assumptions: neither a, b, nor a, c are $\mathcal{A}^*$-similar ultrafilters, and b, c are $\mathcal{A}^*$-similar ultrafilters. In this situation let us say that $\mathcal{A}^*$ satisfies the *$\delta^*(a,b,c)$-conditions*, and put

$$\delta^*(a,b,c)(\mathcal{A}^*) = \#(\{M \mid M \cap \{a,b,c\} = \{b,c\},\ \#(M) = \mathfrak{n}, \text{ and } M \in \mathcal{A}^*\}).$$

Again let us consider our algebras $\mathcal{A}_1, \ldots, \mathcal{A}_{2^m+2}$, and let $\#(X) = 2\mathfrak{n}$.

LEMMA. *If $\mathcal{A}_2$ satisfies the $\delta^*(a,b,c)$-conditions and $m = 6$, then*

$$32 \cdot \delta^*(a,b,c)(\mathcal{A}_2) < \gamma.$$

PROOF. Consider an algebra $\mathcal{A}_0^3$; $\#(\ker \mathcal{A}_0^3) = 6$ and $\ker \mathcal{A}_0^3$ splits into three $\mathcal{A}_0^3$-similar sets, each consisting of two elements. Let $\mathcal{A}_0^3$ satisfy the $\delta^*(a,b,c)$-

[10] For the corresponding algebra $\mathcal{A}^4$, there exists a case where $16 \cdot \delta(a,b,c)(\mathcal{A}^4) > \gamma$.
[11] If $m=5$, then in a similar situation $\delta(a,b,c)(\mathcal{A}_1) = \binom{2\mathfrak{n}-9}{\mathfrak{n}-3} + \binom{2\mathfrak{n}-9}{\mathfrak{n}-4}$ and $32 \cdot \delta(a,b,c)(\mathcal{A}_1) > \gamma$.

conditions. Then it is easy to show that

$$\delta^*(a,b,c)(\mathcal{A}_0^3) \leq 2 \cdot \binom{2\mathfrak{n}-6}{\mathfrak{n}-2}.^{12}$$

This and the arguments presented in 9.17 imply the lemma in the case $\#(\ker \mathcal{A}_2) = 12$.[13] It is clear also that our lemma holds in the case where $\#(\ker \mathcal{A}_2) = 11$ and any $\mathcal{A}_2$-similar set consisting of three elements is disjoint from $\{a,b,c\}$.

Now let us assume that $\#(\ker \mathcal{A}_2) = 11$. As in the proof of Lemma 9.36, we will assume that there is only one $\mathcal{A}_2$-similar set M_0 consisting of three elements. Assume also that

$$M_0 \cap \{a,b,c\} \neq \varnothing.$$

Consider an algebra $\mathcal{A}_{**}^3$; $\#(\ker \mathcal{A}_{**}^3) = 5$ and there are two $\mathcal{A}_{**}^3$-similar sets — M_0^{**}, M_1^{**}; $\#(M_0^{**}) = 3$,

$$\{b,c\} \subset M_0^{**}, \quad M_1^{**} \cap \{a,b,c\} = \varnothing.$$

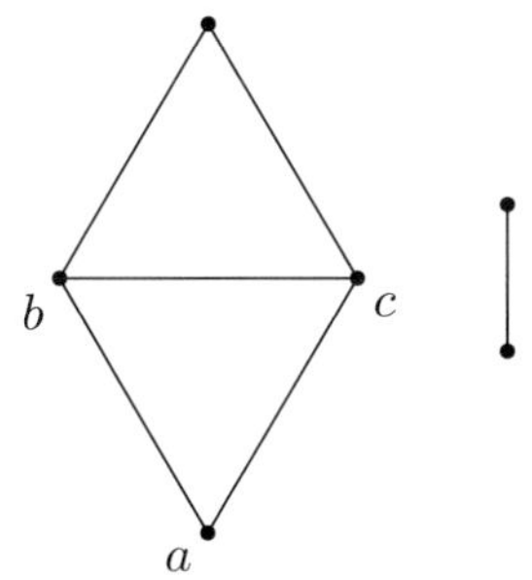

We have

$$\delta^*(a,b,c)(\mathcal{A}_{**}^3) = \binom{2\mathfrak{n}-6}{\mathfrak{n}-1} + \binom{2\mathfrak{n}-6}{\mathfrak{n}-3} < 2 \cdot \binom{2\mathfrak{n}-6}{\mathfrak{n}-2}.$$

This and the arguments from 9.17 imply that if only one $\mathcal{A}_2$-similar set has a nonempty intersection with $\{a,b,c\}$ (this set is necessarily $M_0 \supset \{b,c\}$), then the lemma is true.

Assume now that two $\mathcal{A}_2$-similar sets have nonempty intersection with $\{a,b,c\}$. Here two cases are possible:

1) $\{a,q_1,q_2\}$, $\{b,c\}$, $\{q_3,q_4\}$, $\{q_5,q_6\}$, $\{q_7,q_8\}$ are all $\mathcal{A}_2$-similar sets;
2) $\{a,q_1\}$, $\{b,c,q_2\}$, $\{q_3,q_4\}$, $\{q_5,q_6\}$, $\{q_7,q_8\}$ are all $\mathcal{A}_2$-similar sets.

---

[12] $\delta^*(a,b,c)(\mathcal{A}_0^3) = 2 \cdot \binom{2\mathfrak{n}-6}{\mathfrak{n}-2}$, when a belongs to an $\mathcal{A}_0^3$-similar set.

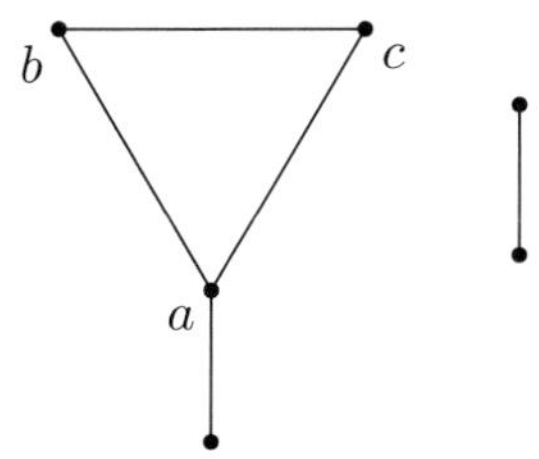

[13] Recall that in this case $\ker \mathcal{A}_2$ splits into six $\mathcal{A}_2$-similar sets, each consisting of two elements.

1) 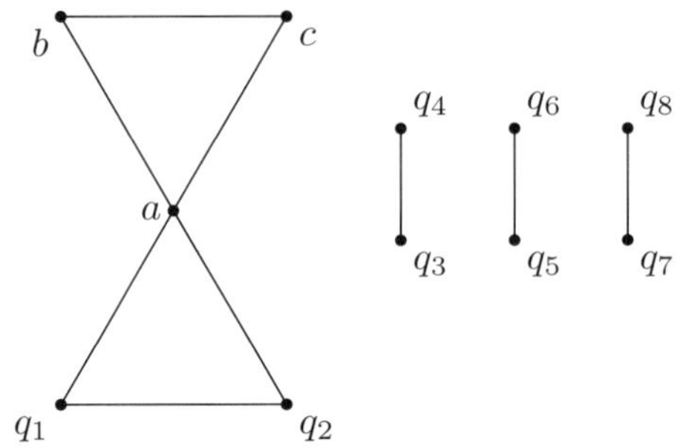

2) 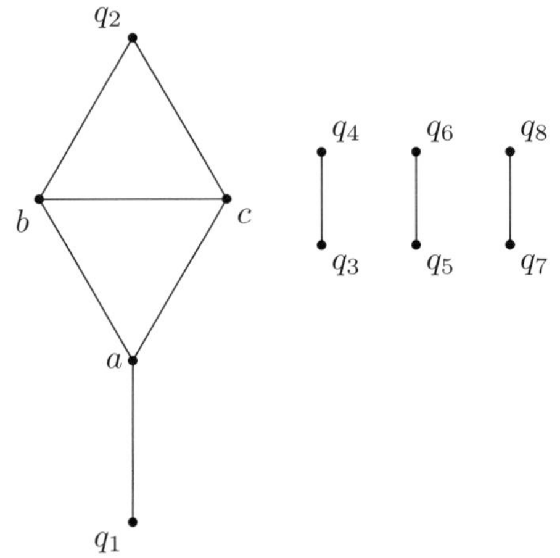

In both cases

$$\delta^*(a,b,c)(\mathcal{A}_2) = \binom{2\mathfrak{n}-11}{\mathfrak{n}-2} + \binom{2\mathfrak{n}-11}{\mathfrak{n}-3} + 3\cdot\binom{2\mathfrak{n}-11}{\mathfrak{n}-4} + 3\cdot\binom{2\mathfrak{n}-11}{\mathfrak{n}-5}$$
$$< 2\left(\binom{2\mathfrak{n}-11}{\mathfrak{n}-3} + \binom{2\mathfrak{n}-11}{\mathfrak{n}-4} + 2\cdot\binom{2\mathfrak{n}-11}{\mathfrak{n}-5}\right);$$

hence $32\cdot\delta^*(a,b,c)(\mathcal{A}_2) < 64\cdot\delta(a,b,c)(\mathcal{A}_1) < \gamma$. □

9.39. Denote by δ_6 the maximal number $\delta(a,b,c)(\mathcal{A}_1)$ of various configurations of $\ker\mathcal{A}_1$, and by δ_6^* the maximal number $\delta^*(a,b,c)(\mathcal{A}_2)$ of various configurations of $\ker\mathcal{A}_2$. If $m > 6$, then let us define the corresponding numbers δ_m, δ_m^*; we have

$$\delta_{m-1} > 2\cdot\delta_m, \quad \delta_{m-1}^* > 2\cdot\delta_m^*.$$

Therefore, if $m \geq 6$, then

$$2^m\cdot\delta_m < \gamma = \binom{2\mathfrak{n}-3}{\mathfrak{n}-1}, \quad 2^{m-1}\cdot\delta_m^* < \gamma.$$

9.40. Remark. Of course, one can define the corresponding numbers δ_m and δ_m^*, when $m \in [3,5]$. But, as was noted in footnote 9.11,

$$32\cdot\delta_5 > \gamma.$$

9.41. The analysis of the proof of Theorem 9.16 implies the following statement for all $m \geq 3$: if there exist ultrafilters a, b and pairwise distinct numbers

$$k_1, k_2, k_3 \in [1, 2^m + 2]$$

such that a, b are $\mathcal{A}_{k_1}$-, $\mathcal{A}_{k_2}$-, $\mathcal{A}_{k_3}$-similar ultrafilters, then

$$\bigcup_{k=1}^{2^m+2} \mathcal{A}_k \neq \mathfrak{P}(X).$$

Hence below we will assume that such ultrafilters a, b and numbers k_1, k_2, k_3 do not exist.

9.42. In this and the next section we have $\#(X) = 2\mathfrak{n}$. *Let* $m \geq 6$. Consider the following situation: there are three ultrafilters a, b, c such that a, b are $\mathcal{A}_1$- and $\mathcal{A}_2$-similar ultrafilters, and a, c are $\mathcal{A}_3$- and $\mathcal{A}_4$-similar ultrafilters. Let us show that in this situation

$$\bigcup_{k=1}^{2^m+2} \mathcal{A}_k \neq \mathfrak{P}(X).$$

Consider the most difficult case: assume that b, c are $\mathcal{A}_5$- and $\mathcal{A}_6$-similar ultrafilters. We have

$$\#\left(\left\{M \mid M \cap \{a, b, c\} = \{b, c\},\ \#(M) = \mathfrak{n},\ \text{and } M \in \bigcup_{k=5}^{2^m+2} \mathcal{A}_k\right\}\right)$$
$$\leq (2^m - 4) \cdot \delta_m + 2 \cdot \delta_m^* < \frac{2^m - 4}{2^m} \cdot \gamma + \frac{2}{2^{m-1}} \cdot \gamma = \gamma. \qquad \square$$

Below we will assume that there do not exist ultrafilters a, b, c and pairwise distinct numbers $k_1, k_2, k_3, k_4 \in [1, 2^m + 2]$ such that a, b are $\mathcal{A}_{k_1}$- and $\mathcal{A}_{k_2}$-similar ultrafilters, and a, c are $\mathcal{A}_{k_3}$- and $\mathcal{A}_{k_4}$-similar ultrafilters.

9.43. Let $m \geq 6$. Consider the following situation: a, b are $\mathcal{A}_1$- and $\mathcal{A}_2$-similar ultrafilters. Let us show that in this situation

$$\bigcup_{k=1}^{2^m+2} \mathcal{A}_k \neq \mathfrak{P}(X).$$

Clearly,

$$a \notin \bigcup_{k=3}^{2^m+2} \ker \mathcal{A}_k \quad \text{implies} \quad \bigcup_{k=1}^{2^m+2} \mathcal{A}_k \neq \mathfrak{P}(X),$$

since $\mathfrak{g}(2^m) \leq 2m - 1$. Hence let us assume that a, c are $\mathcal{A}_3$-similar ultrafilters. Consider the most difficult case: assume that b, c are $\mathcal{A}_4$-similar ultrafilters. We have

$$\#\left(\left\{M \mid M \cap \{a, b, c\} = \{b, c\},\ \#(M) = \mathfrak{n},\ \text{and } M \in \bigcup_{k=4}^{2^m+2} \mathcal{A}_k\right\}\right)$$
$$\leq (2^m - 2) \cdot \delta_m + \delta_m^* < \frac{2^m - 2}{2^m} \cdot \gamma + \frac{\gamma}{2^{m-1}} = \gamma. \qquad \square$$

Below we will assume that there do not exist ultrafilters a, b and pairwise distinct numbers $k_1, k_2 \in [1, 2^m + 2]$ such that a, b are $\mathcal{A}_{k_1}$- and $\mathcal{A}_{k_2}$-similar ultrafilters.

9.44. *Let $m \geq 6$. Consider the following situation: a, b, c are pairwise distinct ultrafilters, a, b are $\mathcal{A}_1$-similar ultrafilters, a, c are $\mathcal{A}_2$-similar ultrafilters, and there does not exist a $k \in [1, 2^m + 2]$ such that b, c are $\mathcal{A}_k$-similar ultrafilters. Then*

$$\bigcup_{k=1}^{2^m+2} \mathcal{A}_k \neq \mathfrak{P}(X).$$

9.45. REMARK. Of course, the statement from 9.41 is true for $m \in [3, 8]$ because $\mathfrak{g}(2^m + 2) \leq 2m - 1$ for $k \in [3, 8]$. Also the statements from 9.42, 9.43, and 9.44 are true for $m \in [6, 8]$, because $\mathfrak{g}(2^m+2) \leq 2m-1$ for $k \in [6, 8]$. It is important that these statements are true for $m \geq 9$.

9.46. Let us summarize our arguments. If $m \geq 9$ and

$$\bigcup_{k=1}^{2^m+2} \mathcal{A}_k = \mathfrak{P}(X),$$

then $\bigcup_{k=1}^{2^m+2} \ker \mathcal{A}_k$ splits into at most two connected components. In addition, if distinct ultrafilters a, b belong to the same connected component, then there exists only one $k \in [1, 2^m + 2]$ such that a, b are $\mathcal{A}_k$-similar ultrafilters.

9.47. LEMMA. *If $m \geq 9$ and*

$$\bigcup_{k=1}^{2^m+2} \mathcal{A}_k = \mathfrak{P}(X),$$

then our family of algebras has only one connected component.

PROOF. Let $\#(X) = 2\mathfrak{n}$. Assume that our family of algebras has two connected components. Since $m(2^m + 2) > (6m - 2) \cdot (6m - 3)$,[14] there exists a connected component K such that $\#(K) > 6m - 2$. Let us take pairwise distinct ultrafilters $a_1, b_1, b_2 \in K$ and pairwise distinct numbers $k_1, k_2, k_3 \in [1, 2^m + 2]$, such that a_1, b_1 are $\mathcal{A}_{k_1}$-similar ultrafilters, a_1, b_2 are $\mathcal{A}_{k_2}$-similar ultrafilters, and b_1, b_2 are $\mathcal{A}_{k_3}$-similar ultrafilters. Let $k_1 = 1$, $k_2 = 4$, $k_3 = 7$. Clearly,

$$\#(\ker \mathcal{A}_1 \cup \ker \mathcal{A}_4 \cup \ker \mathcal{A}_7) \leq 6m - 3.\,^{15}$$

Hence one can take an ultrafilter

$$a_2 \in K \setminus (\ker \mathcal{A}_1 \cup \ker \mathcal{A}_4 \cup \ker \mathcal{A}_7).$$

Assume that a_2, b_2 are $\mathcal{A}_2$-similar ultrafilters, a_2, b_1 are $\mathcal{A}_5$-similar ultrafilters, and a_1, a_2 are $\mathcal{A}_6$-similar ultrafilters. Let a_3, b_3 be distinct ultrafilters from another connected component. Assume that a_3, b_3 are $\mathcal{A}_3$-similar ultrafilters.

---

[14]Here we use the fact that for fixed $k \in [1, 2^m + 2]$ there exist not less than m pairwise distinct pairs of $\mathcal{A}_k$-similar ultrafilters. Let us also note that this inequality does not hold when $m = 8$.

[15]Recall that $\#(\ker \mathcal{A}_k) \leq 2m$.

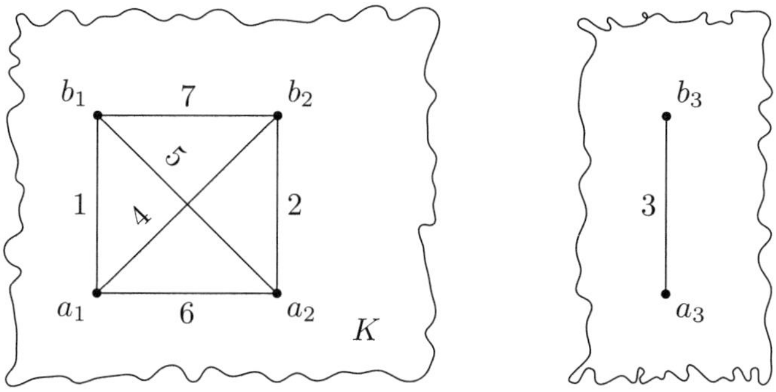

Put

$$\mathcal{M}_1 = \{a_1, a_2, a_3\}, \quad \mathcal{M}_2 = \{b_1, b_2, b_3\}.$$

Then (see 9.20)

$$\#\Big(\Big\{M \mid M \cap (\mathcal{M}_1 \cup \mathcal{M}_2) = \mathcal{M}_1,\ \#(M) = \mathfrak{n},\ \text{and } M \in \bigcup_{k=6}^{2^m+2} \mathcal{A}_k\Big\}\Big)$$
$$< 2 \cdot 2^{1-m} \cdot \gamma_0 + (2^m - 5) \cdot 2^{-m} \cdot \gamma_0 = \gamma_0 - 2^{-m} \cdot \gamma_0$$

up to an infinitely small function with respect to γ_0, a contradiction. On the other hand, if, for example, a_3, b_3 are $\mathcal{A}_4$-similar ultrafilters, then put

$$\mathcal{M}_1 = \{a_1, b_2, a_3\}, \quad \mathcal{M}_2 = \{b_1, a_2, b_3\},$$

and we again come to a contradiction. □

9.48. THEOREM. *$\mathfrak{g}(2^m + 2) \leq 2m - 1$ for $m \in [9, 10]$.*

PROOF. Of course, we consider the algebras $\mathcal{A}_1, \ldots, \mathcal{A}_{2^m+2}$, and $m \in [9, 10]$. Let n'_* be the minimal natural number such that

$$1 + \sqrt{1 + 8kr_m} \leq 4 \cdot n'_*.$$

A verification shows that if $m \in [9, 10]$, then

$$(2^m + 2) \cdot \sum_{k=0}^{m} \binom{m}{k} \cdot \binom{2n'_* - 2m}{n'_* - 2k} < \binom{2n'_*}{n'_*}.$$

Let $\#(X) = 2n'_*$. Since the results of 9.26 and 9.27 hold, we have

$$\bigcup_{k=1}^{2^m+2} \mathcal{A}_k \neq \mathfrak{P}(X).$$

Assume that

$$X = \bigcup_{k=1}^{2^m+2} \ker \mathcal{A}_k \quad \text{and} \quad \#(X) = \widehat{x} \geq 2n'_* + 1.$$ [16]

Further assume that

$$\bigcup_{k=1}^{2^m+2} \mathcal{A}_k = \mathfrak{P}(X).$$

[16] Now one can assume that m is an arbitrary natural number ≥ 9.

We know that for any two distinct ultrafilters $a, b \in X$ there exists a $k \in [1, 2^m + 2]$ such that a, b are $\mathcal{A}_k$-similar ultrafilters. Hence

$$\frac{\widehat{x}(\widehat{x} - 1)}{2} \leq kr_m.$$

But the root of the equation $x(x-1)/2 - kr_m$ is

$$x_0 = \frac{1 + \sqrt{1 + 8kr_m}}{2},$$

and $\widehat{x} > x_0$. Hence $\widehat{x}(\widehat{x} - 1)/2 > kr_m$, a contradiction. □

9.49. THEOREM. $\mathfrak{g}(2^m + 2) \leq 2m - 1$ *if* $m \geq 11$.

PROOF. Let $\#(X) = 2\mathfrak{n}$. Assume that

$$\bigcup_{k=1}^{2^m+2} \mathcal{A}_k = \mathfrak{P}(X).$$

(Clearly, $m \geq 11$.) Then for each pair of distinct ultrafilters

$$a, b \in \bigcup_{k=1}^{2^m+2} \ker \mathcal{A}_k$$

there exists only one $k \in [1, 2^m + 2]$ such that a, b are $\mathcal{A}_k$-similar ultrafilters. We have

$$\#\left(\bigcup_{k=1}^{2^m+2} \ker \mathcal{A}_k\right) > 20m - 14$$

since

$$2m(2^m + 2) > (20m - 14)(20m - 15).$$ [17]

Let us show that one can choose a set of pairwise distinct ultrafilters

$$P = \{a_1, a_2, a_3, b_1, b_2, b_3\} \subset \bigcup_{k=1}^{2^m+2} \ker \mathcal{A}_k$$

such that if a, b are $\mathcal{A}_k$-similar ultrafilters, a', b' are $\mathcal{A}_{k'}$-similar ultrafilters, and

$$\{a, b\} \neq \{a', b'\}, \quad \{a, b\}, \{a', b'\} \subset P,$$

then $k \neq k'$. The construction goes by induction. Suppose that a set consisting of five ultrafilters is constructed. This means that we have constructed

$$P' = \{a_1, a_2, b_1, b_2, b_3\} \subset \bigcup_{k=1}^{2^m+2} \ker \mathcal{A}_k$$

such that if a, b are $\mathcal{A}_k$-similar ultrafilters, a', b' are $\mathcal{A}_{k'}$-similar ultrafilters, and

$$\{a, b\} \neq \{a', b'\}, \quad \{a, b\}, \{a', b'\} \subset P',$$

then $k \neq k'$. Put

$$\Sigma = \big\{k \mid \text{a pair of } \mathcal{A}_k\text{-similar ultrafilters is contained in } P'\big\}.$$

[17] This inequality is not true for $m = 10$.

Clearly, $\#(\Sigma) = 10$ and

$$\#\left(\bigcup_{k\in\Sigma} \ker \mathcal{A}_k\right) \leq 20m - 15.$$

Hence one can take an ultrafilter

$$a_3 \in \bigcup_{k=1}^{2^m+2} \ker \mathcal{A}_k \setminus \bigcup_{k\in\Sigma} \ker \mathcal{A}_k,$$

and put $P = P' \cup \{a_3\}$. Let a_k, b_k be $\mathcal{A}_k$-similar ultrafilters for all $k \in [1,3]$. For each $k \in [4,9]$ let there exist $\mathcal{A}_k$-similar ultrafilters a_{i_k}, b_{j_k}. For each $k \in [10,12]$ let there exist $\mathcal{A}_k$-similar ultrafilters $a_{i'_k}, a_{j'_k}$. For each $k \in [13,15]$ let there exist $\mathcal{A}_k$-similar ultrafilters $b_{i''_k}, b_{j''_k}$.

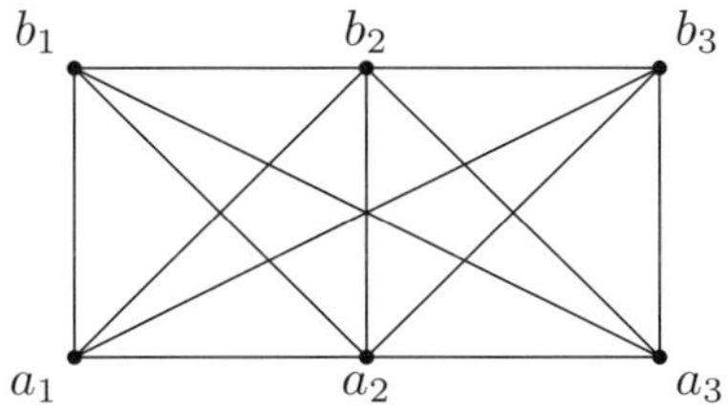

Put $\mathcal{M}_1 = \{a_1, a_2, a_3\}$, $\mathcal{M}_2 = \{b_1, b_2, b_3\}$. Then (see 9.20),

$$\#\Big(\Big\{M \mid M \cap (\mathcal{M}_1 \cup \mathcal{M}_2) = \mathcal{M}_1,\ \#(M) = \mathfrak{n},\ \text{and } M \in \bigcup_{k=10}^{2^m+2} \mathcal{A}_k\Big\}\Big)$$
$$< 6 \cdot 2^{1-m} \cdot \gamma_0 + (2^m - 13) \cdot 2^{-m} \cdot \gamma_0 = \gamma_0 - 2^{-m} \cdot \gamma_0$$

up to an infinitely small function with respect to γ_0, a contradiction. □

So for all $m \geq 3$ the following has been proved:

$$\mathfrak{g}(2^m + 2) \leq 2m - 1.$$

9.50. Let l be a fixed natural number ≥ 3. Just as we considered the family of algebras $\mathcal{A}_1, \ldots, \mathcal{A}_{2^m+2}$ above, we will now consider a family of algebras

$$\mathcal{A}_1, \ldots, \mathcal{A}_{2^m+l},$$

where $2m - 1 \leq \#(\ker \mathcal{A}_k) \leq 2m$, etc. But now we will assume that m is a fixed but sufficiently large integer.

THEOREM. *There exists an $m_l \in \mathbb{N}^+$ such that if $m > m_l$, then*

$$\mathfrak{g}(2^m + l) \leq 2m - 1.$$

PROOF. We must prove that, for m large,

$$\bigcup_{k=1}^{2^m+l} \mathcal{A}_k \neq \mathfrak{P}(X).$$

Suppose the contrary. Similarly as in 2.41, let us assume that there do not exist distinct ultrafilters a, b and pairwise distinct numbers

$$k_1, \ldots, k_{l+1} \in [1, 2^m + l]$$

such that a, b are $\mathcal{A}_{k_i}$-similar ultrafilters for all $i \in [1, l+1]$. Hence we can assume that

$$\#\left(\bigcup_{k=1}^{2^m+l} \ker \mathcal{A}_k\right) > 2^m.$$

Let $p \in \mathbb{N}^+$, and let $P = \{a_1, \dots, a_p\}$ be a sequence of pairwise distinct ultrafilters possessing the following two properties:

(1) each ultrafilter a_i has some $\mathcal{A}_k$-similar ultrafilter a_j (possibly, a_i, a_j are $\mathcal{A}_{k^*}$-similar ultrafilters and $k \neq k^*$);
(2) if (a_{i_1}, a_{i_2}) and (a_{j_1}, a_{j_2}) are distinct pairs of ultrafilters, a_{i_1}, a_{i_2} are $\mathcal{A}_{k'}$-similar ultrafilters, and a_{j_1}, a_{j_2} are $\mathcal{A}_{k''}$-similar ultrafilters, then necessarily $k' \neq k''$.

Put

$$\Sigma = \big\{k \mid \text{there exist } \mathcal{A}_k\text{-similar ultrafilters in } P\big\}.$$

Obviously,

$$\#(\Sigma) \leq l \cdot \binom{p}{2}.$$

Let

$$2m \cdot l \cdot \frac{p(p-1)}{2} < 2^m.$$

Then there exists

$$a_{p+1} \in \bigcup_{k=1}^{2^m+l} \ker \mathcal{A}_k \setminus \bigcup_{k \in \Sigma} \ker \mathcal{A}_k.$$

If a_{p+1} has an $\mathcal{A}_k$-similar ultrafilter in P, then the sequence $a_1, \dots, a_p, a_{p+1}$ satisfies conditions (1) and (2) above. If there does not exist a corresponding ultrafilter a_{p+1}, then let us look at $\mathcal{A}_k$-similar ultrafilters a_{p+1}, a_{p+2} such that

$$a_{p+1}, a_{p+2} \in \bigcup_{k=1}^{2^m+l} \ker \mathcal{A}_k \setminus \bigcup_{k \in \Sigma} \ker \mathcal{A}_k.$$

Assume that such ultrafilters a_{p+1}, a_{p+2} do not exist. Then for m large enough, one can find an ultrafilter

$$b \in \bigcup_{k \in \Sigma} \ker \mathcal{A}_k$$

and pairwise distinct ultrafilters $b_1, \dots, b_{l+1}$ such that b, b_i are $\mathcal{A}_{k^i}$-similar ultrafilters, where $k^1, \dots, k^{l+1}$ are pairwise distinct numbers; and also there do not exist $\mathcal{A}_k$-similar ultrafilters b_i, b_j for any $k \in [1, 2^m + l]$. Put

$$\mathcal{M}_1 = \{b_1, \dots, b_{l+1}\}, \quad \mathcal{M}_2 = \{b\}.$$

By the results of 9.20 we come to a contradiction:

$$\bigcup_{k=1}^{2^m+l} \mathcal{A}_k \neq \mathfrak{P}(X).$$

Hence let us assume the existence of ultrafilters a_{p+1}, a_{p+2}. Clearly, the sequence

$$a_1, \dots, a_p, a_{p+1}, a_{p+2}$$

satisfies conditions (1) and (2). Now it is clear that for m large enough, p can be a large number. Assume that $P^* = \{q_1, \dots, q_l, r_1, \dots, r_l\}$ are pairwise distinct ultrafilters from P, and q_i, r_i are $\mathcal{A}_i$-similar ultrafilters for all $i \in [1, l]$. Let us assume also that if

$$a_j \in P \setminus \bigcup_{k=1}^{l} \ker \mathcal{A}_k,$$

then a_j does not have an $\mathcal{A}_k$-similar ultrafilter in $P \setminus P^*$. Let us argue as above. For p large there exists an ultrafilter $c \in P^*$ and pairwise distinct ultrafilters $c_1, \dots, c_{l+1}$ such that c, c_i are $\mathcal{A}_{k^{(i)}}$-similar ultrafilters. Also, the numbers $k^{(1)}, \dots, k^{(l+1)}$ are pairwise distinct, and there do not exist $\mathcal{A}_k$-similar ultrafilters c_i, c_j for any $k \in [1, 2^m + l]$. We come again to a contradiction:

$$\bigcup_{k=1}^{2^m+l} \mathcal{A}_k \neq \mathfrak{P}(X).$$

Therefore we can assume the existence of a set of pairwise distinct ultrafilters

$$P_* = \{q_1, \dots, q_{l+1}, r_1, \dots, r_{l+1}\}$$

such that the following hold:

(i) q_i, r_i are $\mathcal{A}_i$-similar ultrafilters for all $i \in [1, l+1]$;

(ii) if (a, b) and (c, d) are distinct pairs of ultrafilters in P_*, a, b are $\mathcal{A}_{k'}$-similar ultrafilters, and c, d are $\mathcal{A}_{k''}$-similar ultrafilters, then necessarily $k' \neq k''$.

Denoting some q_i by r_i and r_i by q_i, one can assume that

$$\begin{aligned} &\#(\{k \mid \text{there exist } \mathcal{A}_k\text{-similar ultrafilters } q_i, r_j \text{ and } i \neq j\}) \\ &\quad \geq \#(\{k \mid \text{either there exist } \mathcal{A}_k\text{-similar ultrafilters } q_{i'}, q_{i''}, \\ &\qquad\qquad \text{or there exist } \mathcal{A}_k\text{-similar ultrafilters } r_{j'}, r_{j''}\}). \end{aligned}$$

Now put $\mathcal{M}_1 = \{q_1, \dots, q_{l+1}\}$, $\mathcal{M}_2 = \{r_1, \dots, r_{l+1}\}$. By the results of 9.20 we come to a contradiction:

$$\bigcup_{k=1}^{2^m+l} \mathcal{A}_k \neq \mathfrak{P}(X). \qquad \square$$

CHAPTER 10

A Description of the Class of Functions Ψ_*^7 [1]

10.1. Let us present a rather natural construction.

DEFINITION. Let $\mathcal{A}_1, \ldots, \mathcal{A}_n$ be a finite sequence of algebras, $[1,n] \supset M \neq \varnothing$, and for each $i \in M$ consider either an $\mathcal{A}_i$-special ultrafilter s_i, or $\mathcal{A}_i$-similar ultrafilters s_i, t_i; let $s = s_i$ for all $i \in M$. The set $\{s\} \cup \bigcup\{t_i\}$ is called an $s(M)$-*bush*, or simply an s-*bush*; s is called the *root* of the s-bush, and $\bigcup\{t_i\}$ is called the *crown* of the s-bush.

Let T be the crown of an s-bush. Let us denote an s-bush by (s, T). Below we will consider sequences of bushes $\{s^1, T^1\}, \ldots, \{s^\alpha, T^\alpha\}$ possessing the following properties:

(1) $\{s^1, \ldots, s^\alpha\} \cap \bigcup_{i=1}^{\alpha} T^i = \varnothing$;

(2) if $\{s^1, T^1\}$ is an $s^1(M_1)$-bush, then $M_1 = \{k \in [1,n] \mid s^1 \in \ker \mathcal{A}_k\}$; if $\{s^2, T^2\}$ is an $s^2(M_2)$-bush, then $M_2 = \{k \in [1,n] \setminus M_1 \mid s^2 \in \ker \mathcal{A}_k\}$; if $\{s^3, T^3\}$ is an $s^3(M_3)$-bush, then

$$M_3 = \{k \in [1,n] \setminus (M_1 \cup M_2) \mid s^3 \in \ker \mathcal{A}_k\}, \quad \text{etc.};$$

(3) the maximality property:

$$k \in [1,n] \setminus \bigcup_{i=1}^{\alpha} M_i \quad \text{implies} \quad \ker \mathcal{A}_k \subset \bigcup_{i=1}^{\alpha} T^i.$$

Clearly, $s^1, \ldots, s^\alpha$ are pairwise distinct ultrafilters. Obviously, if

$$\bigcup_{i=1}^{\alpha} M_i = [1,n],$$

then one can put

$$S_n = \{s^1, \ldots, s^\alpha\}, \quad T_n = \bigcup_{i=1}^{\alpha} T^i.$$ [2]

10.2. Let us proceed to the construction of the classes Ψ_*^n for all $n \leq 7$ (these classes are described in the table in 2.13). By Theorem 3.24, we can assume that all algebras under consideration contain X.[3] Below we will use, without explicitly

---

[1] If we know a description of the class Ψ_*^n, then we know a description of the classes Ψ_*^m for $m < n$. But it is natural to construct the class Ψ_*^n, where $n > 1$, by first constructing the class Ψ_*^{n-1}.

[2] Of course, the statement $\bigcup_{i=1}^{\alpha} M_i \neq [1,n]$ does not imply that $\bigcup_{k=1}^{n} \mathcal{A}_k = \mathfrak{P}(X)$.

[3] It is convenient to assume (see 3.25) that if we consider an algebra $\mathcal{A}_k$ with $\#(\ker \mathcal{A}_k) \geq \psi(k)$, $\psi(k) \in \Psi_*^n$, then for $\psi(k)$ odd, we have $\#(\ker \mathcal{A}_k) \leq \psi(k) + 1$, and for $\psi(k)$ even, we have $\#(\ker \mathcal{A}_k) = \psi(k)$. Also it is convenient to assume that any $\mathcal{A}_k$-similar set contains at most three elements.

mentioning it, the following statement from 2.13: the function

$$\psi(k) = \begin{cases} \left[\frac{4}{3}(k-1)\right] + 1 & \text{for } k \neq 2, \\ 1 & \text{for } k = 2, \end{cases}$$

defined on $[1, n]$, belongs to Ψ_*^n. Recall that in the table in 2.13 such functions are marked by a star. We use the numeration of functions from this table. It is clear what the classes Ψ_*^n are for $n \leq 3$.

LEMMA. *The function* (4–1) *belongs to* Ψ_*^4.

PROOF. Let $\mathcal{A}_1, \mathcal{A}_2, \mathcal{A}_3, \mathcal{A}_4$ be algebras such that for each $k \in [1, 4]$

$$\#(\ker \mathcal{A}_k) \geq (4\text{–}1)(k).$$

We know that

$$\bigcup_{k=1}^{3} \mathcal{A}_k \neq \mathfrak{P}(X).$$

Hence for these algebras there exist corresponding sets of ultrafilters S_3 and T_3. Clearly, if $\#(S_3 \cup T_3) > 4$, then one easily constructs S_4 and T_4. This easily implies the following: we can assume that there is a set of ultrafilters Θ and $\#(\Theta) = 4$; $\ker \mathcal{A}_k \subset \Theta$ for all $k \in [1, 4]$. Let a_1, b_1 be $\mathcal{A}_1$-similar ultrafilters. It is natural to assume that $a_1 \notin \ker \mathcal{A}_{k_1}$ and $b_1 \notin \ker \mathcal{A}_{k_2}$. Let $M \subset \Theta$ and $\#(M) = 2$. There exist at most four M's such that for each of them there exists k such that either M is an $\mathcal{A}_k$-similar set, or $M \cap \ker \mathcal{A}_k = \varnothing$. But $\binom{4}{2} = 6$. Hence

$$\bigcup_{k=1}^{4} \mathcal{A}_k \neq \mathfrak{P}(X).$$

It remains to apply Theorem 2.2(2) to conclude that $(4\text{–}1) \in \Psi_*^4$. □

10.3. EXAMPLE. Algebras $\mathcal{A}_1, \mathcal{A}_2, \mathcal{A}_3, \mathcal{A}_4$ such that

$$\#(\ker \mathcal{A}_k) = 2 \text{ for } k \in [1, 2], \quad \#(\ker \mathcal{A}_k) = 4 \text{ for } k \in [3, 4], \quad \bigcup_{k=1}^{4} \mathcal{A}_k = \mathfrak{P}(X).$$

CONSTRUCTION. $\{a_1, b_1\}$ is the $\mathcal{A}_1$-similar set; $\{a_2, b_2\}$ is the $\mathcal{A}_2$-similar set; $\{a_1, a_2\}$, $\{b_1, b_2\}$ are the $\mathcal{A}_3$-similar sets; $\{a_1, b_2\}$, $\{b_1, a_2\}$ are the $\mathcal{A}_4$-similar sets.

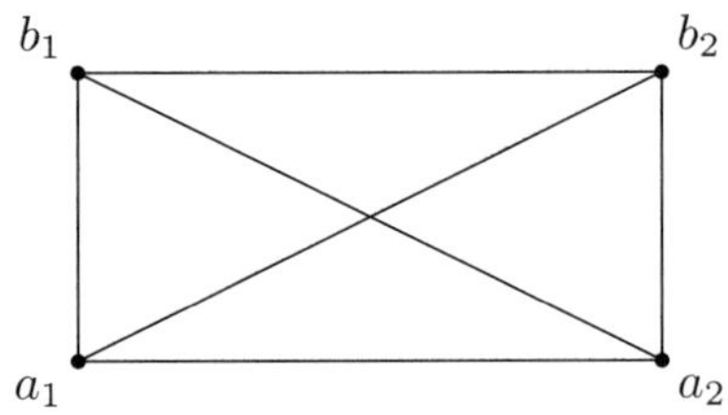

□

10.4. By the results of 10.2 and 10.3, the following theorem is true:

THEOREM. *The class* Ψ_*^4 *consists of the functions* (4–1) *and* (4–2).

10.5. LEMMA. *Let* $\mathcal{A}_1, \ldots, \mathcal{A}_5$ *be algebras such that* $\#(\ker \mathcal{A}_k) \geq (5\text{–}1)(k)$ *for each* $k \in [1, 5]$. *Then*

$$\bigcup_{k=1}^{5} \mathcal{A}_k \neq \mathfrak{P}(X).$$

PROOF. We know that

$$\bigcup_{k=1}^{4} \mathcal{A}_k \neq \mathfrak{P}(X).$$

Therefore for these algebras the corresponding sets of ultrafilters S_4 and T_4 exist. It is not difficult to check that if $\#(S_4 \cup T_4) > 5$, then one can construct S_5 and T_5. This easily implies that we can assume the existence of a set of ultrafilters Θ and $\#(\Theta) = 5$; $\ker \mathcal{A}_k \subset \Theta$ for all $k \in [1, 5]$. Let a_1, b_1 be $\mathcal{A}_1$-similar ultrafilters. It is natural to assume that $a_1 \notin \ker \mathcal{A}_{k_1}$ and $b_1 \notin \ker \mathcal{A}_{k_2}$.[4] Let $L \subset \Theta$ and $\#(L) = 3$. Besides,

$$\#(L \cap \{a_1, b_1\}) = 1.$$

Thus there exist at most five sets L such that for each of them there is a k such that $L \cap \ker \mathcal{A}_k$ is an $\mathcal{A}_k$-similar set. But the total number of such sets L is six. Hence

$$\bigcup_{k=1}^{5} \mathcal{A}_k \neq \mathfrak{P}(X). \qquad \square$$

10.6. LEMMA. *Let $\mathcal{A}_1, \ldots, \mathcal{A}_5$ be algebras such that $\#(\ker \mathcal{A}_k) \geq (5\text{–}2)(k)$ for each $k \in [1, 5]$. Then*

$$\bigcup_{k=1}^{5} \mathcal{A}_k \neq \mathfrak{P}(X).$$

PROOF. If $\ker \mathcal{A}_1 \cap \ker \mathcal{A}_2 \neq \varnothing$, then let us turn to s-bushes; now everything is easy. Assume that $\ker \mathcal{A}_1 \cap \ker \mathcal{A}_2 = \varnothing$. We know that

$$\bigcup_{k=1}^{4} \mathcal{A}_k \neq \mathfrak{P}(X).$$

Therefore for these algebras the corresponding sets of ultrafilters S_4 and T_4 exist. It is not difficult to check that if $\#(S_4 \cup T_4) > 5$, then one can construct S_5 and T_5. This easily implies that we can assume the existence of a set of ultrafilters Θ and $\#(\Theta) = 5$; $\ker \mathcal{A}_k \subset \Theta$ for all $k \in [1, 5]$. Let a_1, b_1 be $\mathcal{A}_1$-similar ultrafilters, and let a_2, b_2 be $\mathcal{A}_2$-similar ultrafilters. We know that

$$\{a_1, b_1\} \cap \{a_2, b_2\} = \varnothing.$$

Among the four sets $\{a_1, a_2\}$, $\{b_1, b_2\}$, $\{a_1, b_2\}$, $\{a_2, b_1\}$, one can find one (denote it by M) such that $M \cap \ker \mathcal{A}_k$ is not an $\mathcal{A}_k$-similar set for any $k \in [3, 5]$. $\square$

10.7. EXAMPLE. Algebras $\mathcal{A}_1, \ldots, \mathcal{A}_5$ such that

$$\#(\ker \mathcal{A}_k) = 2 \text{ for } k \in [1, 2], \quad \#(\ker \mathcal{A}_3) = 4, \quad \#(\ker \mathcal{A}_k) = 5 \text{ for } k \in [4, 5],$$

$$\bigcup_{k=1}^{5} \mathcal{A}_k = \mathfrak{P}(X).$$

CONSTRUCTION. $\{a_1, b_1\}$ is the $\mathcal{A}_1$-similar set; $\{a_2, b_2\}$ is the $\mathcal{A}_2$-similar set; $\{a_1, b_2\}$, $\{a_2, b_1\}$ are the $\mathcal{A}_3$-similar sets; $\{a_1, a_2, q\}$, $\{b_1, b_2\}$ are the $\mathcal{A}_4$-similar sets; $\{a_1, a_2\}$, $\{b_1, b_2, q\}$ are the $\mathcal{A}_5$-similar sets.

[4] Possibly, $k_1 = k_2$.

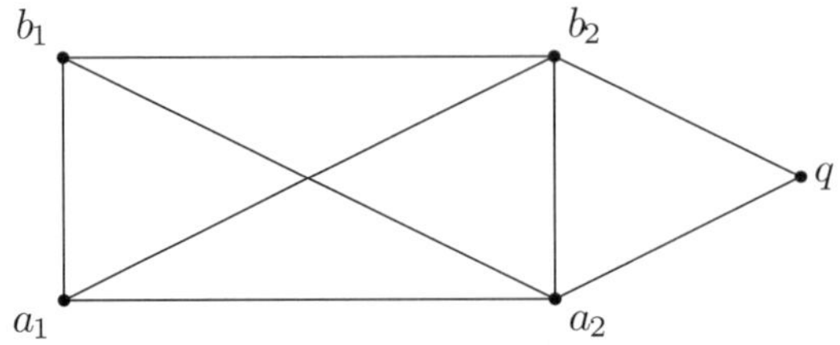

□

10.8. By the results of 10.5, 10.6, 10.7 and by simple arguments (also using the fact that $\mathfrak{g}(5) = 5$) the following theorem is true:

THEOREM. *The class* Ψ^5_* *consists of the functions* (5–1), (5–2), (5–3).

10.9. Let us proceed to the description of the class Ψ^6_*.

EXAMPLE. Algebras $\mathcal{A}_1, \ldots, \mathcal{A}_6$ such that

$$\#(\ker \mathcal{A}_1) = 2, \quad \#(\ker \mathcal{A}_k) = 4 \text{ for } k \in [2,4], \quad \#(\ker \mathcal{A}_k) = 6 \text{ for } k \in [5,6],$$

$$\bigcup_{k=1}^{6} \mathcal{A}_k = \mathfrak{P}(X).$$

CONSTRUCTION. $\{a_1, b_1\}$ is the $\mathcal{A}_1$-similar set; $\{a_1, a_2\}$, $\{b_1, b_2\}$ are the $\mathcal{A}_2$-similar sets; $\{a_1, b_2\}$, $\{a_2, b_1\}$ are the $\mathcal{A}_3$-similar sets; $\{a_2, b_2\}$, $\{a_3, b_3\}$ are the $\mathcal{A}_4$-similar sets; $\{a_1, a_3\}$, $\{a_2, b_2\}$, $\{b_1, b_3\}$ are the $\mathcal{A}_5$-similar sets; $\{a_1, b_3\}$, $\{a_2, b_2\}$, $\{a_3, b_1\}$ are the $\mathcal{A}_6$-similar sets.

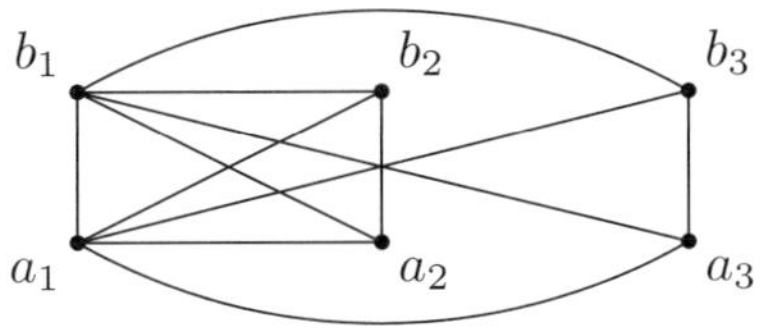

□

Taking into account this example, Theorem 2.1, and the fact that the function (5–1) belongs to Ψ^5_*, we see that the function (6–4) belongs to Ψ^6_*.

10.10. EXAMPLE. Algebras $\mathcal{A}_1, \ldots, \mathcal{A}_6$ such that

$$\#(\ker \mathcal{A}_1) = \#(\ker \mathcal{A}_2) = 2, \quad \#(\ker \mathcal{A}_k) = 5 \text{ for } k \in [3,5], \quad (\# \ker \mathcal{A}_6) = 6,$$

$$\bigcup_{k=1}^{6} \mathcal{A}_k = \mathfrak{P}(X).$$

CONSTRUCTION. $\{s, t_i\}$ is an $\mathcal{A}_i$-similar set for each $i \in [1,4]$; $\{t_1, t_2, t_4\}$ is an $\mathcal{A}_3$-similar set; $\{t_1, t_2, t_3\}$ is an $\mathcal{A}_4$-similar set; $\{t_1, t_2, t_5\}$, $\{t_3, t_4\}$ are the $\mathcal{A}_5$-similar sets; $\{s, t_5\}$, $\{t_1, t_2\}$, $\{t_3, t_4\}$ are the $\mathcal{A}_6$-similar sets.

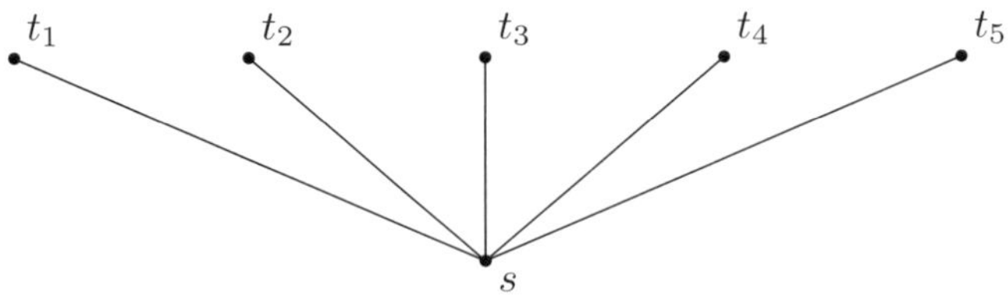

□

Taking into account this example, Theorem 2.1, and the fact that the function (5–2) belongs to Ψ^5_*, we see that the function (6–5) belongs to Ψ^6_*.

10.11. LEMMA. *Let $\mathcal{A}_1, \dots, \mathcal{A}_6$ be algebras such that $\#(\ker \mathcal{A}_k) \geq (6\text{–}3)(k)$ for each $k \in [1,6]$; let $\ker \mathcal{A}_1 \cap \ker \mathcal{A}_2 \neq \varnothing$. Then*

$$\bigcup_{k=1}^{6} \mathcal{A}_k \neq \mathfrak{P}(X).$$

PROOF. Using the technique of s-bushes, one can assert that the only nontrivial situation is similar to the following one: there is a set Θ consisting of six ultrafilters $s, t_1, t_2, t_3, t_5, t_6$, and the following are satisfied:

1) $\{s, t_k\}$ is an $\mathcal{A}_k$-similar set;
2) $\ker \mathcal{A}_k \subset \Theta$ for all $k \in [3,6]$;
3) $\ker \mathcal{A}_4 = \Theta \setminus \{s\}$.

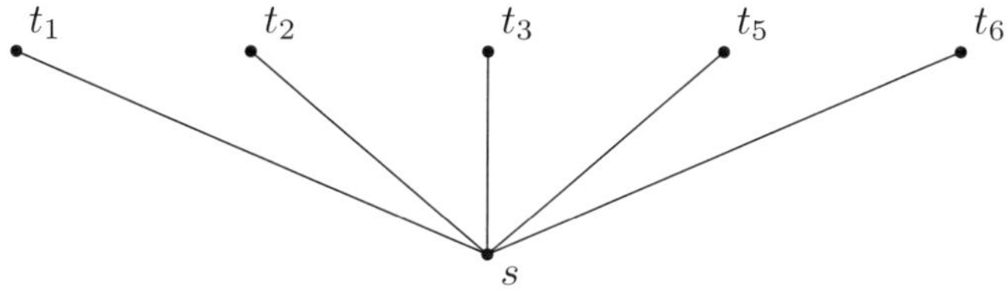

Hence our lemma is true by Lemma 9.15. □

10.12. LEMMA. *Let $\mathcal{A}_1, \dots, \mathcal{A}_6$ be algebras such that $\#(\ker \mathcal{A}_k) \geq (6\text{–}3)(k)$ for each $k \in [1,6]$. Then*

$$\bigcup_{k=1}^{6} \mathcal{A}_k \neq \mathfrak{P}(X).$$

PROOF. By Lemma 10.11, let us assume that $\ker \mathcal{A}_1 \cap \ker \mathcal{A}_2 = \varnothing$. We know that $\bigcup_{k=1}^{5} \mathcal{A}_k \neq \mathfrak{P}(X)$. Hence the corresponding sets of ultrafilters S_5 and T_5 exist. The two situations that are not quite simple are the following:

1) $S_5 \cup T_5$ forms a bush of the fourth kind K and $\#(K) = 6$;
2) $S_5 \cup T_5$ splits into a bush K_1 of the second kind (or of the third kind) and a bush K_2 of the fourth kind ($\#(K_1) = 3$, $\#(K_2) = 4$).

Let us examine the first situation only (the second one is simpler). It is natural to assume that $\ker \mathcal{A}_k \subset K$ for all $k \in [1,6]$. Let a_1, b_1 be $\mathcal{A}_1$-similar ultrafilters, and let a_2, b_2 be $\mathcal{A}_2$-similar ultrafilters. We know that

$$\{a_1, b_1\} \cap \{a_2, b_2\} = \varnothing.$$

Let $L \subset K$, $\#(L) = 3$, and

$$\#(L \cap \{a_1, b_1\}) = 1, \quad \#(L \cap \{a_2, b_2\}) = 1.$$

The number of such sets L is eight. Hence it is natural to assume that if $k \in [3,6]$, then there exists an $\mathcal{A}_k$-similar set consisting of three elements. Among the four sets

$$\{a_1, a_2\}, \ \{b_1, b_2\}, \ \{a_1, b_2\}, \ \{a_2, b_1\}$$

we take a set (denote it by M) such that $M \cap \ker \mathcal{A}_k$ is not an $\mathcal{A}_k$-similar set for $k \in [3,4]$.[5] Obviously, we have $M \cap \ker \mathcal{A}_k \neq \varnothing$ for all $k \in [1,6]$, and M is not an $\mathcal{A}_k$-similar set for any $k \in [1,6]$. □

[5] Clearly, M can be chosen among three sets.

10.13. By Example 10.3, there exist algebras $\mathcal{A}_1, \mathcal{A}_2, \mathcal{A}_3, \mathcal{A}_4$ such that $\#(\ker \mathcal{A}_k) = 2$ for $k \in [1,2]$, $\#(\ker \mathcal{A}_k) = 4$ for $k \in [3,4]$, and

$$\bigcup_{k=1}^{4} \mathcal{A}_k = \mathfrak{P}(X).$$

By Example 1.13, there exist algebras $\mathcal{B}_1, \mathcal{B}_2, \mathcal{B}_3$ such that $\#(\ker \mathcal{B}_k) = 2$ for all $k \in [1,3]$ and

$$\bigcup_{k=1}^{3} \mathcal{B}_k = \mathfrak{P}(X).$$

Using Theorem 9.3 ($n = 4$, $m = 3$, $\nu = 3$, $\mu = 0$), one can construct the following example:

EXAMPLE. Algebras $\mathcal{A}_1, \dots, \mathcal{A}_6$ such that

$$\#(\ker \mathcal{A}_k) = 2 \text{ for } k \in [1,2], \quad \#(\ker \mathcal{A}_3) = 4, \quad \#(\ker \mathcal{A}_k) = 6 \text{ for } k \in [4,6],$$

$$\bigcup_{k=1}^{6} \mathcal{A}_k = \mathfrak{P}(X).$$

From this example, Example 10.10, and Lemma 10.12, we see that the function (6–3) belongs to Ψ_*^6.

10.14. LEMMA. *Let $\mathcal{A}_1, \dots, \mathcal{A}_6$ be algebras such that $\#(\ker \mathcal{A}_k) \geq (6\text{–}2)(k)$ for each $k \in [1,6]$. Let there exist a set of ultrafilters Θ such that $\#(\Theta) = 7$ and $\ker \mathcal{A}_k \subset \Theta$ for all $k \in [1,6]$. Then*

$$\bigcup_{k=1}^{6} \mathcal{A}_k \neq \mathfrak{P}(X).$$

PROOF. Let a_1, b_1 be $\mathcal{A}_1$-similar ultrafilters. Let L be a subset of Θ such that $\#(L) = 3$ and $\#(L \cap \{a_1, b_1\}) = 1$. Let M be a subset of Θ such that $\#(M) = 2$ and $\#(M \cap \{a_1, b_1\}) = 1$. At least one of the following two statements holds:

1) there exists an L such that $L \cap \ker \mathcal{A}_k \neq \varnothing$ and $L \cap \ker \mathcal{A}_k$ is not an $\mathcal{A}_k$-similar set for all $k \in [2,6]$;
2) there exists an M such that $M \cap \ker \mathcal{A}_k \neq \varnothing$ and $M \cap \ker \mathcal{A}_k$ is not an $\mathcal{A}_k$-similar set for all $k \in [2,6]$.

(Here we take into account the fact that there are 20 sets with the properties of L and ten sets with those of M.) □

10.15. LEMMA. *The function* (6–2) *belongs to* Ψ_*^6.

PROOF. Let $\mathcal{A}_1, \dots, \mathcal{A}_6$ be algebras such that $\#(\ker \mathcal{A}_k) \geq (6\text{–}2)(k)$ for each $k \in [1,6]$. Let us prove that

$$\bigcup_{k=1}^{6} \mathcal{A}_k \neq \mathfrak{P}(X).$$

By Lemma 10.14, we can assume that

$$\#\left(\bigcup_{k=1}^{6} \ker \mathcal{A}_k\right) > 7.$$

We know that $\bigcup_{k=1}^{5} \mathcal{A}_k \neq \mathfrak{P}(X)$. Hence for these algebras the corresponding sets of ultrafilters S_5 and T_5 exist. Let us divide $S_5 \cup T_5$ into bushes. If there is no bush of the fourth kind, then everything is easy. If we have a bush K of the fourth kind and $\#(K) = 4$, then everything is also easy. If we have a bush K of the fourth kind, $\#(K) = 5$, and there is one more simple bush,[6] then everything is easy as well. It remains to consider one case: S_5 and T_5 form one bush K of the fourth kind. Since $\#(K) \leq 6$, everything is easy. We have proved that $\bigcup_{k=1}^{6} \mathcal{A}_k \neq \mathfrak{P}(X)$.

It remains to take into account Examples 10.9 and 10.13. □

10.16. LEMMA. *The function* (6–1) *belongs to* Ψ_*^6.

PROOF. Let Θ be a set of ultrafilters and $\#(\Theta) = 7$. For each $k \in [1, 6]$ consider an algebra $\mathcal{A}_k$, $\#(\ker \mathcal{A}_k) \geq (6\text{–}1)(k)$, and $\ker \mathcal{A}_k \subset \Theta$. Let $L \subset \Theta$ and $\#(L) = 3$. If $k \in [1, 4]$, then there exist at most seven sets L such that either $L \cap \ker \mathcal{A}_k$ is an $\mathcal{A}_k$-similar set, or $L \cap \ker \mathcal{A}_k = \varnothing$. If $k \in [5, 6]$, then there exist at most three sets L such that either $L \cap \ker \mathcal{A}_k$ is an $\mathcal{A}_k$-similar set, or $L \cap \ker \mathcal{A}_k = \varnothing$. The total number of such sets L is less than 35. But $\binom{7}{3} = 35$. Hence

$$\bigcup_{k=1}^{6} \mathcal{A}_k \neq \mathfrak{P}(X).$$

To conclude the proof, it remains to mention the arguments from the proof of Lemma 10.15, the fact that $\mathfrak{g}(5) = 5$, and the fact that the function (6–2) belongs to Ψ_*^6. □

10.17. We have shown that the functions (6–1)–(6–6) belong to Ψ_*^6. It is easy to show that Ψ_*^6 does not contain other functions. In other words, we have proved the following theorem.

THEOREM. *The class* Ψ_*^6 *consists of the functions* (6–1)–(6–6).

10.18. Let us proceed to the construction of the class Ψ_*^7.

LEMMA. *Let* $\mathcal{A}_1, \ldots, \mathcal{A}_7$ *be algebras such that* $\#(\ker \mathcal{A}_k) \geq \psi(k)$ *for any* $k \in [1, 7]$, *where* $\psi(k)$ *is one of the four functions* (7–9)–(7–12). *Then*

$$\bigcup_{k=1}^{7} \mathcal{A}_k \neq \mathfrak{P}(X).$$

SKETCH OF THE PROOF. We know that $\bigcup_{k=1}^{6} \mathcal{A}_k \neq \mathfrak{P}(X)$. Hence for these algebras the corresponding sets of ultrafilters S_6 and T_6 exist. Consider the only case which is not absolutely trivial: $S_6 \cup T_6$ splits into two bushes of the fourth kind, K_1 and K_2, and

$$K_1 = \{a_1, a_2, b_1, b_2\}, \quad K_2 = \{a_3, a_4, b_3, b_4\}.$$

Consider six pairs of ultrafilters: (a_1, b_1), (a_1, b_2), (a_2, b_2), (a_3, b_3), (a_3, b_4), (a_4, b_4). There is a one-to-one correspondence between these pairs and the set $[1, 6]$, and the pair corresponding to $k \in [1, 6]$ is a pair of $\mathcal{A}_k$-similar ultrafilters.

[6] See Definition 8.12.

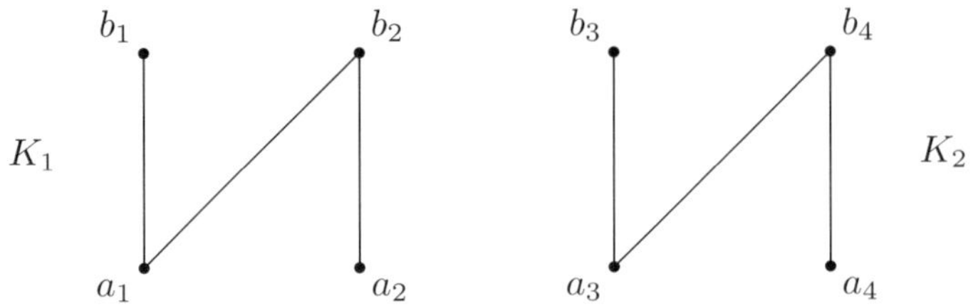

It is natural to assume that $\{a_1, a_2\}$, $\{a_3, a_4\}$, $\{b_1, b_2\}$, $\{b_3, b_4\}$ are $\mathcal{A}_7$-similar sets. Now simple arguments yield the construction of S_7 and T_7. □

10.19. EXAMPLE. Algebras $\mathcal{A}_1, \ldots, \mathcal{A}_7$ such that

$$\#(\ker \mathcal{A}_1) = \#(\ker \mathcal{A}_2) = 2, \quad \#(\ker \mathcal{A}_3) = 4, \quad \#(\ker \mathcal{A}_4) = 5,$$

$$\#(\ker \mathcal{A}_5) = \#(\ker \mathcal{A}_6) = \#(\ker \mathcal{A}_7) = 7, \quad \bigcup_{k=1}^{7} \mathcal{A}_k = \mathfrak{P}(X).$$

CONSTRUCTION. By Example 10.7, let us take algebras $\mathcal{A}_1, \ldots, \mathcal{A}_5$ such that

$$\#(\ker \mathcal{A}_k) = 2 \text{ for } k \in [1, 2], \ \#(\ker \mathcal{A}_3) = 4, \ \#(\ker \mathcal{A}_k) = 5 \text{ for } k \in [4, 5],$$

$$\bigcup_{k=1}^{5} \mathcal{A}_k = \mathfrak{P}(X).$$

By Example 1.13, there exist algebras $\mathcal{B}_1, \mathcal{B}_2, \mathcal{B}_3$ such that $\#(\ker \mathcal{B}_k) = 2$ for all $k \in [1, 3]$ and

$$\bigcup_{k=1}^{3} \mathcal{B}_k = \mathfrak{P}(X).$$

Now we can apply Theorem 9.3 ($n = 5$, $m = 3$, $\nu = 4$, $\mu = 0$). □

By the present example, Lemma 10.18, and Theorem 2.2(2) we see that the function (7–12) belongs to Ψ_*^7.

10.20. EXAMPLE. Algebras $\mathcal{A}_1, \ldots, \mathcal{A}_7$ such that

$$\#(\ker \mathcal{A}_k) = 2 \text{ for } k \in [1, 2], \quad \#(\ker \mathcal{A}_k) = 5 \text{ for } k \in [3, 4], \quad \#(\ker \mathcal{A}_5) = 6,$$

$$\#(\ker \mathcal{A}_k) = 7 \text{ for } k = [6, 7], \quad \bigcup_{k=1}^{7} \mathcal{A}_k = \mathfrak{P}(X).$$

CONSTRUCTION. $\{a, b_1\}$ is the $\mathcal{A}_1$-similar set; $\{a, c_1\}$ is the $\mathcal{A}_2$-similar set; $\{a, c_3\}$, $\{b_1, b_3, c_1\}$ are the $\mathcal{A}_3$-similar sets; $\{a, b_3\}$, $\{b_1, c_1, c_3\}$ are the $\mathcal{A}_4$-similar sets; $\{b_1, c_1\}$, $\{b_2, c_2\}$, $\{b_3, c_3\}$ are the $\mathcal{A}_5$-similar sets; $\{a, c_2\}$, $\{b_1, b_2, c_1\}$, $\{b_3, c_3\}$ are the $\mathcal{A}_6$-similar sets; $\{a, b_2\}$, $\{b_1, c_1, c_2\}$, $\{b_3, c_3\}$ are the $\mathcal{A}_7$-similar sets.

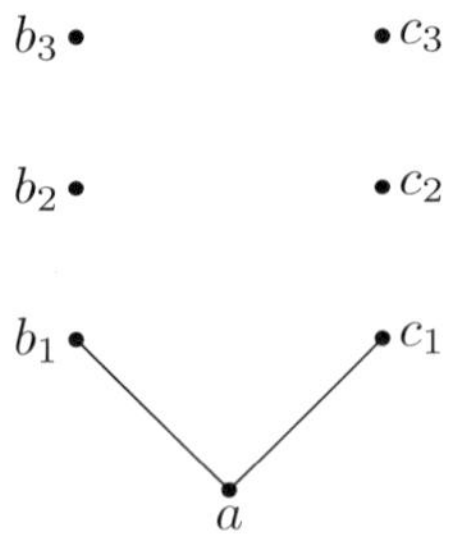

□

By the present example, Lemma 10.18, and the fact that the functions (6–3), (6–5) belong to Ψ_*^6, we see that the functions (7–9), (7–11) belong to Ψ_*^7.

10.21. EXAMPLE. Algebras $\mathcal{A}_1, \dots, \mathcal{A}_7$ such that

$$\#(\ker \mathcal{A}_1) = 2, \quad \#(\ker \mathcal{A}_k) = 4 \text{ for } k \in [2,4], \quad \#(\ker \mathcal{A}_5) = 6,$$
$$\#(\ker \mathcal{A}_k) = 7 \text{ for } k \in [6,7], \quad \bigcup_{k=1}^{7} \mathcal{A}_k = \mathfrak{P}(X).$$

CONSTRUCTION. $\{a_1, b_1\}$ is the $\mathcal{A}_1$-similar set; $\{a_1, a_2\}$, $\{b_1, b_2\}$ are the $\mathcal{A}_2$-similar sets; $\{a_1, b_2\}$, $\{a_2, b_1\}$ are the $\mathcal{A}_3$-similar sets; $\{a_2, b_2\}$, $\{a_3, b_3\}$ are the $\mathcal{A}_4$-similar sets; $\{a_1, a_3\}$, $\{a_2, b_2\}$, $\{b_1, b_3\}$ are the $\mathcal{A}_5$-similar sets; $\{a_1, b_3, c\}$, $\{a_2, b_2\}$, $\{a_3, b_1\}$ are the $\mathcal{A}_6$-similar sets; $\{a_1, b_3\}$, $\{a_2, b_2\}$, $\{a_3, b_1, c\}$ are the $\mathcal{A}_7$-similar sets.

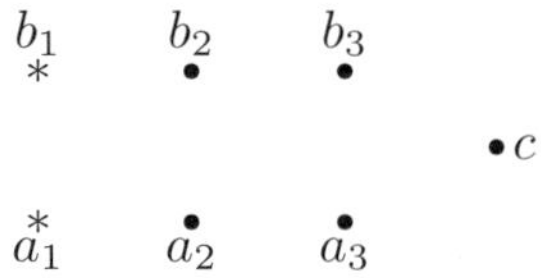

□

By the present example, by Lemma 10.18, and by the fact that the function (6–4) belongs to Ψ_*^6, we see that the function (7–10) belongs to Ψ_*^7.

10.22. LEMMA. *Let $\mathcal{A}_1, \dots, \mathcal{A}_7$ be algebras such that $\#(\ker \mathcal{A}_k) \geq \psi(k)$ for each $k \in [1,7]$, where $\psi(k)$ is one of the six functions (7–3)–(7–8). Let S_6 and T_6 be corresponding sets of ultrafilters for the algebras $\mathcal{A}_1, \dots, \mathcal{A}_6$; let S_6 and T_6 form more than one bush. Then*

$$\bigcup_{k=1}^{7} \mathcal{A}_k \neq \mathfrak{P}(X).$$

SKETCH OF THE PROOF. One must consider four cases:[7]

1)

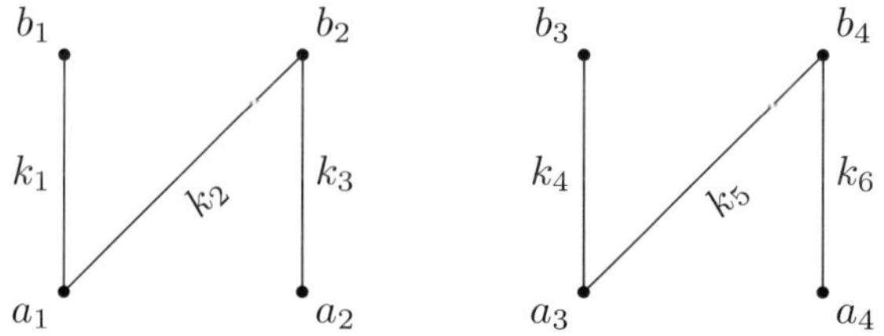

2)

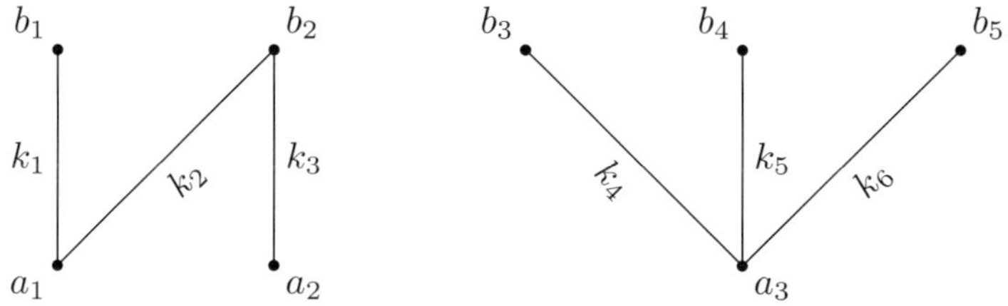

[7] Cases 3) and 4) do not differ, in principle.

3) 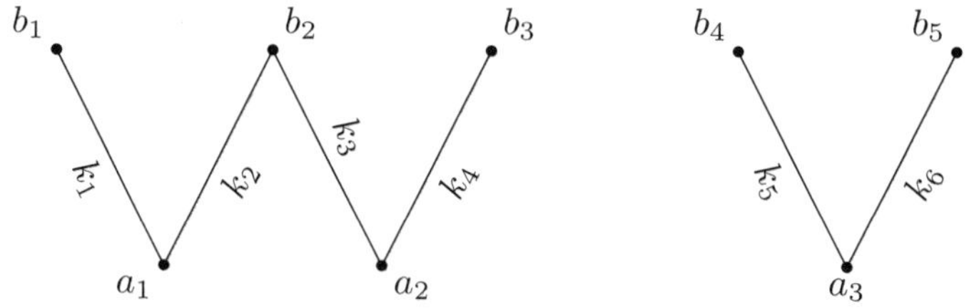

4) 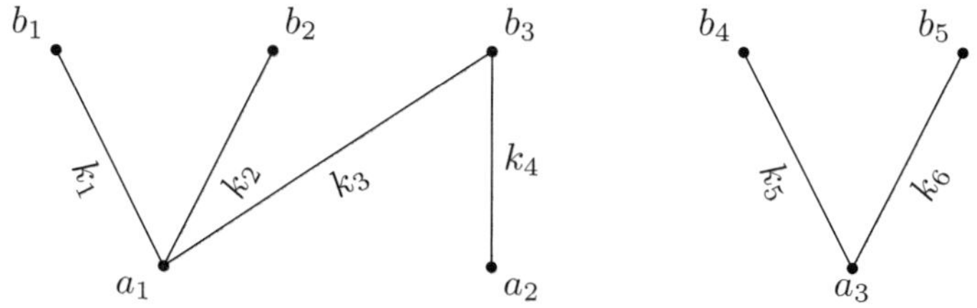

Here $\mathcal{A}_{k_i}$-similar ultrafilters are marked, and $\{k_1, \dots, k_6\} = [1, 6]$. Clearly, in the first case, $\{a_1, a_2\}$, $\{b_1, b_2\}$, $\{a_3, a_4\}$, $\{b_3, b_4\}$ are $\mathcal{A}_7$-similar sets; in the second case, $\{a_1, a_2\}$, $\{b_1, b_2\}$, $\{b_3, b_4, b_5\}$ are $\mathcal{A}_7$-similar sets; and in the third and the fourth cases, $\{a_1, a_2\}$, $\{b_1, b_2, b_3\}$, $\{b_4, b_5\}$ are $\mathcal{A}_7$-similar sets. For any values of $k_1, \dots, k_6$ in $[1, 6]$, one can construct S_7 and T_7.[8] □

10.23. LEMMA. *The function* (7–6) *belongs to* Ψ_*^7.

PROOF. Let $\mathcal{A}_1, \dots, \mathcal{A}_7$ be algebras such that $\#(\ker \mathcal{A}_k) \geq (7\text{–}6)(k)$ for all $k \in [1, 7]$. Let us prove that

$$\bigcup_{k=1}^{7} \mathcal{A}_k \neq \mathfrak{P}(X).$$

By Lemma 10.22, one must consider the situation in which $\Theta = S_6 \cup T_6$ is a bush of the fourth kind, $\#(\Theta) = 7$, and $\ker \mathcal{A}_k \subset \Theta$ for all $k \in [1, 7]$. Let a_1, b_1 be $\mathcal{A}_1$-similar ultrafilters. Consider all sets $L \subset \Theta$ such that $\#(L) = 3$ and

$$\#(L \cap \{a_1, b_1\}) = 1.$$

The number of such sets is 20. In the nontrivial situation there exist $k', k'' \in [2, 4]$ such that

$$(\{a_1, b_1\} \setminus \ker \mathcal{A}_{k'}) \cup (\{a_1, b_1\} \setminus \ker \mathcal{A}_{k''}) = \{a_1, b_1\}$$

(possibly, $k' = k''$). Let us make the count. There is an L such that $L \cap \ker \mathcal{A}_k \neq \varnothing$ if $k \in [2, 4]$, and $L \cap \ker \mathcal{A}_k$ is not an $\mathcal{A}_k$-similar set for any $k \in [2, 7]$.

It remains to take into account Examples 10.3 and 10.21. □

10.24. EXAMPLE. Algebras $\mathcal{A}_1, \dots, \mathcal{A}_7$ such that

$$\#(\ker \mathcal{A}_1) = 2, \quad \#(\ker \mathcal{A}_k) = 4 \text{ for } k \in [2, 3], \quad \#(\ker \mathcal{A}_k) = 6 \text{ for } k \in [4, 7],$$

$$\bigcup_{k=1}^{7} \mathcal{A}_k = \mathfrak{P}(X).$$

---

[8] In some situations arising in the construction of S_7 and T_7, one should take into account the fact that we have already constructed the class Ψ_*^5. For example, let $\psi(k) = (7\text{–}5)(k)$; consider case 3) with $k_5 = 1$, $k_6 = 2$. Then we can use the fact that $(5\text{–}1) \in \Psi_*^5$.

CONSTRUCTION. $\{a_1, b_1\}$ is the $\mathcal{A}_1$-similar set; $\{a_1, a_2\}$, $\{b_1, b_2\}$ are the $\mathcal{A}_2$-similar sets; $\{a_1, b_2\}$, $\{a_2, b_1\}$ are the $\mathcal{A}_3$-similar sets; $\{a_1, a_3\}$, $\{a_2, b_2\}$, $\{b_1, b_3\}$ are the $\mathcal{A}_4$-similar sets; $\{a_1, b_3\}$, $\{a_2, b_2\}$, $\{a_3, b_1\}$ are the $\mathcal{A}_5$-similar sets; $\{a_1, c\}$, $\{a_2, b_2\}$, $\{a_3, b_3\}$ are the $\mathcal{A}_6$-similar sets; $\{a_2, b_2\}$, $\{a_3, b_3\}$, $\{b_1, c\}$ are the $\mathcal{A}_7$-similar sets.

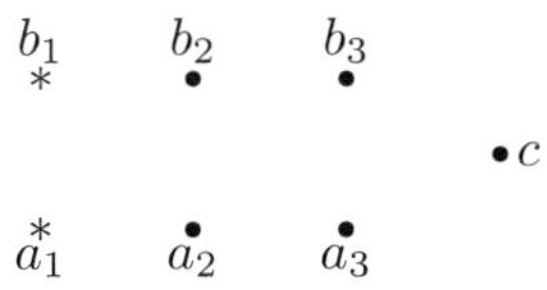

□

10.25. LEMMA. *The function (7–4) belongs to* Ψ^7_*.

PROOF. Let $\mathcal{A}_1, \dots, \mathcal{A}_7$ be algebras such that $\#(\ker \mathcal{A}_k) \geq (7\text{–}4)(k)$ for all $k \in [1, 7]$. Let us prove that

$$\bigcup_{k=1}^{7} \mathcal{A}_k \neq \mathfrak{P}(X).$$

By Lemma 10.22, we can assume that there exists a set of ultrafilters Θ such that $\#(\Theta) = 7$ and $\ker \mathcal{A}_k \subset \Theta$ for all $k \in [1, 7]$. Let a_1, b_1 be $\mathcal{A}_1$-similar ultrafilters. Let L be a subset of Θ such that $\#(L) = 3$ and $\#(L \cap \{a_1, b_1\}) = 1$. Let M be a subset of Θ such that $\#(M) = 2$ and $\#(M \cap \{a_1, b_1\}) = 1$. If there exists a k_0 such that $a_1 \notin \ker \mathcal{A}_{k_0}$, then at least one of the following two statements is true:

(a) there exists an L such that $L \cap \ker \mathcal{A}_k \neq \varnothing$ for each $k \in [2, 3]$ and $L \cap \ker \mathcal{A}_k$ is not an $\mathcal{A}_k$-similar set for any $k \in [2, 7]$;
(b) there exists an M such that $M \cap \ker \mathcal{A}_k \neq \varnothing$ for each $k \in [2, 6]$ and $M \cap \ker \mathcal{A}_k$ is not an $\mathcal{A}_k$ similar set for any $k \in [2, 7]$.

Now it remains to take into account Example 10.24 and the fact that the function (6–2) belongs to Ψ^6_*. □

10.26. EXAMPLE. Algebras $\mathcal{A}_1, \dots, \mathcal{A}_7$ such that

$$\#(\ker \mathcal{A}_1) = \#(\ker \mathcal{A}_2) = 2, \quad \#(\ker \mathcal{A}_3) = 4, \quad \#(\ker \mathcal{A}_4) = \#(\ker \mathcal{A}_5) = 6,$$
$$\#(\ker \mathcal{A}_6) = \#(\ker \mathcal{A}_7) = 8, \quad \bigcup_{k=1}^{7} \mathcal{A}_k = \mathfrak{P}(X).$$

CONSTRUCTION. Using Example 10.3, let us take algebras $\mathcal{A}_1, \dots, \mathcal{A}_4$, $\mathcal{B}_1, \dots, \mathcal{B}_4$ such that

$$\#(\ker \mathcal{A}_1) = \#(\ker \mathcal{A}_2) = \#(\ker \mathcal{B}_1) = \#(\ker \mathcal{B}_2) = 2,$$
$$\#(\ker \mathcal{A}_3) = \#(\ker \mathcal{A}_4) = \#(\ker \mathcal{B}_3) = \#(\ker \mathcal{B}_4) = 4,$$
$$\bigcup_{k=1}^{4} \mathcal{A}_k = \bigcup_{k=1}^{4} \mathcal{B}_k = \mathfrak{P}(X).$$

Let us apply Theorem 9.3 ($n = 4$, $m = 4$, $\nu = 3$, $\mu = 0$). □

10.27. LEMMA. *The function* (7–8) *belongs to* Ψ_*^7.

PROOF. Let $\mathcal{A}_1, \ldots, \mathcal{A}_7$ be algebras such that $\#(\ker \mathcal{A}_k) \geq (7\text{–}8)(k)$ for all $k \in [1,7]$. Assume that $\ker \mathcal{A}_1 \cap \ker \mathcal{A}_2 \cap \ker \mathcal{A}_3 \neq \varnothing$. Using s-bushes and Lemma 9.15, we conclude that

$$\bigcup_{k=1}^{7} \mathcal{A}_k \neq \mathfrak{P}(X).$$

Therefore we can assume that $\ker \mathcal{A}_1 \cap \ker \mathcal{A}_2 \cap \ker \mathcal{A}_3 = \varnothing$; let us prove that

$$\bigcup_{k=1}^{7} \mathcal{A}_k \neq \mathfrak{P}(X).$$

By Lemma 10.22, we must consider only the situation in which $\Theta = S_6 \cup T_6$ is a bush of the fourth kind, $\#(\Theta) = 7$, and $\ker \mathcal{A}_k \subset \Theta$ for all $k \in [1,7]$. We know that

$$\#(\ker \mathcal{A}_1) = \#(\ker \mathcal{A}_2) = 2.$$

We must consider two cases:

Case 1. $\ker \mathcal{A}_1 \cap \ker \mathcal{A}_2 = \varnothing$.

Case 2. $\#(\ker \mathcal{A}_1 \cap \ker \mathcal{A}_2) = 1$.

Let $L \subset \Theta$, $\#(L) = 3$, and $\#(L \cap \ker \mathcal{A}_k) = 1$ for $k \in [1,2]$. In both cases we make the count and see that there exists an L such that $L \cap \ker \mathcal{A}_k \neq \varnothing$ and $L \cap \ker \mathcal{A}_k$ is not an $\mathcal{A}_k$-similar set for any $k \in [3,7]$. It remains to refer to Examples 10.19 and 10.26. □

10.28. LEMMA. *The function* (7–7) *belongs to* Ψ_*^7.

PROOF. Let $\mathcal{A}_1, \ldots, \mathcal{A}_7$ be algebras such that $\#(\ker \mathcal{A}_k) \geq (7\text{–}7)(k)$ for all $k \in [1,7]$. As in the proof of Lemma 10.27, let us take Θ ($\#(\Theta) = 7$), sets L, and consider the same two cases. We make the count, obtaining

$$\bigcup_{k=1}^{7} \mathcal{A}_k \neq \mathfrak{P}(X).$$

It remains to refer to Examples 10.19 and 10.20. □

10.29. EXAMPLE. Algebras $\mathcal{A}_1, \ldots, \mathcal{A}_7$ such that

$$\#(\ker \mathcal{A}_k) = 2 \text{ for } k \in [1,2], \quad \#(\ker \mathcal{A}_k) = 6 \text{ for } k \in [3,7], \quad \bigcup_{k=1}^{7} \mathcal{A}_k = \mathfrak{P}(X).$$

CONSTRUCTION. By Example 1.13, there exist algebras $\mathcal{A}_1, \mathcal{A}_2, \mathcal{A}_3$ such that $\#(\ker \mathcal{A}_k) = 2$ for all $k \in [1,3]$, and $\bigcup_{k=1}^{3} \mathcal{A}_k = \mathfrak{P}(X)$. By the construction in 9.6, there exist algebras $\mathcal{B}_1, \ldots, \mathcal{B}_5$ such that $\#(\ker \mathcal{B}_k) = 4$ for all $k \in [1,5]$, and $\bigcup_{k=1}^{5} \mathcal{B}_k = \mathfrak{P}(X)$. It remains to refer to Theorem 9.3 ($n = 3$, $m = 5$, $\nu = 2$, $\mu = 0$). □

10.30. LEMMA. *Let $\mathcal{A}_1, \dots, \mathcal{A}_7$ be algebras such that $\#(\ker \mathcal{A}_k) \geq (7\text{–}5)(k)$ for all $k \in [1,7]$. If $\ker \mathcal{A}_1 \cap \ker \mathcal{A}_2 \cap \ker \mathcal{A}_3 \neq \varnothing$, then*

$$\bigcup_{k=1}^{7} \mathcal{A}_k \neq \mathfrak{P}(X).$$

PROOF. Using the technique of s-bushes, one can deduce that the only nontrivial situation is similar to the following one: there is a set Θ consisting of seven ultrafilters $s, t_1, t_2, t_3, t_4, t_5, t_7$, and the following conditions are satisfied:

1) $\{s, t_k\}$ is an $\mathcal{A}_k$-similar set;
2) $\ker \mathcal{A}_k \subset \Theta$ for all $k \in [3,7]$;
3) $\ker \mathcal{A}_6 = \Theta \setminus \{s\}$.

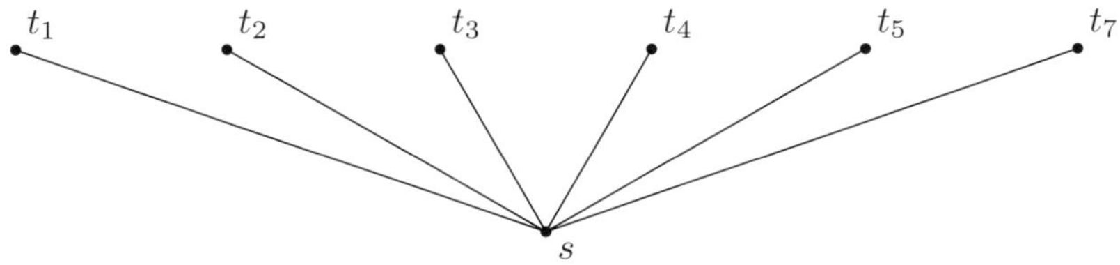

First suppose that $t_7 \in \ker \mathcal{A}_3$. We can assume that $\{t_3, t_7\}$ is an $\mathcal{A}_6$-similar set, because otherwise the sets S_7 and T_7 are constructed immediately by Lemma 9.15. By the same arguments, we can assume that $t_7 \notin \ker \mathcal{A}_4, \ker \mathcal{A}_5$ and $\{t_4, t_5\}$ is an $\mathcal{A}_6$-similar set. Also, by the same arguments, $\{t_1, t_2, t_7\}$ is an $\mathcal{A}_3$-similar set. If one cannot take $S_7 = \{s, t_3, t_4, t_7\}$ (if $\{t_3, t_4\}$ is an $\mathcal{A}_7$-similar set), then it is possible to take $S_7 = \{s, t_3, t_5, t_7\}$.

Now suppose that $t_7 \notin \ker \mathcal{A}_3$, $t_4 \in \ker \mathcal{A}_3$. If $t_3 \notin \ker \mathcal{A}_4$, then it is natural to assume that $\{t_4, t_7\}$ is an $\mathcal{A}_6$-similar set, $t_4 \notin \ker \mathcal{A}_5$, and we can put $S_7 = \{s, t_5, t_7\}$. Now let $t_3 \subset \ker \mathcal{A}_4$. Then it is natural to assume that $\{t_3, t_4\}$ is an $\mathcal{A}_6$-similar set, $t_7 \notin \ker \mathcal{A}_4$, $t_4 \notin \ker \mathcal{A}_5$, $\{t_5, t_7\}$ is an $\mathcal{A}_6$-similar set, and $\{t_1, t_2, t_4\}$ is an $\mathcal{A}_3$-similar set. If we cannot take $S_7 = \{s, t_3, t_4, t_7\}$ (if $\{t_3, t_4\}$ is an $\mathcal{A}_7$-similar set), then it is possible to take $S_7 = \{s, t_4, t_5, t_7\}$. □

10.31. With the proof of Lemma 10.27 and the statement of Lemma 10.30 known, it is easy to prove the following lemma.

LEMMA. *Let $\mathcal{A}_1, \dots, \mathcal{A}_7$ be algebras such that $\#(\ker \mathcal{A}_k) \geq (7\text{–}5)(k)$ for all $k \in [1,7]$. Then*

$$\bigcup_{k=1}^{7} \mathcal{A}_k \neq \mathfrak{P}(X).$$

By this lemma and Examples 10.20, 10.26, and 10.29, we see that the function (7–5) belongs to Ψ^7_*.

10.32. EXAMPLE. Algebras $\mathcal{A}_1, \dots, \mathcal{A}_7$ such that

$$\#(\ker \mathcal{A}_k) = 4 \text{ for } k \in [1,4], \quad \#(\ker \mathcal{A}_k) = 6 \text{ for } k \in [5,7], \quad \bigcup_{k=1}^{7} \mathcal{A}_k = \mathfrak{P}(X).$$

CONSTRUCTION. By the construction in 9.6, there exist algebras $\mathcal{A}_1, \ldots, \mathcal{A}_5$ such that $\#(\ker \mathcal{A}_k) = 4$ for all $k \in [1,5]$, and $\bigcup_{k=1}^{5} \mathcal{A}_k = \mathfrak{P}(X)$. By Example 1.13, there exist algebras $\mathcal{B}_1, \mathcal{B}_2, \mathcal{B}_3$ such that $\#(\ker \mathcal{B}_k) = 2$ for all $k \in [1,3]$, and $\bigcup_{k=1}^{3} \mathcal{B}_k = \mathfrak{P}(X)$. It remains to refer to Theorem 9.3 ($n = 5$, $m = 3$, $\nu = 4$, $\mu = 0$). □

10.33. LEMMA. *The function* (7–3) *belongs to* Ψ_*^7.

PROOF. Let $\mathcal{A}_1, \ldots, \mathcal{A}_7$ be algebras such that $\#(\ker \mathcal{A}_k) \geq (7\text{–}3)(k)$ for all $k \in [1,7]$. Let us prove that

$$\bigcup_{k=1}^{7} \mathcal{A}_k \neq \mathfrak{P}(X).$$

By Lemma 10.22, one can consider the following situation: $\#(\ker \mathcal{A}_7) = 7$ and $\ker \mathcal{A}_k \subset \ker \mathcal{A}_7$ for all $k \in [1,6]$. Let $L \subset \ker \mathcal{A}_7$ and $\#(L) = 3$. The number of such sets L is 35. In the nontrivial situation, $\#(\ker \mathcal{A}_k) = 4$ for all $k \in [1,4]$, and, of course, for each such k there exist two $\mathcal{A}_k$-similar sets, each of which consists of two elements (otherwise we cannot obtain the 35 sets L for each of which there is a $k \in [1,7]$ such that either $L \cap \ker \mathcal{A}_k$ is an $\mathcal{A}_k$-similar set, or $L \cap \ker \mathcal{A}_k = \varnothing$). Let

$$\ker \mathcal{A}_7 = \{a_1, a_2, a_3, b_1, b_2, b_3, c\},$$

and let $\{a_1, a_2\}$, $\{b_1, b_2\}$ be the $\mathcal{A}_1$-similar sets.

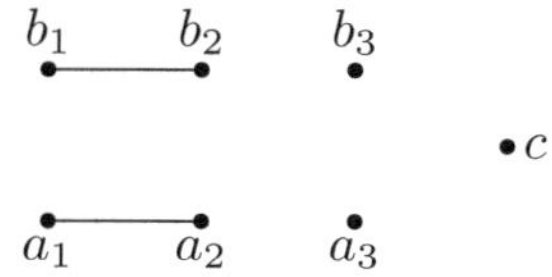

In the nontrivial situation, $\ker \mathcal{A}_2$ can be described as follows (or similarly): $\{a_1, b_1\}$, $\{a_2, a_3\}$ are the $\mathcal{A}_2$-similar sets (otherwise we will not obtain the 35 corresponding sets L).

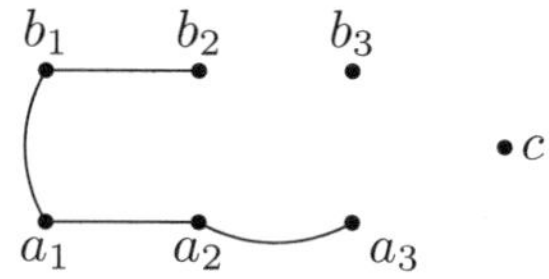

Consider the algebras $\mathcal{A}^1, \mathcal{A}^2, \mathcal{A}^3$; $\#(\ker \mathcal{A}^i) = 4$; $\{a_1, a_3\}$, $\{a_2, b_2\}$ are the $\mathcal{A}^1$-similar sets; $\{a_1, b_2\}$, $\{a_3, b_1\}$ are the $\mathcal{A}^2$-similar sets; $\{a_2, b_1\}$, $\{a_3, b_2\}$ are the $\mathcal{A}^3$-similar sets. In the nontrivial situation, $\mathcal{A}_3 = \mathcal{A}^i$ and $\mathcal{A}_4 = \mathcal{A}^j$, $i \neq j$. Assume that $\mathcal{A}_3 = \mathcal{A}^1$ and $\mathcal{A}_4 = \mathcal{A}^2$.

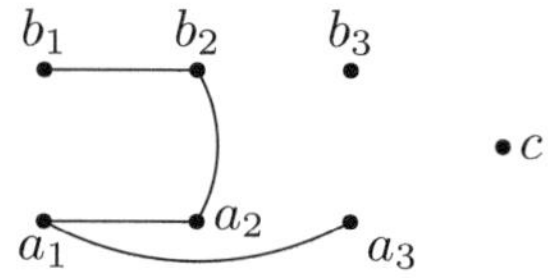

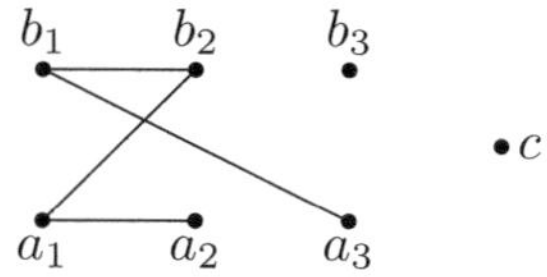

Now consider the sets L such that $L \cap \ker \mathcal{A}_1 \neq \varnothing$ and $L \cap \ker \mathcal{A}_1$ is not an $\mathcal{A}_1$-similar set. The number of such sets L is 28. To avoid confusion, let us denote them by L'. Let us list all L' such that for each of them and for any $k \in [2,4]$, $L' \cap \ker \mathcal{A}_k \neq \varnothing$ and $L' \cap \ker \mathcal{A}_k$ is not an $\mathcal{A}_k$-similar set. The number of such sets is seven:

$$\{a_1,\ a_2,\ b_1\},\ \{a_1,\ a_3,\ b_2\},\ \{a_1,\ b_3,\ c\},\ \{a_2,\ b_1,\ b_3\},$$
$$\{a_2,\ b_1,\ c\},\ \{a_3,\ b_2,\ b_3\},\ \{a_3,\ b_2,\ c\}.$$

Since it is natural to assume that $a_1 \notin \ker \mathcal{A}_5 \cap \ker \mathcal{A}_6$, we can suppose that $a_1 \notin \ker \mathcal{A}_5$.

(i) First let us suppose that $\#(\ker \mathcal{A}_5) = 5$. In our constructions there are two points which play similar roles, namely b_3 and c. If $b_3, c \in \ker \mathcal{A}_5$, we will not have the 35 corresponding sets L. Let $b_3 \notin \ker \mathcal{A}_5$. We will consider two cases.

Case (1). $\{a_2, b_1\}$, $\{a_3, b_2, c\}$ are the $\mathcal{A}_5$-similar sets.

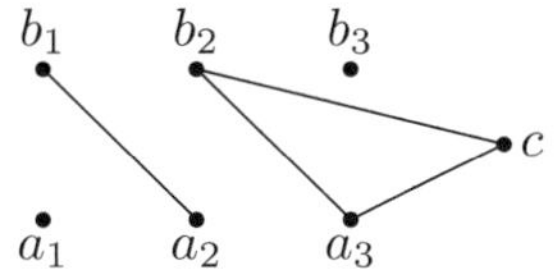

Case (2). $\{a_3, b_2\}$, $\{a_2, b_1, c\}$ are the $\mathcal{A}_5$-similar sets.

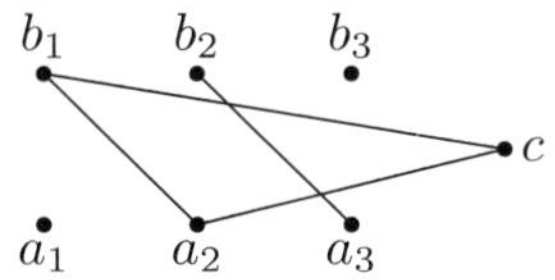

In all other cases we will not have the 35 corresponding sets L.

So, let us consider Case (1). Let us list all L' such that for each of them, for any $k \in [2,5]$, $L' \cap \ker \mathcal{A}_k \neq \varnothing$ and $L' \cap \ker \mathcal{A}_k$ is not an $\mathcal{A}_k$-similar set. There are four such sets L':

$$\{a_1,\ a_3,\ b_2\},\ \{a_1,\ b_3,\ c\},\ \{a_2,\ b_1,\ c\},\ \{a_3,\ b_2,\ b_3\}.$$

Since only one pair in this collection of sets is repeated twice in different L', namely, the pair (a_3, b_2), in the nontrivial situation we must assume that $\{a_3, b_2\}$, $\{a_2, b_1, c\}$ are $\mathcal{A}_6$-similar sets (otherwise we will not have the 35 corresponding sets L). Also, in order to have the 35 corresponding sets L, we must assume that $\{a_1, b_3, c\}$ is an $\mathcal{A}_7$-similar set. Put $S_7 = \{a_1, c\}$.

Now consider Case (2). Let us list all L' such that for each of them and for any $k \in [2,5]$, $L' \cap \ker \mathcal{A}_k \neq \varnothing$ and $L' \cap \ker \mathcal{A}_k$ is not an $\mathcal{A}_k$-similar set. There are four such sets L':

$$\{a_1,\ a_2,\ b_1\},\ \{a_1,\ b_3,\ c\},\ \{a_2,\ b_1,\ b_3\},\ \{a_3,\ b_2,\ c\}.$$

Since only one pair in this collection is repeated twice in different L', namely, the pair (a_2, b_1), in the nontrivial situation we must assume that $\{a_2, b_1\}$, $\{a_3, b_2, c\}$ are $\mathcal{A}_6$-similar sets (otherwise we will not have the 35 corresponding sets L). Also, in order to have the 35 corresponding sets L, we must assume that $\{a_1, b_3, c\}$ is an $\mathcal{A}_7$-similar set. Put $S_7 = \{a_1, c\}$.

(ii) Now let us assume that $\#(\ker \mathcal{A}_5) = 6$. Then we must assume that $\{a_2, b_1\}$, $\{a_3, b_2\}$, $\{b_3, c\}$ are the $\mathcal{A}_5$-similar sets (otherwise we will not have the 35 corresponding sets L).

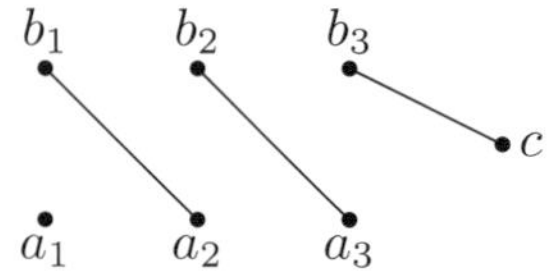

Let us list all L' such that for each of them and for any $k \in [2,5]$ we have $L' \cap \ker \mathcal{A}_k \neq \varnothing$ and $L' \cap \ker \mathcal{A}_k$ is not an $\mathcal{A}_k$-similar set:

$$\{a_2,\ b_1,\ b_3\},\ \{a_2,\ b_1,\ c\},\ \{a_3,\ b_2,\ b_3\},\ \{a_3,\ b_2,\ c\}.$$

None of these four sets contains a_1. But one necessarily has an $\mathcal{A}_6$-similar set which contains a_1. Now it is clear that

$$\bigcup_{k=1}^{7} \mathcal{A}_k \neq \mathfrak{P}(X).$$

The situations in which $\mathcal{A}_3 = \mathcal{A}^1$, $\mathcal{A}_4 = \mathcal{A}^3$ or $\mathcal{A}_3 = \mathcal{A}^2$, $\mathcal{A}_4 = \mathcal{A}^3$ are completely similar to the situation considered above. It remains to take into account the construction in 9.6 and Examples 10.21 and 10.32. □

10.34. LEMMA. *The function* (7–2) *belongs to* Ψ_*^7.

PROOF. Let $\mathcal{A}_1, \dots, \mathcal{A}_7$ be algebras such that $\#(\ker \mathcal{A}_k) \geq (7\text{–}2)(k)$ for all $k \in [1,7]$. Let $n \in \mathbb{N}^+$ be sufficiently large, let Θ be a set of ultrafilters such that $\#(\Theta) = 2n$, and let $\ker \mathcal{A}_k \subset \Theta$ for all $k \in [1,7]$. Let $\{a_1, b_1\}$ be an $\mathcal{A}_1$-similar set. Let $\Pi \subset \Theta$, $\#(\Pi) = n$, and

$$\#(\Pi \cap \{a_1, b_1\}) = 1.$$

Assume that $\ker \mathcal{A}_2 = \{a_1, a_2, b_1, b_2\}$ and $\{a_1, a_2\}$, $\{b_1, b_2\}$ are the $\mathcal{A}_2$-similar sets.

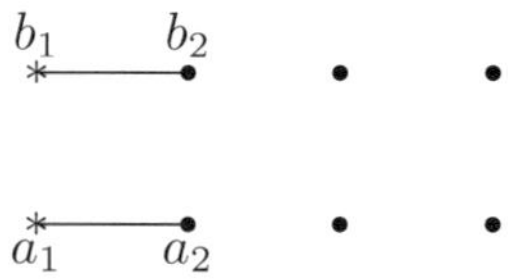

There exist $2 \cdot \binom{2n-4}{n-2}$ sets Π for each of which $\Pi \cap \ker \mathcal{A}_2$ is an $\mathcal{A}_2$-similar set. For an arbitrary algebra $\mathcal{A}_2$ there exist at most $2 \cdot \binom{2n-4}{n-2}$ sets Π for each of which one of the following two conditions holds:

(a) $\Pi \cap \ker \mathcal{A}_2 = \varnothing$;
(b) $\Pi \cap \ker \mathcal{A}_2$ is a union of $\mathcal{A}_2$-similar sets.

Assume that $\ker \mathcal{A}_3 = \{a_1, a_2, a_3, b_1, b_2\}$ and $\{a_1, a_2, a_3\}$, $\{b_1, b_2\}$ are the $\mathcal{A}_3$-similar sets.

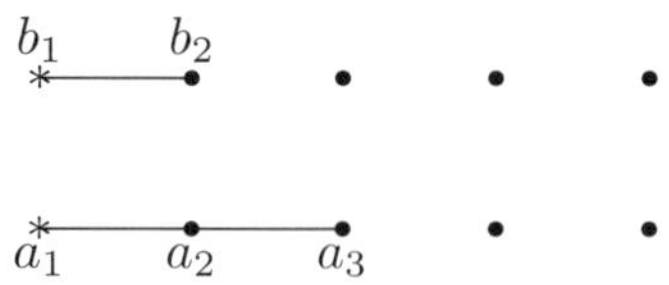

There exist $2 \cdot \binom{2n-5}{n-2}$ sets Π for each of which $\Pi \cap \ker \mathcal{A}_3$ is an $\mathcal{A}_3$-similar set. For an arbitrary algebra $\mathcal{A}_3$ there exist at most $2 \cdot \binom{2n-5}{n-2}$ sets Π for each of which one of the two corresponding conditions (a), (b) holds. The total number of all sets Π is

$$2 \cdot \binom{2n-2}{n-1}.$$

But

$$2 \cdot \binom{2n-2}{n-1} > 2 \cdot \binom{2n-4}{n-2} + 10 \cdot \binom{2n-5}{n-2}.$$

Hence

$$\bigcup_{k=1}^{7} \mathcal{A}_k \neq \mathfrak{P}(X).$$

It remains to take into account that the function (6–5) belongs to Ψ_*^6, and recall Example 10.24. □

10.35. LEMMA. *The function* (7–1) *belongs to* Ψ_*^7.

SKETCH OF THE PROOF. Let $\mathcal{A}_1, \ldots, \mathcal{A}_7$ be algebras such that $\#(\ker \mathcal{A}_k) \geq$ (7–1)(k) for all $k \in [1, 7]$. Let us prove that

$$\bigcup_{k=1}^{7} \mathcal{A}_k \neq \mathfrak{P}(X).$$

Assume that there exists a set of ultrafilters Θ such that $\#(\Theta) = 8$ and $\ker \mathcal{A}_k \subset \Theta$ for all $k \in [1, 7]$. Let $\Pi \subset \Theta$ and $\#(\Pi) = 4$. There can be at most

$$3 \times 14 + 4 \times 6 = 66$$

sets Π such that for each of them there exists $k \in [1, 7]$, with one of the following two conditions holding:

(a) $\Pi \cap \ker \mathcal{A}_k = \varnothing$;
(b) $\Pi \cap \ker \mathcal{A}_k$ is a union of $\mathcal{A}_k$-similar sets.

But since $\binom{8}{4} = 70 > 66$, we have

$$\bigcup_{k=1}^{7} \mathcal{A}_k \neq \mathfrak{P}(X).$$

Here it is convenient to assume that

$$\#\left(\bigcup_{k=1}^{7} \ker \mathcal{A}_k\right) > 8 \quad \text{and} \quad \#(\ker \mathcal{A}_7) = 6.$$

We know that $\bigcup_{k=1}^{6} \mathcal{A}_k \neq \mathfrak{P}(X)$. For these algebras consider the corresponding sets of ultrafilters S_6 and T_6. Let us divide $S_6 \cup T_6$ into bushes. Reviewing all possible configurations of bushes and $\ker \mathcal{A}_7$, we conclude that one can construct S_7 and T_7.

Assume now that Θ is a set of ultrafilters, $\#(\Theta) = 10$, $\ker \mathcal{A}_k \subset \Theta$ for all $k \in [1, 7]$, and $\#(\ker \mathcal{A}_k) = 5$ for all $k \in [4, 7]$. Here let $\Pi \subset \Theta$ and $\#(\Pi) = 5$. There can be at most

$$3 \times 52 + 4 \times 22 = 244$$

sets Π such that for each of them either condition (a) or condition (b) holds (see above). But since $\binom{10}{5} = 252 > 244$, we have

$$\bigcup_{k=1}^{7} \mathcal{A}_k \neq \mathfrak{P}(X).$$

Finally, assume that

$$\#\left(\bigcup_{k=1}^{7} \ker \mathcal{A}_k\right) > 10 \quad \text{and} \quad \#(\ker \mathcal{A}_k) = 5 \text{ for all } k \in [4,7].$$

Again consider the corresponding sets of ultrafilters S_6 and T_6, and divide $S_6 \cup T_6$ into bushes. Considering all possible configurations of bushes and $\ker \mathcal{A}_7$, we conclude that one can construct S_7 and T_7.[9]

By Example 10.32 and by the fact that the function (7–2) belongs to Ψ_*^7, we see that the function (7–1) belongs to Ψ_*^7. □

10.36. We have shown that the functions (7–1)–(7–13) belong to Ψ_*^7. It is easy to show that Ψ_*^7 does not contain other functions. Thus we have proved the following theorem, which summarizes all the results of the present chapter:

THEOREM. *The class Ψ_*^7 consists of the functions* (7–1)–(7–13).

[9]For the proof of our lemma we are using the arguments from 9.22.

CHAPTER 11

The General Problem. Proof of Theorems 2.15 and 2.20

11.1. In this chapter we discuss the results described in 2.14–2.20. Statement 2.14 is Theorem 9.12.

THEOREM. *Let $\mathcal{A}_1, \dots, \mathcal{A}_n$ be a finite sequence of algebras, and for each $k \in [1, n]$ let there exist more than $4 \cdot \log_2 k$ pairwise disjoint sets not belonging to $\mathcal{A}_k$. Then*

$$\bigcup_{k=1}^{n} \mathcal{A}_k \neq \mathfrak{P}(X).$$

PROOF. Put

$$\eta(k) = [4 \cdot \log_2 k] + 1.$$

Here are the first 16 values of this function: $\eta(1) = 1$, $\eta(2) = 5$, $\eta(3) = 7$, $\eta(4) = 9$, $\eta(5) = 10$, $\eta(6) = 11$, $\eta(7) = 12$, $\eta(8) = 13$, $\eta(9) = 13$, $\eta(10) = 14$, $\eta(11) = 14$, $\eta(12) = 15$, $\eta(13) = 15$, $\eta(14) = 16$, $\eta(15) = 16$, $\eta(16) = 17$. Therefore

$$\begin{aligned}\sum_{l=1}^{\infty} 2^{-\left[\frac{\eta(l)+1}{2}\right]} &= \sum_{l=1}^{16} 2^{-\left[\frac{\eta(l)+1}{2}\right]} + \sum_{l=17}^{\infty} 2^{-\left[\frac{\eta(l)+1}{2}\right]} \\ &< \frac{1}{2} + \frac{1}{8} + \frac{1}{16} + \frac{1}{2^5} + \frac{1}{2^5} + \frac{1}{2^6} + \frac{1}{2^6} + \frac{1}{2^7} + \frac{1}{2^7} + \frac{1}{2^7} + \frac{1}{2^7} \\ &\quad + \frac{1}{2^8} + \frac{1}{2^8} + \frac{1}{2^8} + \frac{1}{2^8} + \frac{1}{2^9} + \sum_{l=4}^{\infty} (2^{l+1} - 2^l) \cdot 2^{-\left[\frac{\eta(2^l)+1}{2}\right]} \\ &= \frac{1}{2} + \frac{1}{4} + \frac{1}{16} + \frac{1}{64} + \frac{1}{2^9} + \sum_{l=4}^{\infty} 2^l \cdot 2^{-(2l+1)} \\ &= \frac{1}{2} + \frac{1}{4} + \frac{1}{16} + \frac{1}{64} + \frac{1}{2^9} + \sum_{l=5}^{\infty} 2^{-l} = \frac{1}{2} + \frac{1}{4} + \frac{1}{8} + \frac{1}{64} + \frac{1}{2^9} < 1.\end{aligned}$$

Our theorem is true by Theorem 9.12. □

11.2. Let us try to replace the coefficient 4 in Theorem 11.1 by a smaller number α close to 4. Put $\eta'(k) = [\alpha \cdot \log_2 k] + 1$. The first 16 values of this function differ from the values of $\eta(k)$ only at the numbers $2, 4, 8, 16$; and at these numbers $\eta'(k) = \eta(k) - 1$. Hence

$$\sum_{l=1}^{16} 2^{-\left[\frac{\eta'(l)+1}{2}\right]} = \frac{1}{2} + \frac{1}{4} + \frac{1}{16} + \frac{1}{64} + \frac{1}{2^9} + \frac{1}{8} + \frac{1}{32} + \frac{1}{2^7} + \frac{1}{2^9} = 1 - \frac{1}{2^8}.$$

Further, $\eta'(17) = \eta'(18) = 17$. Therefore

$$\sum_{l=1}^{18} 2^{-\left[\frac{\eta'(l)+1}{2}\right]} = 1.$$

Hence, for $n > 18$,

$$\sum_{l=1}^{n} 2^{-\left[\frac{\eta'(l)+1}{2}\right]} > 1.$$

Let us summarize our arguments. *By our methods, Theorem* 11.1 *cannot be improved by replacing the function* $4 \cdot \log_2 k$ *by some function* $\alpha \cdot \log_2 k$*, where* $\alpha < 4$.

11.3. REMARK. We know a lower estimate for $\mathfrak{g}(n)$, which is approximately $4 \cdot \log_5 n$. Comparing this estimate with the estimate $4 \cdot \log_2 n$, one should recognize Theorem 11.1 as a rather strong one.

11.4. Theorems 1.17 and 11.1 imply the version of Theorem 2.15, by replacing almost σ-algebras by σ-algebras. Theorem 2.15 is a direct corollary of the following statement, which is a particular case of Theorem 2.20 and should be taken into account in the proof of that theorem.

STATEMENT. *Let* $\mathcal{A}_1, \dots, \mathcal{A}_k, \dots$ *be a countable sequence of almost* σ*-algebras, for each* k *let there exist* μ_k *pairwise disjoint sets not belonging to* $\mathcal{A}_k$, $\mu_k \in \mathbb{N}^+$, *and let*

$$\sum_{k=1}^{\infty} 2^{-\left[\frac{\mu_k+1}{2}\right]} \leq 1.$$

Then $\bigcup \mathcal{A}_k \neq \mathfrak{P}(X)$.

PROOF. As in the proof of Theorem 2.7, let us consider the corresponding strictly simple algebras $\mathcal{A}_{l_1}, \dots, \mathcal{A}_{l_m}, \dots$; let the other algebras be ω-saturated. We can assume, without loss of generality, that $m \to \infty$. Put

$$Z = \bigcup \operatorname{sp} \mathcal{A}_k, \quad \alpha = \sum_{m=1}^{\infty} 2^{-\left[\frac{\mu_{l_m}+1}{2}\right]}.$$

Take a countable sequence of positive numbers $\varepsilon_1, \dots, \varepsilon_m, \dots$ such that

$$\sum_{m=1}^{\infty} 2^{\frac{m(m+1)}{2}} \cdot \varepsilon_m \leq \alpha.$$

Further, take an increasing sequence of natural numbers $\gamma_1, \dots, \gamma_m, \dots$ such that

$$\sum_{k=\gamma_m}^{\infty} 2^{-\left[\frac{\mu_k+1}{2}\right]} \leq \varepsilon_m.$$

Let

$$B = \begin{pmatrix} b_1^1 & & \\ \dots & \dots & \dots \\ b_1^m & \dots & b_m^m \\ \dots & \dots & \dots \end{pmatrix}$$

be a matrix of pairwise distinct point of Z such that

1) $b_i^m \in \operatorname{sp} \mathcal{A}_{l_m}$ and b_i^m has an $\mathcal{A}_{l_m}$-similar ultrafilter in $\overline{Z} \setminus Z$;
2) if $\mathcal{A}_k$ is an ω-saturated algebra and $k < \gamma_m$, then $b_i^m \notin \ker \mathcal{A}_k$.

Let $\mathcal{A}_k$ be an ω-saturated algebra. If $k < \gamma_1$, then put $\mathcal{A}'_k = \mathcal{A}_k$; if $\gamma_m \leq k < \gamma_{m+1}$, then let us construct, by our standard way, an algebra $\mathcal{A}'_k > \mathcal{A}_k$ such that

$$\#(\ker \mathcal{A}_k) - m(m+1) \leq \#(\ker \mathcal{A}'_k)$$

and $b_i^p \notin \mathcal{A}'_k$ if $p \leq m$. By Theorem 9.12 and Lemma 7.12, $\bigcup \mathcal{A}'_k \neq \mathfrak{P}(X)$. Besides, $B \cap \bigcup \ker \mathcal{A}'_k = \varnothing$. Now the argument is exactly the same as in the proof of Theorem 2.7. $\square$

11.5. LEMMA. *Let $\psi^1 \in \Psi^m$ and $\psi^2 \in \Psi^{p-m}$. Consider the function $\psi(k)$ on $[1,p]$: $\psi(k) = \psi^1(k)$ for $k \leq m$, and $\psi(k) = 2m + \psi^2(k-m)$ for $k \in [m+1,p]$. Then $\psi \in \Psi^p$.*

PROOF. Let $\mathcal{A}_1, \dots, \mathcal{A}_p$ be algebras such that $\#(\ker \mathcal{A}_k) \geq \psi(k)$ for each $k \in [1,p]$. For the algebras $\mathcal{A}_1, \dots, \mathcal{A}_m$ let us take corresponding sets of ultrafilters S_m and T_m. Put

$$S_m^* = \{s \in S_m \mid \iota_m(s) > 1\},$$
$$T_m^* = \{t \in T_m \mid \text{there exists } s_k \in S_m^* \text{ such that } t_k = t\},$$
$$S_m^{**} = S_m \setminus S_m^*, \quad T_m^{**} = T_m \setminus T_m^*.$$

Clearly,

$$2 \cdot \#(S_m) + \#(T_m^*) \leq 2m.$$

Of course, we can assume that $\#(\ker \mathcal{A}_k) < \aleph_0$ for all $k \in [m+1,p]$. For each $k \in [m+1,p]$ let us construct an algebra $\mathcal{A}'_k$, and let

$$\ker \mathcal{A}'_k \subset G_k = \ker \mathcal{A}_k \setminus (S_m \cup T_m^*).$$

Suppose $q \in G_k$. Let us examine all possible cases:

(a) q is an $\mathcal{A}_k$-special ultrafilter; then q is an $\mathcal{A}'_k$-special ultrafilter;
(b) q, r are $\mathcal{A}_k$-similar ultrafilters and $r \in G_k$; then q, r are $\mathcal{A}'_k$-similar ultrafilters;
(c) q belongs to an $\mathcal{A}_k$-similar set M, $(M \setminus \{q\}) \cap G_k = \varnothing$, and there are two possible cases:
  (c$_1$) $(M \setminus \{q\}) \subset S_m$; then $q \notin \ker \mathcal{A}'_k$;
  (c$_2$) $(M \setminus \{q\}) \cap T_m^* \neq \varnothing$; then q is an $\mathcal{A}'_k$-special ultrafilter.

Clearly,

$$\#(\ker \mathcal{A}'_k) \geq \psi^2(k-m).$$

Hence for the algebras $\mathcal{A}'_{m+1}, \dots, \mathcal{A}'_p$ there exist corresponding sets of ultrafilters S_{p-m}, T_{p-m}; $S_{p-m} = \{s_{m+1}, \dots, s_p\}$, and T_{p-m} is the set of the corresponding ultrafilters t_k. Let us construct the corresponding sets of ultrafilters S_p and T_p for the algebras $\mathcal{A}_1, \dots, \mathcal{A}_p$. If s_k is an $\mathcal{A}'_k$-special but not $\mathcal{A}_k$-special ultrafilter, then let us take $t_k \in T_m^*$ such that s_k, t_k are $\mathcal{A}_k$-similar ultrafilters. Let $t \in T_m^{**}$. It can happen that $t \in S_{p-m}$. In this case, if $s_k \in S_m^{**}$ and $t_k = t$ (clearly, here $k \leq m$), we change the notation and put $t_k = s_k$ and $s_k = t$. It remains to put $S_p = \{s_1, \dots, s_p\}$, and take T_p to be the set of all t_k. $\square$

11.6. LEMMA. *Let* $\mathcal{A}_1, \ldots, \mathcal{A}_n$ *be a finite sequence of algebras,*

$$\bigcup_{k=1}^{n} \mathcal{A}_k \neq \mathfrak{P}(X)$$

(i.e., there exist corresponding sets of ultrafilters S_n *and* T_n*),* $S^* \subset S_n$*, and let*

$$S' = \{s \in S^* \mid s \text{ belongs to a simple bush of the second kind}\}.$$

Then

$$\#(S^* \cup T_n) \leq n + \left[\frac{n}{2}\right] + \left[\frac{\#(S')+1}{2}\right].$$

PROOF. Let us divide S^* into four pairwise disjoint sets:

$$S'; \quad S'' = \{s \in S^* \mid \text{there exists } k \text{ such that } s_k = s \text{ is an } \mathcal{A}_k\text{-special ultrafilter}\};$$
$$S''' = \{s \in S^* \setminus S'' \mid \iota_n(s) > 1\}; \quad S'''' = S^* \setminus (S' \cup S'' \cup S''').$$

We have

$$\#(S''') \leq \frac{n - \#(S^* \setminus S''')}{2} = \frac{n}{2} - \frac{\#(S')}{2} - \frac{\#(S'')}{2} - \frac{\#(S'''')}{2};$$
$$\#(T_n) \leq n - \#(S'') - \frac{\#(S'''')}{2}.$$

Hence

$$\#(S^* \cup T_n) \leq \frac{n}{2} - \frac{\#(S')}{2} - \frac{\#(S'')}{2} - \frac{\#(S'''')}{2} + \#(S') + \#(S'')$$
$$+ \#(S'''') + n - \#(S'') - \frac{\#(S'''')}{2} \leq \frac{3n}{2} + \frac{\#(S')}{2}. \qquad \square$$

11.7. LEMMA. *Let* $\mathcal{A}_1, \ldots, \mathcal{A}_n$ *be a finite sequence of algebras,* $m \in \mathbb{N}^+$*,* $m < n$*, and* $\bigcup_{k=1}^{m} \mathcal{A}_k \neq \mathfrak{P}(X)$*. For each* $k \in [m+1, n]$ *let there exist* $n + \left[\frac{m}{2}\right]$ *pairwise disjoint sets not belonging to* $\mathcal{A}_k$*. Then*

$$\bigcup_{k=1}^{n} \mathcal{A}_k \neq \mathfrak{P}(X).$$

PROOF. Consider, as usual, the sets of ultrafilters S_m and T_m. Among the numbers $m+1, \ldots, n$, let us take the minimal number (if it exists), say $m+1$, such that there exist $\mathcal{A}_{m+1}$-similar ultrafilters $q_{m+1}, r_{m+1} \in S_m$, and at least one of these ultrafilters belongs to a simple bush of the second kind. Turning this bush over, let us define S_{m+1}, T_{m+1} in the natural way. Among the numbers $m+2, \ldots, n$, let us take the minimal number (if it exists), say $m+2$, such that there exist $\mathcal{A}_{m+2}$-similar ultrafilters $q_{m+2}, r_{m+2} \in S_{m+1}$, and at least one of these ultrafilters belongs to a simple bush of the second kind. Turning this bush over, let us define S_{m+2}, T_{m+2} in the natural way. Assume that, continuing this way, we constructed algebras $\mathcal{A}_{m+1}, \ldots, \mathcal{A}_{m'}$ and cannot continue the construction further. If $m' = n$, the proof is over.[1] If $m' < n$, denote

$$M_0 = \{k \in [m'+1, n] \mid \text{either there exists an } \mathcal{A}_k\text{-special ultrafilter } s_k \in S_{m'},$$
$$\text{or there exist } \mathcal{A}_k\text{-similar ultrafilters } s_k, t_k$$
$$\text{such that } s_k \in S_{m'},\ t_k \notin S_{m'}\}.$$

[1] Note that $m' < 2m$.

Assume that $M_0 = [m'+1, m'']$. For the algebras $\mathcal{A}_1, \dots, \mathcal{A}_{m''}$, let us construct $S_{m''}, T_{m''}$ in the natural way. Let $k_1 \in [m''+1, n]$ and

$$S^*_{k_1} = S_{m''} \cap \ker \mathcal{A}_{k_1}.$$

Clearly, there cannot exist a simple bush of the second kind $\{s,t\}$ of (S_m, T_m) such that $s \in S^*_{k_1}$ which has not been turned over yet. Hence, by Lemma 11.6,

$$\#(S^*_{k_1} \cup T_{m'}) \le m + \left[\frac{m}{2}\right] + \left[\frac{2(m'-m)+1}{2}\right] = m' + \left[\frac{m}{2}\right].$$

Therefore we have

$$\#(S^*_{k_1} \cup T_{m''}) \le m'' + \left[\frac{m}{2}\right].$$

So there exists an ultrafilter $s^1 \in \ker \mathcal{A}_{k_1} \setminus (S_{m''} \cup T_{m''})$. Put

$$M_1 = \{k \in [m''+1, n] \mid s^1 \in \ker \mathcal{A}_k\},$$

and let us construct an $s^1(M_1)$-bush $\{s^1, T^1\}$. If $k_2 \in [m''+1, n] \setminus M_1$, then there exists an ultrafilter

$$s^2 \in \ker \mathcal{A}_{k_2} \setminus (S_{m''} \cup T_{m''} \cup \{s^1, T^1\}).$$

We can now construct an s^2-bush $\{s^2, T^2\}$ in the natural way, and so on. Having constructed the bushes $\{s^i, T^i\}$ (let there be α such bushes), put

$$S_n = S_{m''} \cup \{s^1, \dots, s^\alpha\}, \quad T_n = T_{m''} \cup \bigcup_{i=1}^{\alpha} T^i. \qquad \square$$

11.8. By Lemmas 11.5 and 11.7, Statement 2.17 is true.[2] We have proved that one can construct functions by the inductive process described in 2.18. Is this a good process? On one hand, we have the possibility of constructing many functions from the classes Ψ^n. But on the other hand, none of the functions from Ψ^n_* for $n \in [4,7]$ not marked by a star in the table from 2.13 can be obtained by this process.

11.9. After introducing the notions of functions $\xi(l;k)$ and $\xi^*(k)$ (see 2.14, 2.19), it is natural to introduce the following notion. Let us fix a natural number $l > 1$, and let $k \in [1,l]$ be a positive integer variable. By definition, $\xi^*(l;k)$ *is any function defined on* $[1,l]$, *taking values in* $\mathbb{N}^+$, *and such that*

$$\sum_{k=1}^{l} 2^{-\left[\frac{\xi^*(l;k)+1}{2}\right]} = 1.$$

[2] It is clear that under the conditions of Statement 2.17, the function $\psi_2(m, n+1; k)$ should be used if and only if

$$[m/2] + n + 1 < \min([4n/3] + 1, 2m + \mathfrak{g}(n+1-m)).$$

11.10. Now let us study the classes of functions described in 2.19. Corollaries of the arguments presented in this and the following section will be used, without special reference, in the proof of Statement 2.19. Let $\varphi_1 \in \widehat{\Psi}^m$ be a function defined on $[1, m]$, let $\varphi_2 \in \widehat{\Psi}^p$ be a function defined on $[1, p]$, and let us define the following function from $\widehat{\Psi}^{m+p}$:

$$\varphi(k) = \begin{cases} \varphi_1(k) & \text{if } k \leq m, \\ 2m + \varphi_2(k - m) & \text{if } k \in [m+1, m+p]. \end{cases}$$

Assume that $\varphi_2(k) = \psi_2(\tau, p; k)$ if $k \in [\tau + 1, p]$. Clearly, we always have

$$\left[\frac{m+\tau}{2}\right] + m + p \leq 2m + \left[\frac{\tau}{2}\right] + p.$$

Strict inequality can happen only if at least one of the following two conditions is satisfied:

(a) $m > 1$;

(b) τ is even.

Consider the following function defined on $[1, m + p]$:

$$\varphi'(k) = \begin{cases} \varphi(k) & \text{if } k \leq m + \tau, \\ \left[\frac{m+\tau}{2}\right] + m + p & \text{if } k \in [m + \tau + 1, m + p]. \end{cases}$$

By Lemma 11.7, there exists a function $\varphi''(k)$ on $[1, m+p]$ such that $\varphi''(k) \leq \varphi'(k)$ for all $k \leq m+p$, and $\varphi'' \in \widehat{\Psi}^{m+p}$. Let at least one of the conditions (a), (b) above be satisfied. Since $\varphi''(k) \leq \varphi(k)$ for all $k \in [1, m + \tau]$ and $\varphi''(k) < \varphi(k)$ for all $k \in [m + \tau + 1, m + p]$, we come to a contradiction.

Let us now assume that

$$\varphi_2(p) = \left[\frac{4}{3}(p-1)\right] + 1.$$

Clearly, we always have

$$\left[\frac{4}{3}(m + p - 1)\right] + 1 \leq 2m + \left[\frac{4}{3}(p-1)\right] + 1.$$

Here the equality is possible only for $m = 1$ and $p \equiv 0 \pmod 3$.

Let $p > 1$, and assume that all functions from $\widehat{\Psi}^{p-1}$ as well as from $\widehat{\Psi}^p$ are constructed by means of our inductive process. Due to the above, for φ_2 one of the following properties is true:

1) $\varphi_2(k) = \mathfrak{g}(p-1)$ if $k < p$; here

$$m = 1, \quad \varphi_2(p) = \left[\frac{4}{3}(p-1)\right] + 1, \quad p \equiv 0 \pmod 3;$$

2) $\varphi_2(k) = \mathfrak{g}(p)$ if $k \leq p$;

3) $\varphi_2(k) = \xi^*(p-1; k)$ if $k < p$; here

$$m = 1, \quad \varphi_2(p) = \left[\frac{4}{3}(p-1)\right] + 1, \quad p \equiv 0 \pmod 3;$$

4) $\varphi_2(k) = \xi^*(p; k)$ if $k \leq p$;

5) there exists a natural number $\tau < p - 1$ and a function $\widehat{\varphi}_2 \in \widehat{\Psi}^\tau$ such that

$$\varphi_2(k) = \begin{cases} \widehat{\varphi}_2(k) & \text{if } k \leq \tau, \\ \psi_1(\tau, p-1; k); \end{cases}$$

here

$$m = 1, \quad \varphi_2(p) = \left[\frac{4}{3}(p-1)\right] + 1, \quad p \equiv 0 \pmod 3;$$

6) there exist a natural number $\tau < p$ and a function $\widehat{\varphi}_2 \in \widehat{\Psi}^\tau$ such that

$$\varphi_2(k) = \begin{cases} \widehat{\varphi}_2(k) & \text{if } k \le \tau, \\ \psi_1(\tau, p; k); \end{cases}$$

7) there exist a natural number $\tau < p - 1$ and a function $\widehat{\varphi}_2 \in \widehat{\Psi}^\tau$ such that

$$\varphi_2(k) = \begin{cases} \widehat{\varphi}_2(k) & \text{if } k \le \tau, \\ \psi_2(\tau, p-1; k); \end{cases}$$

here

$$m = 1, \quad \tau \text{ is an odd integer}, \quad \varphi_2(p) = \left[\frac{4}{3}(p-1)\right] + 1, \quad p \equiv 0 \pmod 3;$$

8) there exist a natural number $\tau < p$ and a function $\widehat{\varphi}_2 \in \Psi^\tau$ such that

$$\varphi_2(k) = \begin{cases} \widehat{\varphi}_2(k) & \text{if } k \le \tau, \\ \psi_2(\tau, p; k); \end{cases}$$

here $m = 1$, τ is an odd integer.

In cases 1), 3), 5), 7)

$$\varphi(p+1) = \left[\frac{4}{3}p\right] + 1$$

and $\varphi(k)$ belongs to $\widehat{\Psi}^p$ on $[1, p]$. If 6) holds, then

$$\varphi(k) = \begin{cases} \varphi_1^*(k) & \text{if } k \le m^*, \\ \psi_1(m^*, m+p; k), \end{cases}$$

where $\varphi_1^* \in \widehat{\Psi}^{m^*}$ and $m^* = m + \tau$. If 8) holds, then

$$\varphi(k) = \begin{cases} \varphi_1^*(k) & \text{if } k \le m^*, \\ \psi_2(m^*, p+1; k), \end{cases}$$

where $\varphi_1^* \in \widehat{\Psi}^{m^*}$ and $m^* = \tau + 1$.

11.11. Consider the natural numbers $m^1 < m^2 < l^1 < l^2$, and let $\psi_{l^2} \in \widehat{\Psi}^{l^2}$ be a function on $[1, l^2]$. Assume that

(i) the function ψ_{l^2}, regarded as defined on $[1, m^1]$, $[1, m^2]$, $[1, l^1]$, belongs respectively to $\widehat{\Psi}^{m^1}, \widehat{\Psi}^{m^2}, \widehat{\Psi}^{l^1}$, and is denoted respectively by $\psi_{m^1}, \psi_{m^2}, \psi_{l^1}$;

(ii)

$$\psi_{l^1}(k) = \begin{cases} \psi_{m^1}(k) & \text{if } k \le m^1, \\ 2m^1 + \psi'(k - m^1) & \text{if } k \in [m^1 + 1, l^1], \end{cases}$$

where $\psi' \in \widehat{\Psi}^{l^1 - m^1}$;

(iii)

$$\psi_{l^2}(k) = \begin{cases} \psi_{m^2}(k) & \text{if } k \le m^2, \\ 2m^2 + \psi''(k - m^2) & \text{if } k \in [m^2 + 1, l^2], \end{cases}$$

where $\psi'' \in \widehat{\Psi}^{l^2 - m^2}$.

Consider the following function on $[1, m^2 - m^1]$: $\varepsilon_1(k) = \psi_{m^2}(m^1 + k) - 2m^1$, and the following function on $[1, l^1 - m^2]$: $\varepsilon_2(k) = \psi_{l^1}(m^2 + k) - 2m^2$. There exists a function $\varepsilon_2'(k)$, defined on $[1, l^1 - m^2]$, such that $\varepsilon_2'(k) \leq \varepsilon_2(k)$ for all $k \in [1, l^1 - m^2]$, and $\varepsilon_2' \in \widehat{\Psi}^{l^1 - m^2}$. Since the function

$$\psi'(k) = \begin{cases} \varepsilon_1(k) & \text{if } k \in [1, m^2 - m^1], \\ 2(m^2 - m^1) + \varepsilon_2(k + m^1 - m^2) & \text{if } k \in [m^2 - m^1 + 1, l^1 - m^1] \end{cases}$$

belongs to $\widehat{\Psi}^{l^1 - m^1}$, we have $\varepsilon_2(k) \equiv \varepsilon_2'(k)$. Now let us state a corollary necessary to us:

$$\psi_{l^1}(k) = \begin{cases} \psi_{m^2}(k) & \text{if } k \leq m^2, \\ 2m^2 + \varepsilon_2(k - m^2) & \text{if } k \in [m^2 + 1, l^1], \end{cases}$$

where $\varepsilon_2 \in \widehat{\Psi}^{l^1 - m^2}$.

11.12. PROOF OF STATEMENT 2.19.[3] Let $\psi \in \widehat{\Psi}_*$. The function ψ regarded as defined on $[1, n]$ will be denoted by ψ_n. Let us fix $n > 2$. For each k such that $\psi_n(k) > 1$, let us find a natural number $\tau_k \geq n$ such that the following function, defined on $[1, \tau_k]$,

$$\widetilde{\psi}_k(l) = \begin{cases} \psi(l) & \text{if } l \neq k, \\ \psi(k) - 1 & \text{if } l = k, \end{cases}$$

does not belong to Ψ^{τ_k}. Put $\mathfrak{z}^* = \max\{\tau_k\}$. Consider a function $\eta^* \in \widehat{\Psi}^{\mathfrak{z}^*}$ on $[1, \mathfrak{z}^*]$ such that $\eta^*(k) \leq \psi(k)$ for all $k \leq \mathfrak{z}^*$. Clearly, $\psi_n(k) = \eta^*(k)$ for all $k \leq n$. Therefore there exists the minimal number $\mathfrak{z} \geq n$ and a function $\eta \in \widehat{\Psi}^{\mathfrak{z}}$ such that $\eta(k) \leq \psi(k)$ for all $k \leq \mathfrak{z}$ and $\psi_n(k) = \eta(k)$ for all $k \leq n$. If $\mathfrak{z} = n$, then $\psi_n \in \widehat{\Psi}^n$. Let $\mathfrak{z} > n$. Assume that η is constructed by means of our inductive process. Let us also assume that ψ_n is not a constant. There are the following possible cases:

1) η is a function of the type of $\xi^*(\mathfrak{z}; k)$, so that ψ_n is a function of type $\xi(n; k)$;
2) there exists a natural number $m < n$ such that $\psi_m \in \widehat{\Psi}^m$ and
$$\eta(k) = \psi_1(m, \mathfrak{z}; k) \quad \text{on} \quad [m + 1, \mathfrak{z}];$$
3) there exists a natural number $m < n$ such that $\psi_m \in \widehat{\Psi}^m$ and
$$\eta(k) = \psi_2(m, \mathfrak{z}; k) \quad \text{on} \quad [m + 1, \mathfrak{z}].$$

This implies that one of the following two cases holds:

I. There exists a countable sequence of natural numbers $p_1 < \cdots < p_r < \cdots$ such that $\psi_{p_r} \in \widehat{\Psi}^{p_r}$.
II. There exists a nonnegative integer m_0 and a function $\psi'(k)$ of the type of $\xi^*(k)$ such that
$$\psi(k) = 2m_0 + \psi'(k - m_0) \quad \text{if } k > m_0;$$
if $m_0 > 0$, then $\psi_{m_0} \in \widehat{\Psi}^{m_0}$.

Case I. We will assume that $p_1 > n_0$ (see 2.19); p_2 is the minimal number among all numbers p greater than p_1 such that $\psi_p \in \widehat{\Psi}^p$; p_3 is the minimal number among all numbers p greater than p_2 such that $\psi_p \in \widehat{\Psi}^p$, etc. Assume that

$$\psi(k) = \psi_2(m, p_r; k) \quad \text{if } k \in [m + 1, p_r] \text{ and } r > 1.$$

[3] Recall that in the proof we use, without explicit reference, corollaries of the arguments from the two previous sections.

It is easy to show that $m = p_{r-1}$. Define the subset $\mathbb{N}^1 \subset \mathbb{N}^+$ as follows: $r \in \mathbb{N}^1$ if and only if

$$\psi(p_r) = \left[\frac{4}{3}(p_r - 1)\right] + 1.$$

Define the subset $\mathbb{N}^2 \subset \mathbb{N}^+$ as follows: $r \in \mathbb{N}^2$ if and only if

$$\psi(k) = \psi_2(m, p_r; k)$$

for $k \in [m+1, p_r]$. If $\#(\mathbb{N}^+ \setminus (\mathbb{N}^1 \cup \mathbb{N}^2)) < \aleph_0$, then ψ satisfies condition (1) of our statement. Assume the opposite. Then there exists the maximal countable increasing sequence

$$r_1, \dots, r_l, \dots \subset \mathbb{N}^+ \setminus (\mathbb{N}^1 \cup \mathbb{N}^2),$$

and for each l we have

$$\psi(k) = \psi_1(m_l, p_{r_l}; k)$$

for $k \in [m_l + 1, p_{r_l}]$ and $m_l = p_{v(l)}$. Among all such numbers $v(l)$ for a fixed l, let us take the maximal number $v'(l)$. Among all the numbers $v'(l)$, let us take the minimal number v_1. Put $\pi_1 = p_{v_1}$. On $[\pi_0 + 1, \pi_1]$, where $\pi_0 = 0$, we have $\psi(k) = \psi_1(\pi_0, \pi_1; k)$. Among all the numbers l for which $v'(l) = v_1$, let us take the maximal number l_2. Put $\pi_2 = p_{r_{l_2}}$. On $[\pi_1 + 1, \pi_2]$ we have $\psi(k) = \psi_1(\pi_1, \pi_2; k)$. There are two possible cases:

a) there exists an $l > l_2$ such that $v'(l) = r_{l_2}$;
b) such an l does not exist.

In case a), let us take the maximal number $l_3 > l_2$ such that $v'(l_3) = r_{l_2}$. Put $\pi_3 = p_{r_{l_3}}$. On $[\pi_2 + 1, \pi_3]$ we have $\psi(k) = \psi_1(\pi_2, \pi_3; k)$.

In case b), let us take the minimal natural number $l_2^* > r_{l_2}$ such that there exists $l > l_2$ with $v'(l) = l_2^*$. Among all the numbers $l > l_2$ for which $v'(l) = l_2^*$, let us take the maximal number l_3. Consider the numbers $p_{r_{l_2}}, p_{r_{l_2}+1}, \dots, p_{l_2^*}$ and put

$$\pi_3 = p_{r_{l_2}+1}, \quad \pi_4 = p_{r_{l_2}+2}, \quad \dots, \pi_\gamma = p_{l_2^*}, \quad \pi_{\gamma+1} = p_{r_{l_3}}.$$

If $i \in [3, \gamma]$ and $\pi_i - \pi_{i-1} = 1$, then

$$\psi(\pi_i) = \left[\frac{4}{3}\pi_{i-1}\right] + 1.$$

If $i \in [3, \gamma]$ and $\pi_i - \pi_{i-1} > 1$, then

$$\psi(k) = \psi_2(\pi_{i-1}, \pi_i; k) \quad \text{for all } k \in [\pi_{i-1} + 1, \pi_i].$$

On $[\pi_\gamma + 1, \pi_{\gamma+1}]$ we have

$$\psi(k) = \psi_1(\pi_\gamma, \pi_{\gamma+1}; k).$$

Continuing our process, we will construct a countable sequence of numbers π_i. □

11.13. For the proof of Theorem 2.20 one uses the following lemma, which is a refinement of Lemma 11.6. In the proof of this lemma, one uses the proof of Lemma 11.6 and of the statement in 4.11.

LEMMA. *Let $\mathcal{A}_1, \dots, \mathcal{A}_n$ be a finite sequence of algebras, $\bigcup_{k=1}^n \mathcal{A}_k \neq \mathfrak{P}(X)$ (i.e., there exist corresponding sets of ultrafilters S_n and T_n), $S^* \subset S_n$, and let*

1) *$S' = \{s \in S^* \mid s$ belongs to a simple bush of the second kind$\}$;*
2) *$\alpha = \|K\|$, and K is a bush of (S_n, T_n);*

3) $\beta = \#(\{s \in S^* \setminus K \mid$ *there exists* k *such that* $s_k = s$ *is an* $\mathcal{A}_k$*-special ultrafilter*$\})$;

4) $\mathbb{N}_* = \{k \in [1, n] \mid \mathcal{A}_k$*-special ultrafilter* $s_k \in S_n \setminus (S^* \cup K)\}$.

Then

$$\#(S^* \cup T_n) \leq n + \left[\frac{n}{2}\right] + \frac{\#(S') - \alpha - \beta - 3 \cdot \#(\mathbb{N}_*) + 3}{2}.$$[4]

PROOF. Put

$$S_* = \{s \in S_n \setminus (S^* \cup K) \mid \{s\} \text{ is a bush of the first kind}\}.$$

Consider the algebras $\mathcal{A}_k$, where $s_k \notin K$ and $k \notin \mathbb{N}_*$, and consider the sets $S_n \setminus (K \cup S_*)$, $T_n \setminus K$, $S^* \setminus K = S_0^*$ corresponding to the algebras in question. Let us divide S_0^* into four pairwise disjoint sets $S_0', S_0'', S_0''', S_0''''$ as in the proof of Lemma 11.6 and obtain:

$$\#(S_0''') \leq \frac{n}{2} - \frac{\alpha + \#(\mathbb{N}_*)}{2} - \frac{\#(S_0')}{2} - \frac{\beta}{2} - \frac{\#(S_0'''')}{2},$$

$$\#(T_n \setminus K) \leq n - \alpha - \beta - \#(\mathbb{N}_*) - \frac{\#(S_0'''')}{2}.$$

Hence

$$\begin{aligned}\#((S^* \cup T_n) \setminus K) &\leq \frac{n}{2} - \frac{\alpha + \#(\mathbb{N}_*)}{2} - \frac{\#(S_0')}{2} - \frac{\beta}{2} - \frac{\#(S_0'''')}{2} + \#(S_0') + \beta \\ &\quad + \#(S_0'''') + n - \alpha - \beta - \#(\mathbb{N}_*) - \frac{\#(S_0'''')}{2} \\ &= \frac{3n}{2} - \frac{3(\alpha + \#(\mathbb{N}_*))}{2} + \frac{\#(S_0')}{2} - \frac{\beta}{2} \leq \frac{3n}{2} + \frac{\#(S')}{2} - \frac{3(\alpha + \#(\mathbb{N}_*))}{2} - \frac{\beta}{2}.\end{aligned}$$

Therefore,

$$\begin{aligned}\#(S^* \cup T_n) &\leq \frac{3n}{2} + \frac{\#(S')}{2} - \frac{3(\alpha + \#(\mathbb{N}_*))}{2} - \frac{\beta}{2} + \#(K) \\ &\leq \frac{3n}{2} + \frac{\#(S')}{2} - \frac{3(\alpha + \#(\mathbb{N}_*))}{2} - \frac{\beta}{2} + \alpha + 1.\end{aligned}$$ □

11.14. ABOUT THE PROOF OF THEOREM 2.20. Because of the methods that we have worked out above, the proof of this theorem is not difficult, but technically it is somewhat cumbersome; therefore, it is not sensible to present it here. As in the proofs of Theorem 2.7 and 2.8, we should consider three cases. Case I is almost obvious. Case II is not too difficult. More difficult is Case III.

11.15. REMARK. Here it is appropriate to note that for each function $\psi \in \Psi$ which we can construct "concretely" using the described methods, the following holds: if $\mathcal{A}_1, \ldots, \mathcal{A}_k, \ldots$ is a countable sequence of almost σ-algebras and for each k there exist $\psi(k)$ pairwise disjoint sets not belonging to $\mathcal{A}_k$, then $\bigcup \mathcal{A}_k \neq \mathfrak{P}(X)$.

---

[4]Formally our lemma is a refinement of Lemma 11.6 if $\alpha + \beta + 3 \cdot \#(\mathbb{N}_*) > 2$.

11.16. The time has come to look back on our exposition. We have finished studying the "general question" stated in 1.15, in which the conditions imposed on the algebra $\mathcal{A}_k$ are autonomous, unlike the conditions on the algebras from Theorems 2.21, 2.23, and 2.24. It remains to give an example that justifies what we asserted in footnote 1.7.

STATEMENT. *Let* $\mathcal{A}_1, \dots, \mathcal{A}_5$ *be algebras such that* $\#(\ker \mathcal{A}_k) = 3$ *for all* $k \in [1,5]$. *Then*

$$\bigcup_{k=1}^{5} \mathcal{A}_k \neq \mathfrak{P}(X).$$

PROOF. First assume that $X \in \mathcal{A}_k$ for all $k \in [1,5]$. Let Θ be a set of ultrafilters, $\#(\Theta) = 14$, and $\ker \mathcal{A}_k \subset \Theta$ for all $k \in [1,5]$. Let $\Pi \subset \Theta$ and $\#(\Pi) = 7$. For each $k \in [1,5]$ there exist $2 \cdot \binom{11}{4}$ sets Π such that for each of them either $\ker \mathcal{A}_k \subset \Pi$ or $\ker \mathcal{A}_k \cap \Pi = \varnothing$. The number of all sets Π equals $\binom{14}{7}$. But since $\binom{14}{7} > 10 \cdot \binom{11}{4}$, we have $\bigcup_{k=1}^{5} \mathcal{A}_k \neq \mathfrak{P}(X)$. Now if $\#\left(\bigcup_{k=1}^{5} \ker \mathcal{A}_k\right) = 15$, then, obviously, $\bigcup_{k=1}^{5} \mathcal{A}_k \neq \mathfrak{P}(X)$.

Let us now assume that s is an $\mathcal{A}_1$-special ultrafilter. Putting $s^1 = s$, we can start constructing s-bushes, and take into account Lemma 9.15. □

It remains to recall that the function $\varphi(k) \equiv 3$ on $[1,5]$ does not belong to Ψ^5.

CHAPTER 12

Proof of Theorems 2.21(1,3), 2.24[1]

12.1. Proof of Theorem 2.21(1). Let us assign to each set U_i^k either an $\mathcal{A}_k$-special ultrafilter $s_i^k \ni U_i^k$, or $\mathcal{A}_k$-similar ultrafilters s_i^k, t_i^k; $U_i^k \in s_i^k$, $U_i^k \notin t_i^k$. Without loss of generality, we will consider the ultrafilters t_1^1, t_1^2. The sets $U_1^k, \dots, U_{m_k}^k$ $(k > 2)$ are said to be *neighboring*. In the course of the proof each set U_i^k will be marked either by 0 or by 1, and it will be clear that

$$\bigcup\{U_i^k \mid U_i^k \text{ is marked by } 1\} = U \notin \bigcup \mathcal{A}_k.$$

For each $k > 2$ let us choose, if possible, an $\mathcal{A}_k$-special ultrafilter among the ultrafilters $s_1^k, \dots, s_{m_k}^k$, say $s_{i_0}^k$, and let us mark $U_{i_0}^k$ by 1; let us mark each of the other sets neighboring with $U_{i_0}^k$ by 0.

A finite or countable sequence of sets $U_{i_1}^{k_1}, \dots, U_{i_n}^{k_n}, \dots$ is called a *bunch* if for each n one considers the ultrafilter $t_{i_n}^{k_n}$ and the following conditions hold:

(1) if $U_{i_n}^{k_n}$ is not the last element, then $t_{i_n}^{k_n} \ni U_{i'_{n+1}}^{k_{n+1}}$; moreover, if $k_{n+1} > 2$, then $i_{n+1} \neq i'_{n+1}$;
(2) all the sets $U_{i_n}^{k_n}$ are pairwise distinct;
(3) among the sets $U_{i_n}^{k_n}$ there are no pairs of neighboring sets;
(4) among the sets $U_{i_n}^{k_n}$ and $U_{i'_{n+1}}^{k_{n+1}}$ there are no marked sets;
(5) the maximality property: if $U_{i_l}^{k_l}$ is the last element, then there does not exist a set $U_{i_{l+1}}^{k_{l+1}}$ such that the sets $U_{i_n}^{k_n}$, where $n \in [1, l+1]$, satisfy conditions (1)–(4).

First let us construct a bunch whose first element is $U_{i_1}^{k_1} = U_1^1$.

Case I. *One can construct an infinite bunch whose first element is* $U_{i_1}^{k_1}$. If $U_{i_n}^{k_n} \neq U_1^2$, then let us mark $U_{i_n}^{k_n}$ by 1 and mark $U_{i'_{n+1}}^{k_{n+1}}$ by 0. If $U_{i_n}^{k_n} = U_1^2$, then let us mark $U_{i_n}^{k_n}$ by 0 and mark $U_{i'_{n+1}}^{k_{n+1}}$ by 1.

If it is impossible to construct an infinite bunch, then let us construct a finite bunch, whose last element is $U_{i_l}^{k_l}$. If

$$U_{i_n}^{k_n} \notin \{U_1^2, U_{i_{l-1}}^{k_{l-1}}, U_{i_l}^{k_l}\}$$

(this case and, in general, elements $U_{i_n}^{k_n}$ distinct from $U_{i_l}^{k_l}$, are considered only when $l > 1$), then let us mark $U_{i_n}^{k_n}$ by 1 and mark $U_{i'_{n+1}}^{k_{n+1}}$ by 0. If $U_{i_n}^{k_n} = U_1^2 \neq U_{i_l}^{k_l}$, then let us mark $U_{i_n}^{k_n}$ by 0 and mark $U_{i'_{n+1}}^{k_{n+1}}$ by 1. If $U_1^2 \notin \{U_{i_{l-1}}^{k_{l-1}}, U_{i_l}^{k_l}\}$, then let us mark $U_{i_{l-1}}^{k_{l-1}}$ by 1 and mark $U_{i'_l}^{k_l}$ by 0.

[1] Recall what was said in 2.22. Theorem 2.21(2) is already proved in 1.13. Also, note that the arguments of this chapter are ideologically related to the arguments in 3.16 and 3.18.

Case II. $t_{i_l}^{k_l} \ni U_\beta^\alpha$. If U_β^α is marked by 1, then let us mark $U_{i_l}^{k_l}$ by 0. Otherwise let us mark $U_{i_l}^{k_l}$ by 1. Let $U_{i_l}^{k_l} = U_1^2$. Then, if U_1^2 was marked by 0 (respectively, by 1), then let us mark $U_{i_{l-1}}^{k_{l-1}}$ by 1 (respectively, by 0).

Case III. $t_{i_l}^{k_l} \not\ni U_\beta^\alpha$. If there exists a sequence of sets marked by 1,

$$U_{\beta_1}^{\alpha_1}, \ldots, U_{\beta_n}^{\alpha_n}, \ldots \quad (\alpha_1 < \alpha_2 < \cdots < \alpha_n < \cdots) \quad \text{and} \quad t_{i_l}^{k_l} \ni \bigcup_n U_{\beta_n}^{\alpha_n},$$

then let us mark $U_{i_l}^{k_l}$ by 0. Otherwise let us mark $U_{i_l}^{k_l}$ by 1. If there exists a sequence of nonmarked sets

$$U_{\delta_1}^{\gamma_1}, \ldots, U_{\delta_n}^{\gamma_n}, \ldots \quad (\gamma_1 < \gamma_2 < \cdots < \gamma_n < \cdots) \quad \text{and} \quad t_{i_l}^{k_l} \ni \bigcup_n U_{\delta_n}^{\gamma_n},$$

then let us assume that for each $U_{\delta_n}^{\gamma_n}$ there exist two other nonmarked sets neighboring with it. (Clearly, if there exists one sequence of sets $U_{\delta_n}^{\gamma_n}$, then there exist many such sequences; we consider only one such sequence.) Let us mark each of the sets $U_{\delta_n}^{\gamma_n}$ by 0. If $U_{i_l}^{k_l} = U_1^2$, then we argue in the same way as in Case II.

In all cases, if a nonmarked set U_i^k is neighboring with some set of the constructed bunch, then let us mark U_i^k by 0.[2] If all the sets U_i^k have been marked, then $U \notin \bigcup \mathcal{A}_k$. If all the sets U_i^k have not been marked, then let us take the minimal number $k_* \in \mathbb{N}^+$ such that among the sets $U_i^{k_*}$ some are not marked. Further, let us construct a bunch whose first element is some nonmarked set $U_i^{k_*}$. If $k_* = 2$ and U_1^2 is not the only element of the bunch, then let us mark U_1^2 by 1. In general, if $U_{i_n}^{k_n}$ is not the last element of the new bunch, then let us mark $U_{i_n}^{k_n}$ by 1, and let us mark $U_{i'_{n+1}}^{k_{n+1}}$ by 0, etc. If all the sets U_i^k have not been marked, then let us construct one more bunch, and so on. □

12.2. Remark. Assume that in the proof of Theorem 2.21(1) always either $t_{i_l}^{k_l} \ni U_\beta^\alpha$, or $t_{i_l}^{k_l} \not\ni \bigcup_{\alpha,\beta} U_\beta^\alpha$. Then the statement of Theorem 2.21(1) is true if one puts $m_k = 2$ for all $k > 2$.

12.3. Remark. Theorem 2.24 now follows by Remark 12.2 and Lemma 7.4.

12.4. Example. Let $\#(X) = \aleph_0$, $n \in \mathbb{N}^+$, and

$$X = \begin{pmatrix} a_1^1 & \ldots & a_n^1 \\ \ldots & \ldots & \ldots \\ a_1^k & \ldots & a_n^k \\ \ldots & \ldots & \ldots \end{pmatrix}$$

(the points a_i^k are pairwise distinct). For each $i \in [1, n]$ consider the set $X_i = \{a_i^1, \ldots, a_i^k, \ldots\}$ and a point $a_i \in \beta X_i \setminus X_i$. There exists a natural similarity relation between the points of sets $\beta X_1, \ldots, \beta X_n$. Assume that all points $a_1, \ldots, a_n$ are similar to one another. For each k let us define an ω-saturated algebra $\mathcal{A}_k$:

$$\ker \mathcal{A}_k = \bigcup_{i=1}^n \{a_i^k, a_i\},$$

[2] After this procedure we can assert that if $U_{i_l}^{k_l}$ is the last element of the constructed bunch and it is marked by 1 and $t_{i_l}^{k_l} \ni U_\beta^\alpha$, then U_β^α is marked by 0.

and assume there are only n pairs of $\mathcal{A}_k$-similar ultrafilters, $(a_i^k, a_i)_{i\leq n}$. Clearly,

$$\bigcup \mathcal{A}_k = \mathfrak{P}(X).$$ [3]

12.5. REMARK. For the proof of Theorem 2.21(3) one must slightly modify Example 12.4.

12.6. REMARK. If in the assumptions of Theorem 2.24 one does not impose any restrictions on the algebras $\mathcal{A}_k$, such as σ-additivity or almost σ-additivity, then, clearly, one can assert only the following: for any real number $\rho < 1$ there exists a subset $\mathbb{N}^* \subset \mathbb{N}^+$ such that $p_a(\mathbb{N}^*) > \rho$ and $\bigcup_{k\in\mathbb{N}^*} \mathcal{A}_k \neq \mathfrak{P}(X)$.

[3] Clearly, $\mathcal{A}_k$ is not an almost σ-algebra.

CHAPTER 13

The Inverse Problem

13.1. DEFINITION. Consider an at most countable family of algebras $\mathfrak{A}$. Let us define the *rank* of $\mathfrak{A}$, denoted by $\mathfrak{r}(\mathfrak{A})$, as follows. If $\bigcup \mathfrak{A} \neq \mathfrak{P}(X)$, then $\mathfrak{r}(\mathfrak{A}) = 0$. If $\bigcup \mathfrak{A} = \mathfrak{P}(X)$, then

$$\mathfrak{r}(\mathfrak{A}) = \min\left\{\#(\mathfrak{B}) \mid \mathfrak{B} \subset \mathfrak{A} \text{ and } \bigcup \mathfrak{B} = \mathfrak{P}(X)\right\}.$$

13.2. Obviously, $\mathfrak{r}(\mathfrak{A}) = 1$ if and only if $\mathfrak{A}$ contains the algebra $\mathfrak{P}(X)$. Clearly, $\mathfrak{r}(\mathfrak{A})$ can be equal to 0, $\aleph_0$, or any natural number except 2 (see Theorem 2.2(2)). We are interested in the following question: If $2 < \mathfrak{r}(\mathfrak{A}) \leq \aleph_0$, what can one say about $\mathfrak{A}$?

13.3. THEOREM. *Let $2 < \mathfrak{r}(\mathfrak{A}) < \aleph_0$. Then there exists a family $\mathfrak{B}$ of ω-saturated algebras such that $\mathfrak{B} \subset \mathfrak{A}$ and $\#(\mathfrak{B}) = \mathfrak{r}(\mathfrak{B}) = \mathfrak{r}(\mathfrak{A})$.*

PROOF. By definition, there exists a family $\mathfrak{B} \subset \mathfrak{A}$ such that

$$\#(\mathfrak{B}) = \mathfrak{r}(\mathfrak{B}) = \mathfrak{r}(\mathfrak{A}).$$

Let $\mathfrak{B} = \{\mathcal{A}_1, \dots, \mathcal{A}_n\}$. Assume that the algebra $\mathcal{A}_n$ is not ω-saturated. Then $\bigcup_{i=1}^{n-1} \mathcal{A}_i \neq \mathfrak{P}(X)$, and there exist $\aleph_0$ pairwise disjoint sets not belonging to $\mathcal{A}_n$. By Theorem 2.1, $\bigcup_{i=1}^{n} \mathcal{A}_i \neq \mathfrak{P}(X)$ (this also follows from Theorem 3.2, whose proof is easier than the proof of Theorem 2.1). But since $\mathfrak{r}(\mathfrak{B}) = \mathfrak{r}(\mathfrak{A})$, we have $\bigcup_{i=1}^{n} \mathcal{A}_i = \mathfrak{P}(X)$, a contradiction. □

13.4. THEOREM. *Let $\mathfrak{r}(\mathfrak{A}) = \aleph_0$, and let $\mathfrak{A}$ contain only almost σ-algebras. Then $\mathfrak{A}$ contains $\aleph_0$ ω-saturated algebras and at least one strictly simple algebra.*

PROOF. By Theorem 5.17, let us assume that all algebras in $\mathfrak{A}$ are simple. Let

$$\mathfrak{A} = \{\mathcal{A}_1, \dots, \mathcal{A}_k, \dots\}.$$

Assume that $\mathcal{A}_1, \dots, \mathcal{A}_{k_0}$ are ω-saturated algebras, and if $k > k_0$, then $\mathcal{A}_k$ is a strictly simple algebra. It is rather easy to construct a set W and an infinite matrix of pairwise disjoint sets

$$\begin{pmatrix} U_1^1 & & \\ U_1^2 & U_2^2 & \\ \dots & \dots & \dots \\ U_1^k & \dots & U_k^k \\ \dots & \dots & \dots \end{pmatrix}$$

such that $W \notin \mathcal{A}_m$ if $m \leq k_0$, $U_i^k \notin \mathcal{A}_{k_0+k}$, $W \cap \bigcup_{k,i} U_i^k = \varnothing$, and

$$\overline{\bigcup_{k,i} U_i^k} \cap \ker \mathcal{A}_m = \varnothing \text{ if } m \leq k_0.$$

By Theorem 5.2, we have

$$\bigcup_{k=1}^{\infty} \mathcal{A}_k \neq \mathfrak{P}(X).$$

But this contradicts the fact that $\mathfrak{r}(\mathfrak{A}) = \aleph_0$. Hence $\mathfrak{A}$ contains not a finite but an infinite number of ω-saturated algebras. Assume that all algebras in $\mathfrak{A}$ are ω-saturated. Then, by Lemma 7.12, $\bigcup \mathfrak{A} \neq \mathfrak{P}(X)$. This contradiction implies that $\mathfrak{A}$ contains at least one strictly simple algebra. □

13.5. EXAMPLE. A family of σ-algebras $\mathfrak{A} = \{\mathcal{A}_0\} \cup \mathfrak{A}'$, $\mathfrak{A}' = \{\mathcal{A}^j_{n,i}\}_{j,n,i}$; the algebra $\mathcal{A}_0$ is strictly simple, all the algebras $\mathcal{A}^j_{n,i}$ are ω-saturated, $\mathfrak{r}(\mathfrak{A}) = \aleph_0$, $\mathfrak{r}(\mathfrak{A}') = 0$.

CONSTRUCTION. All algebras in the family $\mathfrak{A}$ contain X. Consider an infinite matrix of pairwise distinct irregular points of βX:

$$\begin{pmatrix} a_1^1 & & \\ \dots & \dots & \dots \\ a_1^n & \dots & a_{2^n}^n \\ \dots & \dots & \dots \end{pmatrix}.$$

Let us assign to each point a_i^n two algebras $\mathcal{A}^1_{n,i}$ and $\mathcal{A}^2_{n,i}$:

$$\ker \mathcal{A}^1_{n,i} = \{a_i^n, a_{2i-1}^{n+1}\}, \quad \ker \mathcal{A}^2_{n,i} = \{a_i^n, a_{2i}^{n+1}\}.$$

The family $\mathfrak{A}'$ consists of all these algebras. Now let us define the algebra $\mathcal{A}_0$:

$$\operatorname{sp} \mathcal{A}_0 = \{a_i^n \mid n \text{ is odd}\},$$

and any two distinct ultrafilters from $\ker \mathcal{A}_0$ are $\mathcal{A}_0$-similar. For $Q \notin \bigcup \mathcal{A}^j_{n,i}$ it is necessary and sufficient that one of the following two statements be true:

(1) $a_i^n \ni Q$ for all odd n, and $a_i^n \not\ni Q$ for all even n;
(2) $a_i^n \not\ni Q$ for all odd n, and $a_i^n \ni Q$ for all even n.

But such a Q belongs to $\mathcal{A}_0$. So we have shown that $\mathfrak{r}(\mathfrak{A}') = 0$ and $\mathfrak{r}(\mathfrak{A}) \neq 0$. But

$$\mathcal{A}_0 \cup \bigcup_{n \leq k} \mathcal{A}^j_{n,i} \neq \mathfrak{P}(X) \quad \text{for any } k \in \mathbb{N}^+.$$

Hence $\mathfrak{r}(\mathfrak{A}) = \aleph_0$. □

13.6. REMARK. In Example 12.4, a family of ω-saturated algebras $\mathfrak{A}$ was constructed such that $\mathfrak{r}(\mathfrak{A}) = \aleph_0$. But there $\mathfrak{A}$ does not contain almost σ-algebras.

CHAPTER 14

Finite Sequences of Algebras (3). Proof of Theorems 2.27, 2.31, 2.36, 2.38

14.1. Let $\mathcal{A}_1, \ldots, \mathcal{A}_k, \ldots$ be an at most countable sequence of algebras, and let there exist pairwise disjoint sets $U_1, \ldots, U_k, \ldots$ such that if $U_k \subset Q \subset \bigcup U_i$, then $Q \notin \mathcal{A}_k$. Assume that among the ultrafilters from $\overline{U}_k$ there are no $\mathcal{A}_k$-special ultrafilters. Then, by the compactness of $\overline{U}_k$, there exists a $D \supset U_k$ such that $D \in \mathcal{A}_k$. Clearly, $D' = D \setminus \bigcup U_i \neq \varnothing$. If there do not exist $\mathcal{A}_k$-similar ultrafilters s_k, t_k such that $s_k \ni U_k$ and $t_k \ni D'$, then, by the compactness of $\overline{U}_k$ and $\overline{D'}$, there exists a $D'' \supset D'$ such that $D'' \cap U_k = \varnothing$, $D'' \in \mathcal{A}_k$. By the assumptions, $D \setminus D'' \notin \mathcal{A}_k$ since $U_k \subset D \setminus D'' \subset \bigcup U_i$. But $D \setminus D'' \in \mathcal{A}_k$, since $D, D'' \in \mathcal{A}_k$. The contradiction thus obtained implies that there exist $\mathcal{A}_k$-similar ultrafilters s_k, t_k such that $s_k \ni U_k$ and $t_k \ni D'$.

14.2. THEOREM. *Let $\mathcal{A}_1, \ldots, \mathcal{A}_n, \mathcal{A}_{n+1}$ be a sequence of algebras, $n \in \mathbb{N}^+$, and let there exist pairwise disjoint sets $U_1, \ldots, U_n$ such that if*

$$U_k \subset Q \subset \bigcup_{i=1}^{n} U_i \quad \text{and} \quad k \leq n,$$

then $Q \notin \mathcal{A}_k$. Assume also that there exist more than $5n/2$ pairwise disjoint sets not belonging to $\mathcal{A}_{n+1}$. Then there exist pairwise disjoint sets $U'_1, \ldots, U'_n, U'_{n+1}$ such that if

$$U'_k \subset Q' \subset \bigcup_{i=1}^{n+1} U'_i \quad \text{and} \quad k \leq n+1,$$

then $Q' \notin \mathcal{A}_k$.

PROOF. For each $k \in [1, n]$ consider either an $\mathcal{A}_k$-special ultrafilter $s_k \ni U_k$, or $\mathcal{A}_k$-similar ultrafilters s_k, t_k; $s_k \ni U_k$, $t_k \not\ni \bigcup_{i=1}^{n} U_i$. Put $S_n = \{s_1, \ldots, s_n\}$. Denote by T_n the set of all t_k. Put

$$K = \{s_i \in S_n \mid \text{either } s_i \text{ is an } \mathcal{A}_i\text{-special ultrafilter, or there exists } t_j = t_i \text{ and } i \neq j\}.$$

Clearly,

$$\#(S_n \cup T_n) \leq \frac{3 \cdot \#(K)}{2} + 2(n - \#(K)).$$

Since $\#(\ker \mathcal{A}_{n+1}) > 5n/2$, there exists an $L \subset \ker \mathcal{A}_{n+1}$ such that $L \cap (S_n \cup T_n) = \varnothing$ and

$$\#(L) > \frac{5n}{2} - \frac{3 \cdot \#(K)}{2} - 2n + 2 \cdot \#(K) = \frac{n + \#(K)}{2} \geq \#(K).$$

If L contains an $\mathcal{A}_{n+1}$-special ultrafilter, we denote this ultrafilter by s_{n+1}. If each ultrafilter from L has an $\mathcal{A}_{n+1}$-similar ultrafilter in K, then, by an argument from 3.8, L contains $\mathcal{A}_{n+1}$-similar ultrafilters q, r (recall that $\#(L) > \#(K)$); put $s_{n+1} = q$, $t_{n+1} = r$. It remains to consider the case in which there exist $\mathcal{A}_{n+1}$-similar ultrafilters q, r such that $q \in L$ and $r \notin K$. Put $s_{n+1} = q$, $t_{n+1} = r$; if $r = s_{i_0}$, then put $s_{i_0} = t_{i_0}$, $t_{i_0} = r$. Summarizing, we have constructed a sequence of pairwise distinct ultrafilters $S_{n+1} = \{s_1, \dots, s_n, s_{n+1}\}$. If the ultrafilter s_k ($k \le n+1$) is not $\mathcal{A}_k$-special, then one considers an ultrafilter t_k which is $\mathcal{A}_k$-similar to s_k. Denote the set of all t_k ($k \le n+1$) by T_{n+1}; $S_{n+1} \cap T_{n+1} = \varnothing$. This obviously implies the existence of required sets $U_1', \dots, U_n', U_{n+1}'$. □

14.3. An obvious corollary of Theorem 14.2 is Theorem 2.27(1). The refinement of Theorem 2.27(1) given in 2.28 is true for $n = 1, 2$. By Theorem 14.2, it is true for all $n > 2$.

14.4. The following theorem, which is a refinement of Theorem 2.27(1), can be proved exactly like Theorem 4.19.

THEOREM. *Let $\mathcal{A}_1, \dots, \mathcal{A}_n$ be a finite sequence of algebras, and for each $k \in [1, n]$ let there exist a finite sequence of pairwise disjoint sets $U_1^k, \dots, U_{m_k}^k \notin \mathcal{A}_k$ with $m_k > \frac{5}{2}(k-1)$. Then for each U_i^k one can take $\widehat{U}_i^k \subset U_i^k$ such that $\widehat{U}_i^k \notin \mathcal{A}_k$; either $\widehat{U}_i^k = \widehat{U}_j^l$, or $\widehat{U}_i^k \cap \widehat{U}_j^l = \varnothing$; there exists a sequence of pairwise disjoint sets $\widehat{U}_{j_1}^{\gamma_1}, \dots, \widehat{U}_{j_n}^{\gamma_n}$ such that*

$$\widehat{U}_{j_k}^{\gamma_k} \subset Q \subset \bigcup_{i=1}^{n} \widehat{U}_{j_i}^{\gamma_i} \quad \textit{implies} \quad Q \notin \mathcal{A}_k.$$

14.5. PROOF OF THEOREM 2.27(2). Let $\ker \mathcal{A}_1 = \{a_1, q_1\}$, and let a_1, q_1 be $\mathcal{A}_1$-similar ultrafilters. Put $\mathcal{B}_2^* = \mathcal{A}_1$. The algebras $\mathcal{A}_1, \mathcal{A}_2 = \mathcal{B}_2^*$ satisfy all the conditions of the theorem for $n = 1$.

Let $\ker \mathcal{B}_2 = \{a_2, q_1\}$ and $\ker \mathcal{B}_3^* = \{a_1, a_2, b_1, b_2, q_1\}$, let a_2, q_1 be $\mathcal{B}_2$-similar ultrafilters, let $\{a_1, b_1\}$ and $\{a_2, b_2\}$ be the $\mathcal{B}_3^*$-similar sets, and let q_1 be the $\mathcal{B}_3^*$-special ultrafilter.

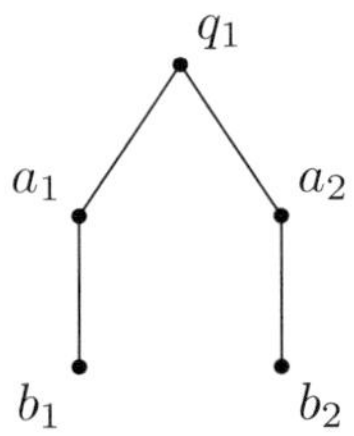

The algebras $\mathcal{A}_1, \mathcal{A}_2 = \mathcal{B}_2, \mathcal{A}_3 = \mathcal{B}_3^*$ satisfy all the conditions of the theorem for $n = 2$. Note that all algebras which appear in the proof of our theorem, except $\mathcal{B}_3^*$ ($\mathcal{A}_3$ for $n = 2$), contain X.

Let us define the algebras $\mathcal{B}_3, \mathcal{B}_4^*$. The $\mathcal{B}_3$-similar sets are: $\{a_1, b_1\}$, $\{a_2, b_2\}$, $\{a_3, q_2\}$; the $\mathcal{B}_4^*$-similar sets are: $\{a_1, b_1\}$, $\{a_2, b_2\}$, $\{a_3, q_1, q_2\}$.

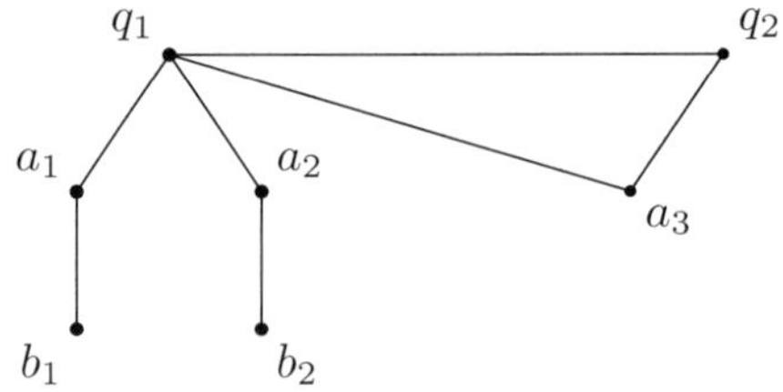

The algebras $\mathcal{A}_1, \mathcal{A}_2 = \mathcal{B}_2, \mathcal{A}_3 = \mathcal{B}_3, \mathcal{A}_4 = \mathcal{B}_4^*$ satisfy all the conditions of the theorem for $n = 3$.

Let us define the algebras $\mathcal{B}_4, \mathcal{B}_5^*$. The $\mathcal{B}_4$-similar sets are: $\{a_1, b_1\}$, $\{a_2, b_2\}$, $\{a_3, q_1\}$, $\{a_4, q_2\}$; the $\mathcal{B}_5^*$-similar sets are: $\{a_1, b_1\}$, $\{a_2, b_2\}$, $\{a_3, b_3\}$, $\{a_4, b_4\}$, $\{q_1, q_2\}$.

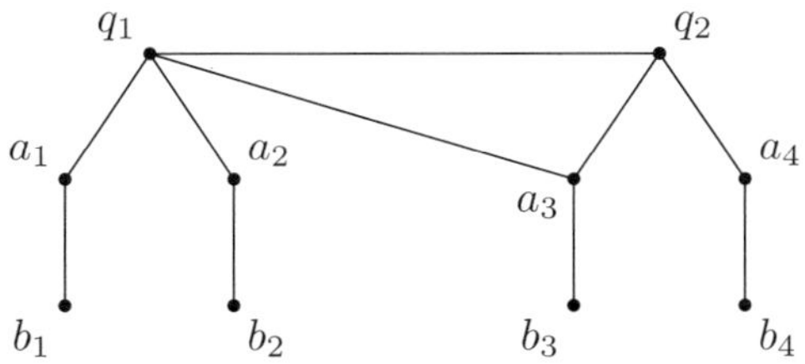

The algebras $\mathcal{A}_1, \mathcal{A}_2 = \mathcal{B}_2, \mathcal{A}_3 = \mathcal{B}_3, \mathcal{A}_4 = \mathcal{B}_4, \mathcal{A}_5 = \mathcal{B}_5^*$ satisfy all the conditions of the theorem for $n = 4$.

The algebra $\mathcal{B}_5 = \mathcal{B}_6^*$. The $\mathcal{B}_5$-similar sets are: $\{a_1, b_1\}$, $\{a_2, b_2\}$, $\{a_3, b_3\}$, $\{a_4, b_4\}$, $\{q_1, q_2\}$, $\{a_5, q_3\}$.

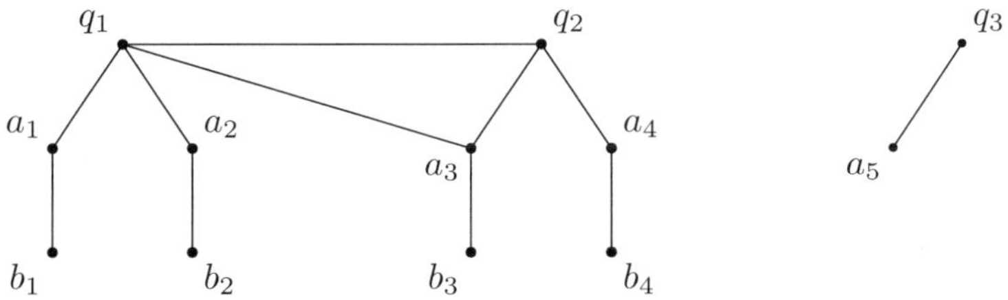

The algebras $\mathcal{A}_1, \mathcal{A}_2 = \mathcal{B}_2, \mathcal{A}_3 = \mathcal{B}_3, \mathcal{A}_4 = \mathcal{B}_4, \mathcal{A}_5 = \mathcal{B}_5$, $\mathcal{A}_6 = \mathcal{B}_6^*$ satisfy all the conditions of the theorem for $n = 5$.

Let us define the algebras $\mathcal{B}_6, \mathcal{B}_7^*$. The $\mathcal{B}_6$-similar sets are: $\{a_1, b_1\}$, $\{a_2, b_2\}$, $\{a_3, b_3\}$, $\{a_4, b_4\}$, $\{a_5, q_1, q_2\}$, $\{a_6, q_3\}$; the $\mathcal{B}_7^*$-similar sets are: $\{a_1, b_1\}$, $\{a_2, b_2\}$, $\{a_3, b_3\}$, $\{a_4, b_4\}$, $\{a_5, b_5\}$, $\{a_6, b_6\}$, $\{q_1, q_2, q_3\}$.

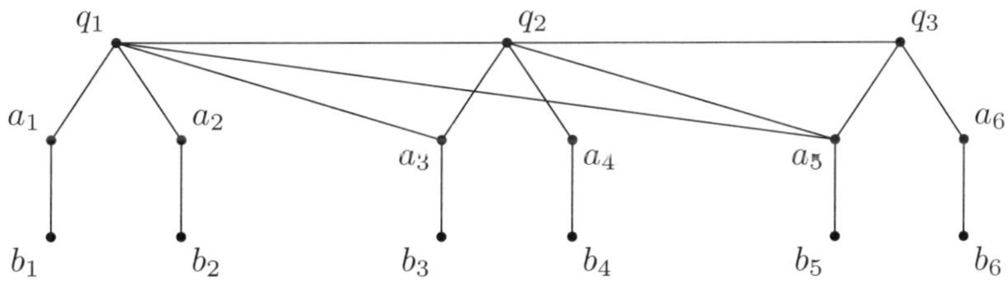

The algebras $\mathcal{A}_1, \mathcal{A}_2 = \mathcal{B}_2, \mathcal{A}_3 = \mathcal{B}_3, \mathcal{A}_4 = \mathcal{B}_4, \mathcal{A}_5 = \mathcal{B}_5, \mathcal{A}_6 = \mathcal{B}_6, \mathcal{A}_7 = \mathcal{B}_7^*$ satisfy all the conditions of the theorem for $n = 6$.

For any $n > 6$ one can construct algebras $\mathcal{A}_1, \ldots, \mathcal{A}_n, \mathcal{A}_{n+1}$, satisfying all the conditions of our theorem, in a similar way. □

14.6. This and the next section are devoted to the results discussed in 2.30.

THEOREM. *Let $\mathcal{A}_1, \ldots, \mathcal{A}_n$ be a finite sequence of algebras, and for each $k \in [1, n]$ let there exist a finite sequence of pairwise disjoint sets $U_1^k, \ldots, U_{m_k}^k \notin \mathcal{A}_k$ with $m_k > 3(k-1)$. Then for each U_i^k one can choose a subset $\widehat{U}_i^k \notin \mathcal{A}_k$ so that*

(i) *either $\widehat{U}_i^k = \widehat{U}_j^l$, or $\widehat{U}_i^k \cap \widehat{U}_j^l = \varnothing$;*

(ii) *there exist pairwise disjoint sets* $\widehat{U}^1_{\psi_1}, \ldots, \widehat{U}^n_{\psi_n}$ *such that*

$$\widehat{U}^k_{\psi_k} \subset Q \subset \bigcup_{i=1}^{n} \widehat{U}^i_{\psi_i} \quad \text{implies} \quad Q \notin \mathcal{A}_k.$$

PROOF. For each U^k_i let us choose a subset $\widehat{U}^k_i \subset U^k_i$ and either an $\mathcal{A}_k$-special ultrafilter s^k_i, or $\mathcal{A}_k$-similar ultrafilters s^k_i, t^k_i so that

1) $\widehat{U}^k_i \in s^k_i$;
2) if the ultrafilter t^k_i is considered, then either $t^k_i = s^l_j$, or $t^k_i \not\ni \bigcup_{l,j} \widehat{U}^l_j$;
3) either $\widehat{U}^k_i = \widehat{U}^l_j$ (then $s^k_i = s^l_j$), or $\widehat{U}^k_i \cap \widehat{U}^l_j = \varnothing$.[1]

Below the situation in which $t^k_i = t^k_j$ $(i \neq j)$ can arise, and then we must put $t^k_i = s^k_j$, $t^k_j = s^k_i$. If $n = 1$, then we consider the set $\widehat{U}^1_1$. We know that if the ultrafilter t^1_1 is considered, then $t^1_1 \not\ni \widehat{U}^1_1$. Let $n = p$. Assume that there exist pairwise disjoint sets $\widehat{U}^1_{\psi_1}, \ldots, \widehat{U}^p_{\psi_p}$ such that if the ultrafilter $t^i_{\psi_i}$ is considered, then

$$t^i_{\psi_i} \not\ni \bigcup_{j=1}^{p} \widehat{U}^j_{\psi_j}.$$

Let us prove that a similar statement holds for $n = p+1$. Take the sets $\widehat{U}^1_{\psi_1}, \ldots, \widehat{U}^p_{\psi_p}$, which exist by assumption. Since $m_{p+1} > 3p$, there exists an ultrafilter $s^{p+1}_{\psi_{p+1}}$ possessing the following properties:

a) $s^{p+1}_{\psi_{p+1}} \neq s^i_{\psi_i}$, where $i \in [1, p]$;
b) $s^{p+1}_{\psi_{p+1}} \neq t^i_{\psi_i}$;
c) if the ultrafilter $t^{p+1}_{\psi_{p+1}}$ is considered, then $t^{p+1}_{\psi_{p+1}} \neq s^i_{\psi_i}$.[2]

Clearly, the sets $\widehat{U}^1_{\psi_1}, \ldots, \widehat{U}^p_{\psi_p}, \widehat{U}^{p+1}_{\psi_{p+1}}$ are pairwise disjoint, and if the ultrafilter $t^i_{\psi_i}$ is considered $(i \in [1, p+1])$, then

$$t^i_{\psi_i} \not\ni \bigcup_{j=1}^{p+1} \widehat{U}^j_{\psi_j}.$$

By induction, the theorem is proved. □

14.7. THEOREM. *For any* $n \in \mathbb{N}^+$ *it is possible to construct algebras* $\mathcal{A}_1, \ldots, \mathcal{A}_n, \mathcal{A}_{n+1}$, *satisfying the following conditions*:

(i) *there exist pairwise disjoint sets* $U^k_1, \ldots, U^k_{m_k} \notin \mathcal{A}_k$ *with* $m_k > 3(k-1)$ *for* $k \in [1, n]$;
(ii) *there exist pairwise disjoint sets* $U^{n+1}_1, \ldots, U^{n+1}_{3n} \notin \mathcal{A}_{n+1}$;

[1] The statement about the existence of the sets $\widehat{U}^k_i$ is trivial when one considers a finite sequence of algebras. We present it in the theorem because a similar statement for a countable sequence of σ-algebras is true but nontrivial (it suffices to say that it uses the nontrivial Gitik–Shelah theorem), and it will be proved later. Of course, similar remarks make sense for Theorems 4.19 and 14.4 above.

[2] For the proof of condition c) one should use, if necessary, the arguments in 3.8, and introduce two new ultrafilters of the type of t^{p+1}_i, as mentioned above in our proof.

(iii) *there do not exist pairwise disjoint sets* $U_1, \dots, U_{n+1}$, *where* $U_k \subset U_{j_k}^{\gamma_k}$ *and* $\gamma_k \le k$ *for which the following is always true*:

$$U_k \subset Q \subset \bigcup_{i=1}^{n+1} U_i \quad \textit{implies} \quad Q \notin \mathcal{A}_k.$$

PROOF. Let us construct the algebras $\mathcal{A}_1, \mathcal{B}_2^*$:

$\ker \mathcal{A}_1 = \{a_1, b_1\}$, and a_1, b_1 are $\mathcal{A}_1$-similar ultrafilters;

$\ker \mathcal{B}_2^* = \{a_1, b_1, c_1\}$, the ultrafilters a_1, c_1 are $\mathcal{B}_2^*$-similar, and the ultrafilter b_1 is $\mathcal{B}_2^*$-special.

Put $\mathcal{A}_2 = \mathcal{B}_2^*$, $m_1 = 1$, $m_2 = 3$, $U_1^1 \in a_1$, $U_1^2 \in a_1$, $U_2^2 \in b_1$, $U_3^2 \in c_1$. The theorem is proved for $n = 1$. Let us construct the algebras $\mathcal{B}_2, \dots, \mathcal{B}_n, \mathcal{B}_{n+1}^*$ when $n = 4$:

$\ker \mathcal{B}_2 = \{a_1, a_2, b_1, b_2, c_1\}$; $\{a_1, c_1\}$, $\{a_2, b_2\}$ are the $\mathcal{B}_2$-similar sets, and b_1 is the $\mathcal{B}_2$-special ultrafilter;

$\ker \mathcal{B}_3 = \{a_1, a_2, a_3, b_1, b_2, b_3, c_1, c_2\}$, and $\{a_1, c_1\}$, $\{a_2, c_2\}$, $\{a_3, b_3\}$, $\{b_1, b_2\}$ are the $\mathcal{B}_3$-similar sets;

$\ker \mathcal{B}_4 = \{a_1, a_2, a_3, a_4, b_1, b_2, b_3, b_4, c_1, c_2, c_3\}$, and $\{a_1, c_1\}$, $\{a_2, c_2\}$, $\{a_3, c_3\}$, $\{a_4, b_4\}$, $\{b_1, b_2, b_3\}$ are the $\mathcal{B}_4$-similar sets;

$\ker \mathcal{B}_5^* = \{a_1, a_2, a_3, a_4, b_1, b_2, b_3, b_4, c_1, c_2, c_3, c_4\}$; $\{a_i, c_i\}$, where $i \in [1, 4]$, and $\{b_1, b_2, b_3, b_4\}$ are the $\mathcal{B}_5^*$-similar sets.

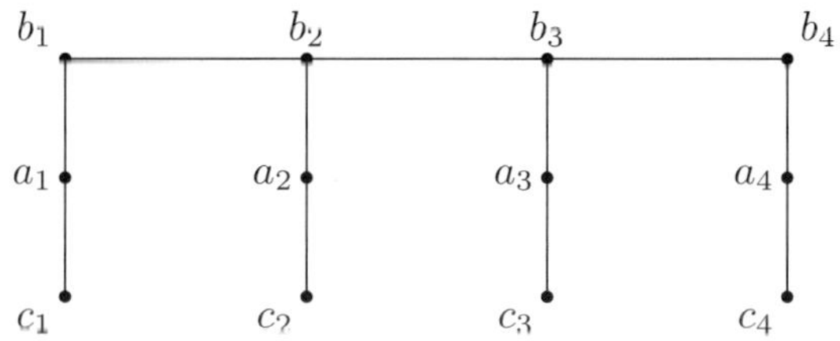

The algebra $\mathcal{A}_1$ was defined above. Put $\mathcal{A}_2 = \mathcal{B}_2$, $\mathcal{A}_3 = \mathcal{B}_3$, $\mathcal{A}_4 = \mathcal{B}_4$, $\mathcal{A}_5 = \mathcal{B}_5^*$, $m_1 = 1$, $m_2 = 4$, $m_3 = 7$, $m_4 = 10$, $m_5 = 12$. For each $k \in [1, 4]$ consider a one-to-one correspondence between the set $\{U_1^k, \dots, U_{m_k}^k\}$ and the set $\ker \mathcal{A}_k \setminus \{b_k\}$. The set U_i^k belongs to the corresponding ultrafilter. Consider a one-to-one correspondence between the set $\{U_1^5, \dots, U_{12}^5\}$ and the set $\ker \mathcal{A}_5$. The set U_i^5 belongs to the corresponding ultrafilter. It is easy to check that the theorem is now proved for $n = 4$. After this example, the construction proving the theorem in the general case is clear. □

14.8. Let us prove a theorem whose obvious corollary is Theorem 2.31(1).

THEOREM. *Let* $\mathcal{A}_1, \dots, \mathcal{A}_n, \mathcal{A}_{n+1}$ *be a sequence of algebras*, $n \in \mathbb{N}^+$, *and let there exist pairwise disjoint sets* $U_1, \dots, U_n, V_1, \dots, V_n$ *such that if* $k \in [1, n]$, $U_k \subset Q$, *and* $Q \cap V_k = \varnothing$, *then* $Q \notin \mathcal{A}_k$. *Assume also that there exist more than* $4n$ *pairwise disjoint sets not belonging to* $\mathcal{A}_{n+1}$. *Then there exist pairwise disjoint sets* $U'_1, \dots, U'_n, U'_{n+1}, V'_1, \dots, V'_n, V'_{n+1}$ *such that if* $k \in [1, n+1]$, $U'_k \subset Q'$, *and* $Q' \cap V'_k = \varnothing$, *then* $Q' \notin \mathcal{A}_k$.

PROOF. If $k \in [1, n]$ and $s_k \ni U_k$ is an $\mathcal{A}_k$-special ultrafilter, then we put $V_k = \varnothing$. Otherwise there exist $\mathcal{A}_k$-similar ultrafilters s_k and t_k such that $s_k \ni U_k$, $t_k \ni V_k$. Denote by Z the set of all ultrafilters s_k and t_k. Clearly, $\#(Z) \leq 2n$. Our usual arguments from 3.8 yield the following conclusion: if there does not exist an $\mathcal{A}_{n+1}$-special ultrafilter $s_{n+1} \notin Z$, then there exist $\mathcal{A}_{n+1}$-similar ultrafilters $s_{n+1}, t_{n+1} \notin Z$. This obviously implies the existence of the required sets $U'_1, \dots, U'_n, U'_{n+1}, V'_1, \dots, V'_n, V'_{n+1}$. □

14.9. REMARK. By the assumption of Theorem 2.31(1), for each $k \in [1, n]$ there exist pairwise disjoint sets $U_1^k, \dots, U_{m_k}^k \notin \mathcal{A}_k$ with $m_k > 4(k-1)$. The proof of Theorem 14.8 shows that the sets $U_1, \dots, U_n$ can be chosen so that $U_k \subset U_{i_k}^k$.

14.10. PROOF OF THEOREM 2.31(2). Let $n = 4$. Let us construct algebras $\mathcal{A}_1, \mathcal{A}_2, \mathcal{A}_3, \mathcal{A}_4, \mathcal{A}_5$, each of which contains X:

the $\mathcal{A}_1$-similar sets are: $\{a_1, b_1\}$;

the $\mathcal{A}_2$-similar sets are: $\{a_1, c_1\}$, $\{b_1, c_2\}$, $\{a_2, b_2\}$;

the $\mathcal{A}_3$-similar sets are: $\{a_1, c_1\}$, $\{b_1, c_2\}$, $\{a_2, c_3\}$, $\{b_2, c_4\}$, $\{a_3, b_3\}$;

the $\mathcal{A}_4$-similar sets are: $\{a_1, c_1\}$, $\{b_1, c_2\}$, $\{a_2, c_3\}$, $\{b_2, c_4\}$, $\{a_3, c_5\}$, $\{b_3, c_6\}$, $\{a_4, b_4\}$;

the $\mathcal{A}_5$-similar sets are: $\{a_1, c_1\}$, $\{b_1, c_2\}$, $\{a_2, c_3\}$, $\{b_2, c_4\}$, $\{a_3, c_5\}$, $\{b_3, c_6\}$, $\{a_4, c_7\}$, $\{b_4, c_8\}$.

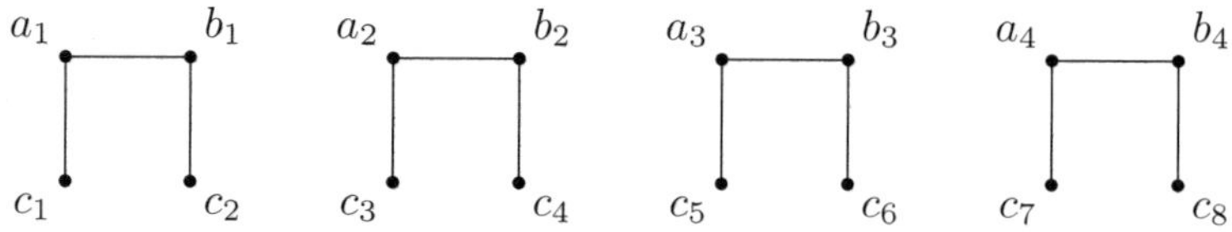

Similarly one constructs algebras $\mathcal{A}_1, \dots, \mathcal{A}_n, \mathcal{A}_{n+1}$ for any n. □

14.11. This section opens the second part of the present chapter. This and the next section are devoted to questions which naturally arise in connection with Theorem 3.24.

Let $\psi\colon [1, n] \to \mathbb{N}^+$ be a function such that if $\mathcal{A}_1, \dots, \mathcal{A}_n$ is a sequence of algebras and for each $k \in [1, n]$ we have $X \in \mathcal{A}_k$ and there exist $\psi(k)$ pairwise disjoint sets not belonging to $\mathcal{A}_k$, then there exist pairwise disjoint sets $U_1, \dots, U_n$, for which the following is true:

$$U_k \subset Q \subset \bigcup_{i=1}^{n} U_i \quad \text{implies} \quad Q \notin \mathcal{A}_k.$$

The following question arises: if $\mathcal{B}_1, \dots, \mathcal{B}_n$ is a sequence of algebras such that for each $k \in [1, n]$ there exist $\psi(k)$ pairwise disjoint sets not belonging to $\mathcal{B}_k$, then do there exist pairwise disjoint sets $V_1, \dots, V_n$, for which

$$V_k \subset Q \subset \bigcup_{i=1}^{n} V_i \quad \text{implies} \quad Q \notin \mathcal{B}_k?$$

The answer is that in general this is not true.

It is easy to check that if $\mathcal{A}_1, \mathcal{A}_2, \mathcal{A}_3$ are algebras such that $X \in \mathcal{A}_k$ and $\#(\ker \mathcal{A}_k) \geq 2k - 1$ for all $k \in [1, 3]$, then the corresponding sets U_1, U_2, U_3 exist.

On the other hand, one can construct algebras $\mathcal{B}_1, \mathcal{B}_2, \mathcal{B}_3$ so as to have $\#(\ker \mathcal{B}_k) \geq 2k-1$ for all $k \in [1,3]$ and no corresponding sets V_1, V_2, V_3 exist.

CONSTRUCTION. $\ker \mathcal{B}_1 = \{a_1, c\}$, and a_1, c are $\mathcal{B}_1$-similar ultrafilters;

$\ker \mathcal{B}_2 = \{a_1, a_2, c\}$, a_1 is a $\mathcal{B}_2$-special ultrafilter, and a_2, c are $\mathcal{B}_2$-similar ultrafilters;

$\ker \mathcal{B}_3 = \{a_1, a_2, b_1, b_2, c\}$, c is a $\mathcal{B}_3$-special ultrafilter, and $\{a_1, b_1\}$, $\{a_2, b_2\}$ are the $\mathcal{B}_3$-similar sets.

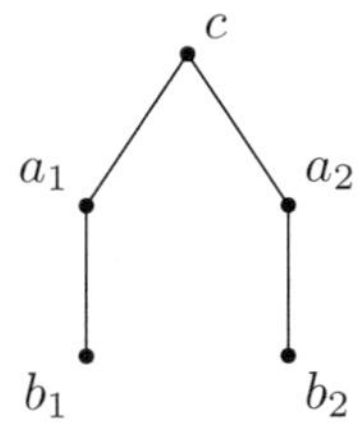

□

14.12. THEOREM. *Let $\psi\colon [1,n] \to \mathbb{N}^+$ be a function such that if $\mathcal{A}_1, \ldots, \mathcal{A}_n$ is a sequence of algebras such that for each $k \in [1,n]$ we have $X \in \mathcal{A}_k$ and there exist $\psi(k)$ pairwise disjoint sets not belonging to $\mathcal{A}_k$, then there exist pairwise disjoint sets $U_1, \ldots, U_n, V_1, \ldots, V_n$, for which the following is true: if $U_k \subset Q$ and $V_k \cap Q = \varnothing$, then $Q \notin \mathcal{A}_k$. Then, if $\mathcal{B}_1, \ldots, \mathcal{B}_n$ is a sequence of algebras such that for each $k \in [1,n]$ there exist $\psi(k)$ pairwise disjoint sets not belonging to $\mathcal{B}_k$, then there exist pairwise disjoint sets $U'_1, \ldots, U'_n, V'_1, \ldots, V'_n$, for which the following is true: if $U'_k \subset Q$ and $V'_k \cap Q = \varnothing$, then $Q \notin \mathcal{B}_k$.*

PROOF. One can assume that $\#(X) < \aleph_0$ and there exist pairwise distinct points

$$q_1, \ldots, q_n \in X \setminus \bigcup_{k=1}^{n} \ker \mathcal{B}_k.$$

If there does not exist a $\mathcal{B}_k$-special ultrafilter, then we put $\mathcal{B}'_k = \mathcal{B}_k$. If there exists a $\mathcal{B}_k$-special ultrafilter r_k (of course, we assume that there are no other $\mathcal{B}_k$-special ultrafilters), then let us construct an algebra $\mathcal{B}'_k \ni X$: each $\mathcal{B}_k$-similar set is a $\mathcal{B}'_k$-similar set; and there is one more $\mathcal{B}'_k$-similar set $\{q_k, r_k\}$. There exist pairwise distinct ultrafilters $s_1, \ldots, s_n, t_1, \ldots, t_n$ such that s_k, t_k are $\mathcal{B}'_k$ similar ultrafilters. Hence the corresponding sets $U'_1, \ldots, U'_n, V'_1, \ldots, V'_n$ exist. □

14.13. PROOF OF THEOREM 2.36. **1)** Let us prove that $\mathfrak{h}(n) \leq 2n-1$. Let $\mathcal{A}_1, \ldots, \mathcal{A}_n$ be a sequence of algebras, and let $\#(\ker \mathcal{A}_k) \geq 2n-1$ for each $k \in [1,n]$. We must prove that the corresponding sets $U_1, \ldots, U_n$ exist. Let us take a maximal set $L_0 \subset [1,n]$, for which there exists a one-to-one map from L_0 to βX: any $k \in L_0$ corresponds to an $\mathcal{A}_k$-special ultrafilter t_k^0. It is possible that $L_0 = \varnothing$. If $L_0 = [1,n]$, then the statement is proved. Otherwise, put $P_k^1 = \ker \mathcal{A}_k \setminus (T_0 \cup T_0^k)$ for each $k \in [1,n] \setminus L_0$, where

$$T_0 = \{t_l^0\}_{l \in L_0},$$

$$T_0^k = \{q \in \ker \mathcal{A}_k \setminus T_0 \mid \text{there exists } q' \in T_0\ \mathcal{A}_k\text{-similar to } q,$$
$$\text{and there exist no } q'' \notin T_0\ \mathcal{A}_k\text{-similar to } q\}.$$

Clearly, $P_k^1 \neq \varnothing$ since

$$\#(T_0 \cup T_0^k) \leq 2 \cdot \#(T_0) \leq 2n - 2.$$

Let us take some ultrafilter $s^1 \in \bigcup P_k^1$. By the maximality of L_0, for any $k \in [1,n] \setminus L_0$ the ultrafilter s^1 is not $\mathcal{A}_k$-special. Let us take a maximal set $L_1 \subset [1,n] \setminus L_0$ for which there exists a one-to-one map from L_1 to βX: any $k \in L_1$ corresponds to an ultrafilter $t_k^1 \in P_k^1$ which is $\mathcal{A}_k$-similar to s^1. If $L_0 \cup L_1 = [1,n]$, then the statement is proved. Otherwise, for each $k \in [1,n] \setminus (L_0 \cup L_1)$ put

$$P_k^2 = P_k^1 \setminus (\{s^1\} \cup T_1 \cup T_1^k),$$

where

$$\begin{aligned} T_1 &= \{t_l^1\}_{l \in L_1}, \\ T_1^k &= \big\{q \in P_k^1 \setminus (\{s^1\} \cup T_1) \mid \text{there exists } q' \in T_1 \ \mathcal{A}_k\text{-similar to } q, \\ &\qquad\qquad \text{and there exist no } q'' \in P_k^1 \setminus T_1 \ \mathcal{A}_k\text{-similar to } q\big\}. \end{aligned}$$

Let us show that $P_k^2 \neq \varnothing$. If $s^1 \notin \ker \mathcal{A}_k$, then

$$\#(T_0 \cup T_0^k \cup T_1 \cup T_1^k) \leq 2 \cdot \#(T_0) + 2 \cdot \#(T_1) \leq 2n - 2.$$

Therefore, $P_k^2 \neq \varnothing$. Let $s^1 \in \ker \mathcal{A}_k$. Then, by the maximality of L_1, s^1 has an $\mathcal{A}_k$-similar ultrafilter $t^{(k)}$ in T_1. Hence

$$\begin{aligned} \#(T_0 \cup T_0^k \cup \{s^1\} \cup T_1 \cup T_1^k) &\leq 2 \cdot \#(T_0) + 1 + \#(T_1) + \#(T_1 \setminus \{t^{(k)}\}) \\ &\leq 2 \cdot \#(T_0) + 2 \cdot \#(T_1) \leq 2n - 2. \end{aligned}$$

Hence $P_k^2 \neq \varnothing$. Let us take some ultrafilter $s^2 \in \bigcup P_k^2$. Let us take a maximal set $L_2 \subset [1,n] \setminus (L_0 \cup L_1)$, for which there exists a one-to-one map from L_2 to βX: any $k \in L_2$ corresponds to an ultrafilter $t_k^2 \in P_k^2$ which is $\mathcal{A}_k$-similar to s^2. If $L_0 \cup L_1 \cup L_2 = [1,n]$, then the statement is proved, etc. Finally we will construct pairwise disjoint sets of ultrafilters $T_0, T_1, \ldots, T_m$; and for each $i \in [1,m]$ an ultrafilter $s^i \notin T = \bigcup_{i=0}^m T_i$ will be considered. We also have

(i) $T = \{t_1, \ldots, t_n\}$;
(ii) $t_k \neq t_j$ if $k \neq j$;
(iii) the ultrafilter $t_k \in T_0$ is $\mathcal{A}_k$-special;
(iv) the ultrafilter $t_k \in T_i$, where $i > 0$, is $\mathcal{A}_k$-similar to s^i.

The corresponding sets $U_1, \ldots, U_n$ are constructed in the obvious way.

2) Let us now prove that $\mathfrak{h}(n) \geq 2n - 1$. Let $n > 1$. It suffices to construct algebras $\mathcal{A}_1, \ldots, \mathcal{A}_n$ such that $\#(\ker \mathcal{A}_k) = 2n - 2$ for each $k \leq n$ and the corresponding sets $U_1, \ldots, U_n$ do not exist. Indeed, let us construct algebras

$$\mathcal{A}_1 = \mathcal{A}_2 = \cdots = \mathcal{A}_n;$$

$\ker \mathcal{A}_k = \{a_1, \ldots, a_{n-1}, b_1, \ldots, b_{n-1}\}$, and $\{a_i, b_i\}$ is an $\mathcal{A}_k$-similar set for each $i \in [1, n-1]$.

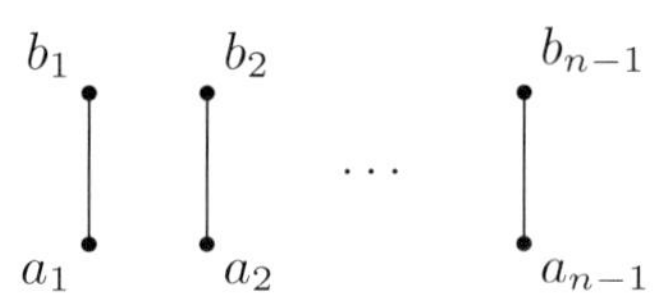

It is clear that one cannot construct the corresponding sets $U_1, \ldots, U_n$. $\square$

14.14. Let us proceed to the study of the function $\mathfrak{v}(n)$ defined in 2.37. By Theorem 2.31, we have $\mathfrak{v}(n) \le 4n-3$. It is rather easy to show that $\mathfrak{v}(n) = 4n-3$ for $n \le 3$. For $n=1$ it is obvious. Let a_1, a_2, b_1, b_2 be pairwise distinct ultrafilters. Let

$$\ker \mathcal{A}_1 = \ker \mathcal{A}_2 = \{a_1, a_2, b_1, b_2\};$$
$$\{a_1, b_1\}, \{a_2, b_2\} \text{ be the } \mathcal{A}_1\text{-similar sets;}$$
$$\{a_1, a_2\}, \{b_1, b_2\} \text{ be the } \mathcal{A}_2\text{-similar sets.}$$

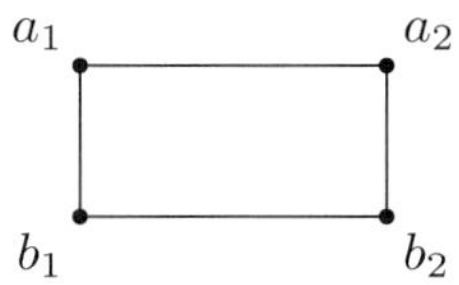

Obviously, for $\mathcal{A}_1, \mathcal{A}_2$ the corresponding sets U_1, U_2, V_1, V_2 do not exist. Since $\#(\ker \mathcal{A}_1) = \#(\ker \mathcal{A}_2) = 4$, by Theorem 2.31 we have $\mathfrak{v}(2) = 5$. □

Let $a_1, a_2, a_3, a_4, b_1, b_2, b_3, b_4$ be pairwise distinct ultrafilters. Let

$$\ker \mathcal{A}_1 = \ker \mathcal{A}_2 = \ker \mathcal{A}_3 = \{a_1, a_2, a_3, a_4, b_1, b_2, b_3, b_4\};$$
$$\{a_1, b_1\},\ \{a_2, b_2\},\ \{a_3, b_3\},\ \{a_4, b_4\} \text{ be the } \mathcal{A}_1\text{-similar sets;}$$
$$\{a_1, a_2\},\ \{b_1, b_2\},\ \{a_3, a_4\},\ \{b_3, b_4\} \text{ be the } \mathcal{A}_2\text{-similar sets;}$$
$$\{a_1, b_2\},\ \{a_2, b_1\},\ \{a_3, b_4\},\ \{a_4, b_3\} \text{ be the } \mathcal{A}_3\text{-similar sets.}$$

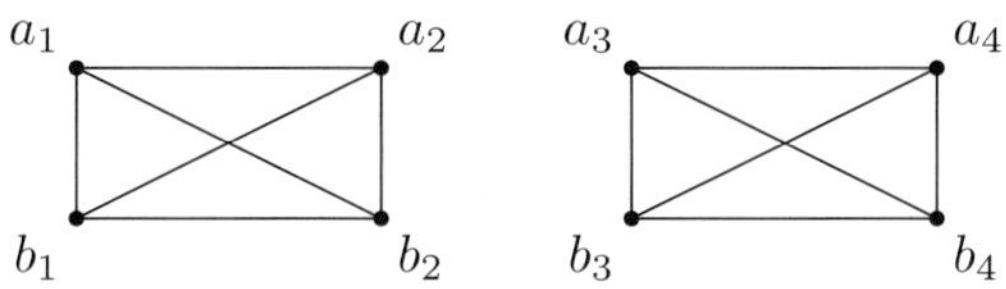

Obviously, for $\mathcal{A}_1, \mathcal{A}_2, \mathcal{A}_3$ the corresponding sets $U_1, U_2, U_3, V_1, V_2, V_3$ do not exist. Now since

$$\#(\ker \mathcal{A}_1) = \#(\ker \mathcal{A}_2) = \#(\ker \mathcal{A}_3) = 8,$$

by Theorem 2.31 we have $\mathfrak{v}(3) = 9$. □

14.15. Let us show that $\mathfrak{v}(4) \le 12$. Indeed, let $\mathcal{A}_1, \mathcal{A}_2, \mathcal{A}_3, \mathcal{A}_4$ be algebras with $\#(\ker \mathcal{A}_k) = 12$ for all $k \in [1,4]$.[3] By Theorem 14.12, let us assume that $X \in \mathcal{A}_k$ for all $k \in [1,4]$. Let us take pairwise distinct ultrafilters $a_1, a_2, a_3, b_1, b_2, b_3$ such that a_i, b_i are $\mathcal{A}_i$-similar ultrafilters for all $i \le 3$. Let $q_1, q_2, q_3, r_1, r_2, r_3$ be pairwise distinct ultrafilters from $\ker \mathcal{A}_4$ and

$$\{a_1, a_2, a_3, b_1, b_2, b_3\} \cap \{q_1, q_2, q_3, r_1, r_2, r_3\} = \varnothing.$$

In the more complicated situation the picture should be, for example, the following: (a_1, q_1), (b_1, r_1), (a_2, q_2), (b_2, r_2), (a_3, q_3), (b_3, r_3) are pairs of $\mathcal{A}_4$-similar ultrafilters.

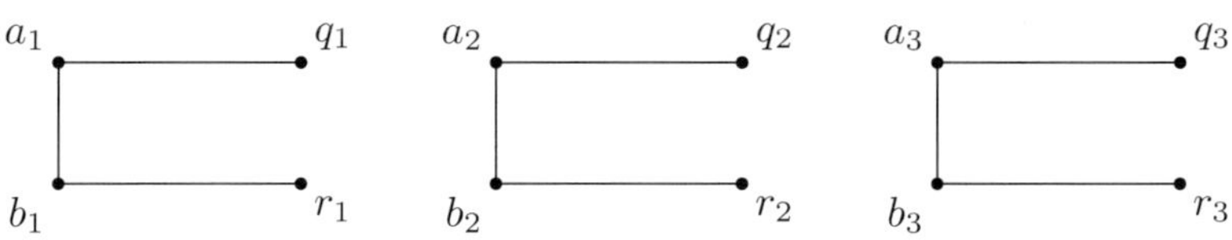

[3]Here one should recall Remark 3.25, which is related to problems discussed in this chapter.

Now consider the $\mathcal{A}_4$-similar ultrafilters a_1 and q_1, the $\mathcal{A}_2$-similar ultrafilters a_2 and b_2, and the $\mathcal{A}_3$-similar ultrafilters a_3 and b_3. Put

$$M = \ker \mathcal{A}_1 \setminus \{a_1, q_1, a_2, b_2, a_3, b_3\}.$$

Since $\#(M) \geq 6$, in the more complicated situation there exists an ultrafilter $r \in M$ which is $\mathcal{A}_1$-similar to q_1. One of the following three cases holds:

1) $r = r_1$;
2) $r = b_1$;
3) $r \notin \{b_1, r_1\}$.

In cases 2) and 3) one readily constructs the corresponding sets

$$U_1, U_2, U_3, U_4, V_1, V_2, V_3, V_4.$$

So let us assume that q_1, r_1 are $\mathcal{A}_1$-similar ultrafilters. Similarly, let us assume that q_i, r_i are $\mathcal{A}_i$-similar ultrafilters if $i \in [2,3]$. The only situation where one cannot immediately construct the corresponding sets $U_1, U_2, U_3, U_4, V_1, V_2, V_3, V_4$ is the situation where for distinct $i, j \in [1,3]$ one of the following two cases holds:

a) (a_j, q_j), (b_j, r_j) are pairs of $\mathcal{A}_i$-similar ultrafilters;
b) (a_j, r_j), (b_j, q_j) are pairs of $\mathcal{A}_i$-similar ultrafilters.

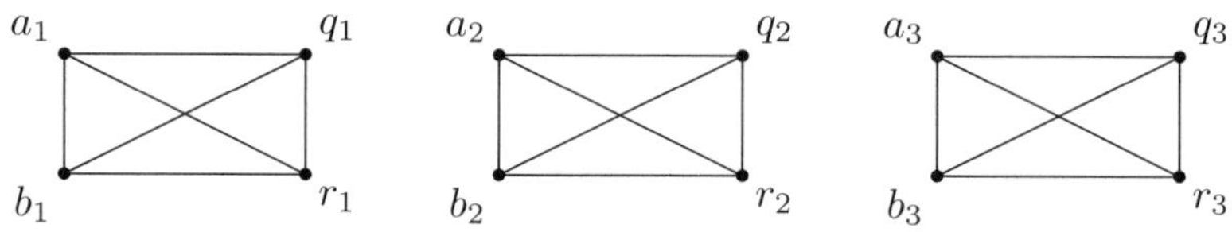

But now again it is very easy to construct the corresponding sets U_1, U_2, U_3, U_4, V_1, V_2, V_3, V_4. □

14.16. One can actually show that $\mathfrak{v}(4) \leq 11$. Since $\mathfrak{v}(4) \leq 12$, it suffices to consider algebras $\mathcal{A}_1, \mathcal{A}_2, \mathcal{A}_3, \mathcal{A}_4$ such that $11 \leq \#(\ker \mathcal{A}_k) \leq 12$ for all $k \in [1,3]$, $\#(\ker \mathcal{A}_4) = 11$, and $X \in \mathcal{A}_k$ for all $k \in [1,4]$. In principle, there are two possible more complicated situations. In the first one the picture should be, for example, the following:

$$a_1, a_2, a_3, b_1, b_2, b_3, q_1, q_2, q_3, r_1, r_2$$

are pairwise distinct ultrafilters, a_i, b_i are $\mathcal{A}_i$-similar ultrafilters if $i \in [1,3]$; $\{a_1, q_1\}$, $\{b_1, r_1\}$, $\{a_2, q_2\}$, $\{b_2, r_2\}$, $\{a_3, b_3, q_3\}$ are the $\mathcal{A}_4$-similar sets.

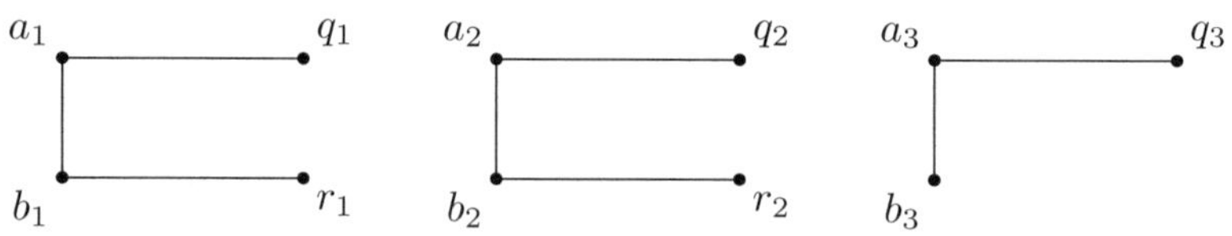

In the second situation only the structure of $\ker \mathcal{A}_4$ is changed: $\{a_1, q_1\}$, $\{b_1, r_1\}$, $\{a_2, q_2\}$, $\{a_3, q_3\}$, $\{b_2, r_2, b_3\}$ are the $\mathcal{A}_4$-similar sets.

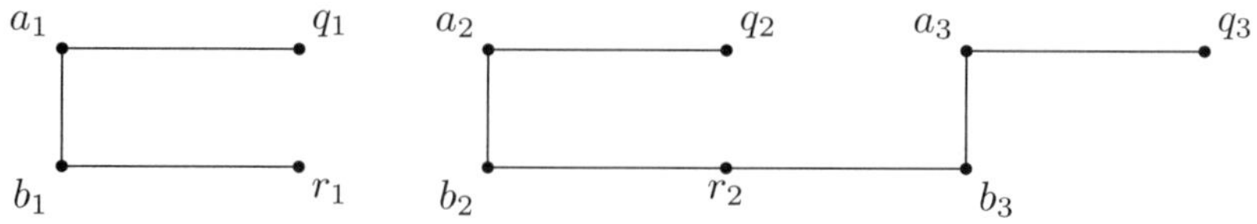

Both in the first and in the second situation one can construct the corresponding sets $U_1, U_2, U_3, U_4, V_1, V_2, V_3, V_4$. □

14.17. It is rather easy to show that $\mathfrak{v}(n+1) - \mathfrak{v}(n) \leq 4$. This property and the inequality $\mathfrak{v}(4) \leq 11$ imply that $\mathfrak{v}(n) \leq 4n - 5$ for all $n \geq 4$.

14.18. STATEMENT. $\mathfrak{v}(n) \leq 4n - \left[\frac{n+3}{2}\right]$.

PROOF. **1)** From the above, it follows that the statement is true for all $n \leq 8$. Let $m \in \mathbb{N}^+$ and $m \geq 6$. Put $n' = 2m - 3$. Let us assume that

$$4(n'-1) - \mathfrak{v}(n'-1) \geq m - 1.$$

Our goal is to prove that $4n' - \mathfrak{v}(n') \geq m$. Suppose the contrary. Consider algebras $\mathcal{A}_1, \ldots, \mathcal{A}_{n'}$; $X \in \mathcal{A}_k$,[4]

$$\#(\ker \mathcal{A}_k) \geq 4n' - m \quad \text{for all } k \in [1, n'],$$

and there do not exist pairwise distinct ultrafilters $a_1^*, \ldots, a_{n'}^*, b_1^*, \ldots, b_{n'}^*$ such that a_k^*, b_k^* are $\mathcal{A}_k$-similar ultrafilters for each $k \in [1, n']$. We know that there exist pairwise distinct ultrafilters $a_1, \ldots, a_{n'-1}, b_1, \ldots, b_{n'-1}$ such that a_k, b_k are $\mathcal{A}_k$-similar ultrafilters for each $k \in [1, n'-1]$. Put $r = m - 2$. Let us show that one can choose ultrafilters $c_1, \ldots, c_{r-1}, c_{n'}, d_1, \ldots, d_{r-1}, d_{n'}$ so that $(a_1, c_{n'})$ and $(b_1, d_{n'})$ are pairs of $\mathcal{A}_{n'}$-similar ultrafilters, and (a_{k+1}, c_k), (b_{k+1}, d_k) are pairs of $\mathcal{A}_k$-similar ultrafilters for each $k \in [1, r-1]$; moreover,

$$Z = \{a_1, \ldots, a_{n'-1}, b_1, \ldots, b_{n'-1}, c_1, \ldots, c_{r-1}, c_{n'}, d_1, \ldots, d_{r-1}, d_{n'}\}$$

is a set of pairwise distinct ultrafilters. (To obtain this, perhaps the indices of the algebras must be changed.) It suffices to show the following: if one can find the corresponding ultrafilters

$$c_1, \ldots, c_{r-2}, c_{n'}, d_1, \ldots, d_{r-2}, d_{n'},$$

then one can also find ultrafilters c_{r-1}, d_{r-1}. Put

$$K = \ker \mathcal{A}_{r-1} \setminus \{a_1, \ldots, a_{n'-1}, b_1, \ldots, b_{n'-1}, c_1, \ldots, c_{r-2}, c_{n'}, d_1, \ldots, d_{r-2}, d_{n'}\}.$$

A simple count shows that $\#(K) \geq m + 2$. If in K there exists an ultrafilter that is $\mathcal{A}_{r-1}$-similar to one of the ultrafilters in the set

$$\{a_1, \ldots, a_{r-1}, b_1, \ldots, b_{r-1}, c_1, \ldots, c_{r-2}, c_{n'}, d_1, \ldots, d_{r-2}, d_{n'}\},$$

then it is easy to construct the corresponding ultrafilters $a_1^*, \ldots, a_{n'}^*, b_1^*, \ldots, b_{n'}^*$ (and this contradicts our assumption). Hence each ultrafilter in K has an $\mathcal{A}_{r-1}$-similar ultrafilter in the set

$$\{a_r, \ldots, a_{n'-1}, b_r, \ldots, b_{n'-1}\}.$$

But since

$$\#([r, n'-1]) = \#([m-2, 2m-4]) = m - 1,$$

there exists $k \in [r, n'-1]$ and distinct ultrafilters $c_{r-1}, d_{r-1} \in K$ such that (a_k, c_{r-1}) and (b_k, d_{r-1}) are pairs of $\mathcal{A}_{r-1}$-similar ultrafilters. If $k > r$, then one should exchange the indices of the algebras $\mathcal{A}_r$ and $\mathcal{A}_k$.

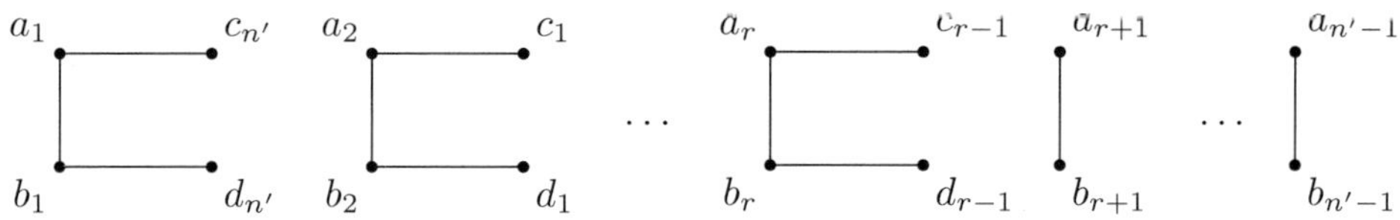

[4]By Theorem 14.12.

Put $L_r = \ker \mathcal{A}_r \setminus Z$. Clearly, $\#(L_r) \geq m$. If in L_r there is an ultrafilter $\mathcal{A}_r$-similar to one of the ultrafilters in the set

$$\{a_1, \dots, a_r, b_1, \dots, b_r, c_1, \dots, c_{r-1}, c_{n'},\ d_1, \dots, d_{r-1}, d_{n'}\},$$

then it is easy to construct the corresponding ultrafilters $a_1^*, \dots, a_{n'}^*, b_1^*, \dots, b_{n'}^*$ (and this contradicts our assumption). Hence each ultrafilter from L_r has an $\mathcal{A}_r$-similar ultrafilter in the set $\{a_{r+1}, \dots, a_{n'-1}, b_{r+1}, \dots, b_{n'-1}\}$. Put

$$I_r = \big\{k \in [r+1, n'-1] \mid \text{there exists an ultrafilter } q_k \in L_r \\ \text{which has an } \mathcal{A}_r\text{-similar ultrafilter in } \{a_k, b_k\}\big\}.$$

Clearly,

$$\#(I_r) \geq \left[\frac{m+1}{2}\right],$$

and $q_{k_1} \neq q_{k_2}$ if $k_1 \neq k_2$. Put $L_{n'} = \ker \mathcal{A}_{n'} \setminus Z$. We know that $\#(L_{n'}) \geq m$. If in $L_{n'}$ there exists an ultrafilter $\mathcal{A}_{n'}$-similar to one of the ultrafilters from the set

$$\{a_1, b_1, c_1, \dots, c_{r-1}, c_{n'}, d_1, \dots, d_{r-1}, d_{n'}\},$$

then it is easy to construct the corresponding ultrafilters $a_1^*, \dots, a_{n'}^*, b_1^*, \dots, b_{n'}^*$ (and this contradicts our assumption). Let $q \in L_{n'}$, and let q, a_2 be $\mathcal{A}_{n'}$-similar ultrafilters. Then for d_1 there are four possible cases:

1) $d_1 \notin \ker \mathcal{A}_{n'}$;
2) b_2, d_1 are $\mathcal{A}_{n'}$-similar ultrafilters;
3) d_1 has an $\mathcal{A}_{n'}$-similar ultrafilter in $\{a_3, \dots, a_r, b_3, \dots, b_r\}$;
4) d_1 has an $\mathcal{A}_{n'}$-similar ultrafilter in $\{a_{r+1}, \dots, a_{n'-1}, b_{r+1}, \dots, b_{n'-1}\}$.

In case 2) let us consider c_1. For this ultrafilter the possible corresponding cases are:

1) $c_1 \notin \ker \mathcal{A}_{n'}$;
3) c_1 has an $\mathcal{A}_{n'}$-similar ultrafilter in $\{a_3, \dots, a_r, b_3, \dots, b_r\}$;
4) c_1 has an $\mathcal{A}_{n'}$-similar ultrafilter in $\{a_{r+1}, \dots, a_{n'-1}, b_{r+1}, \dots, b_{n'-1}\}$.

Consider case 3) for d_1. Let d_1, b_3 be $\mathcal{A}_{n'}$-similar ultrafilters. Then for c_2 there are four possible cases:

1) $c_2 \notin \ker \mathcal{A}_{n'}$;
2) a_3, c_2 are $\mathcal{A}_{n'}$-similar ultrafilters;
3) c_2 has an $\mathcal{A}_{n'}$-similar ultrafilter in $\{a_4, \dots, a_r, b_2, b_4, \dots, b_r\}$;
4) c_2 has an $\mathcal{A}_{n'}$-similar ultrafilter in $\{a_{r+1}, \dots, a_{n'-1}, b_{r+1}, \dots, b_{n'-1}\}$.

Continuing these arguments in an obvious way, we find an ultrafilter

$$q^* \in \{c_1, \dots, c_{r-1}, d_1, \dots, d_{r-1}\}$$

such that one of the following two statements is true:

(a) $q^* \notin \ker \mathcal{A}_{n'}$;
(b) q^* has an $\mathcal{A}_{n'}$-similar ultrafilter in $\{a_{r+1}, \dots, a_{n'-1}, b_{r+1}, \dots, b_{n'-1}\}$.

Now let us put

$$\alpha = \#\big(\{q \in L_{n'} \mid q \text{ has an } \mathcal{A}_{n'}\text{-similar ultrafilter in } \{a_2, \dots, a_r, b_2, \dots, b_r\}\}\big),$$

$$\beta = \#\big(\{c_1, \dots, c_{r-1}, d_1, \dots, d_{r-1}\} \setminus \ker \mathcal{A}_{n'}\big),$$

$$\gamma = \#\big(\{q^* \in \{c_1, \dots, c_{r-1}, d_1, \dots, d_{r-1}\} \mid q^* \text{ has} \\ \text{an } \mathcal{A}_{n'}\text{-similar ultrafilter in } \{a_{r+1}, \dots, a_{n'-1}, b_{r+1}, \dots, b_{n'-1}\}\}\big).$$

The above constructions clearly show that $\alpha \le \beta + \gamma$. Put

$$\widehat{L} = L_{n'} \cup \{c_1, \dots, c_{r-1}, d_1, \dots, d_{r-1}\},$$

$$\widehat{I} = \big\{k \in [r+1, n'-1] \mid \text{there exists an ultrafilter } q'_k \in \widehat{L} \text{ which has an } \mathcal{A}_{n'}\text{-similar ultrafilter in } \{a_k, b_k\}\big\}.$$

Clearly,

$$\#(\widehat{I}) \ge \left[\frac{m+1}{2}\right],$$

and $q'_{k_1} \ne q'_{k_2}$ if $k_1 \ne k_2$. Recall that

$$\#([r+1, n'-1]) = m - 2.$$

Hence $\#(I_r \cap \widehat{I}) \ge 2$. Let $I_r \cap \widehat{I} = \{k_1, k_2\}$. Then there exist pairwise distinct ultrafilters $q_1^a, q_1^b, q_2^a, q_2^b \in L_r$ and pairwise distinct ultrafilters $\widehat{q}_1^a, \widehat{q}_1^b, \widehat{q}_2^a, \widehat{q}_2^b \in \widehat{L}$ such that

$$(a_{k_1}, q_1^a),\ (b_{k_1}, q_1^b),\ (a_{k_2}, q_2^a),\ (b_{k_2}, q_2^b)$$

are pairs of $\mathcal{A}_r$-similar ultrafilters, and

$$(a_{k_1}, \widehat{q}_1^a),\ (b_{k_1}, \widehat{q}_1^b),\ (a_{k_2}, \widehat{q}_2^a),\ (b_{k_2}, \widehat{q}_2^b)$$

are pairs of $\mathcal{A}_{n'}$-similar ultrafilters. We will consider two cases:

Case 1. $\#(I_r \cap \widehat{I}) = 2$; let us assume that $\{r+1, r+2\} = I_r \cap \widehat{I}$.

Case 2. $\#(I_r \cap \widehat{I}) > 2$; let us assume that $\{r+1, r+2, r+3\} \subset I_r \cap \widehat{I}$.

We know that $q'_{r+1} \in \widehat{L}$, and q'_{r+1} has an $\mathcal{A}_{n'}$-similar ultrafilter in $\{a_{r+1}, b_{r+1}\}$. Let a_{r+1}, q'_{r+1} be $\mathcal{A}_{n'}$-similar ultrafilters. Consider the algebras

$$\mathcal{A}_1, \dots, \mathcal{A}_r, \mathcal{A}_{r+2}, \dots, \mathcal{A}_{n'}$$

and the set of pairwise distinct ultrafilters Z^*, which we define by listing the pairs of its elements:

$$(a_1, b_1),\ \dots,\ (a_r, b_r),\ (a_{r+2}, b_{r+2}),\ \dots,\ (a_{n'-1}, b_{n'-1}),\ (a_{r+1}, q'_{r+1})$$

(clearly, each algebra $\mathcal{A}_k$ corresponds to a pair of $\mathcal{A}_k$-similar ultrafilters).

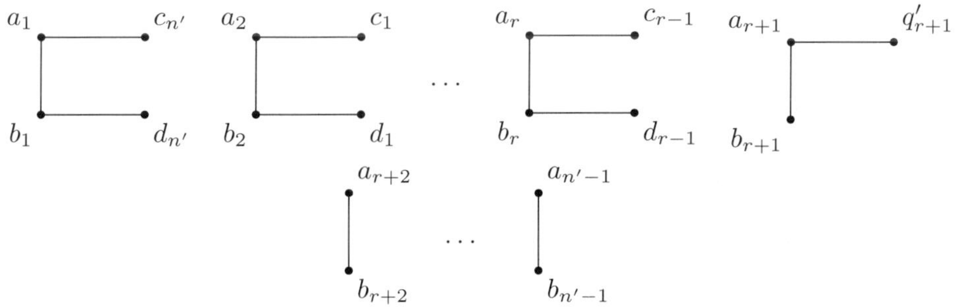

(In the figure we have $q'_{r+1} \in L_{n'}$ although it can happen, for example, that $q'_{r+1} = c_{r-1}$.)

2) Let $s \in \ker \mathcal{A}_{r+1} \backslash Z^*$, and let s have an $\mathcal{A}_{r+1}$-similar ultrafilter in $\{a_1, \dots, a_r, b_1, \dots, b_r\}$. Let a_2, s be $\mathcal{A}_{r+1}$-similar ultrafilters. Let $s \ne d_1$. In Case 1 we come to a contradiction, since we construct the corresponding ultrafilters

$$a_1^*, \dots, a_{n'}^*, b_1^*, \dots, b_{n'}^*.$$

Consider Case 2. We know that $q_{r+1} \in L_r$ and q_{r+1} has an $\mathcal{A}_r$-similar ultrafilter in $\{a_{r+1}, b_{r+1}\}$. Assume that in L_r there does not exist another ultrafilter which has an $\mathcal{A}_r$-similar ultrafilter in $\{a_{r+1}, b_{r+1}\}$. The difficulty (i.e., the impossibility to construct the corresponding ultrafilters $a_1^*, \ldots, a_{n'}^*, b_1^*, \ldots, b_{n'}^*$ immediately) arises if and only if $s = q_{r+1}$.

3) Obviously, in Case 1 we can assume that $q_{r+2} \neq q'_{r+1}$. In Case 2 there exist ultrafilters q_{r+2}, q_{r+3}, and at least one of them does not coincide with q'_{r+1}. Suppose that $q_{r+2} \neq q'_{r+1}$.

Let us imagine the following situation: $q'_{r+1} = c_{r-1}$ (which means that c_{r-1}, a_{r+1} are $\mathcal{A}_{n'}$-similar ultrafilters) and (a_r, b_r), (c_{r-1}, d_{r-1}) are two pairs of $\mathcal{A}_{r+1}$-similar ultrafilters. (Our assumptions are meaningful when we have Case 2 and a_{r+1}, q_{r+1} are $\mathcal{A}_r$-similar ultrafilters.) Then there exist pairwise distinct ultrafilters

$$q_1^*, \ldots, q_p^*, q_*^1, \ldots, q_*^p \in \ker \mathcal{A}_{r+1} \setminus Z^*$$

such that $(a_{k(l)}, q_l^*)$ and $(b_{k(l)}, q_*^l)$ are pairs of $\mathcal{A}_{r+1}$-similar ultrafilters; also $k(l) \in [1, r-1]$, $p \geq \frac{m-2}{2} = \frac{r}{2}$,

$$q_l^* = \begin{cases} d_{n'} & \text{if } k(l) = 1, \\ d_{k(l)-1} & \text{if } k(l) > 1, \end{cases}$$

$$q_*^l = \begin{cases} c_{n'} & \text{if } k(l) = 1, \\ c_{k(l)-1} & \text{if } k(l) > 1. \end{cases}$$

Moreover (see the previous part of the proof) it is possible that there exists $k_* \in [1, r-1]$ such that either (a_{k_*}, q_{r+1}) and (b_{k_*}, c_l) or (a_{k_*}, d_l) and (b_{k_*}, q_{r+1}) is a pair of $\mathcal{A}_{r+1}$-similar ultrafilters; here

$$l = \begin{cases} n' & \text{if } k_* = 1, \\ k_* - 1 & \text{if } k_* > 1. \end{cases}$$

If such a k_* does not exist, we can take $p \geq \frac{m-1}{2} > \frac{r}{2}$. (Let us recall our assumption: there do not exist the corresponding ultrafilters $a_1^*, \ldots, a_{n'}, b_1^*, \ldots, b_{n'}$.) Analyzing other possible situations, we come to the same conclusion: $p \geq \frac{r}{2}$ and if k_* does not exist, we can always take $p > \frac{r}{2}$. (If $q'_{r+1} \in L_{n'}$, then $p > \frac{r}{2}$ and it does not make sense to consider k_*; also, we are obviously assuming that $k(l) \in [1, r]$. If $q'_{r+1} \in \{c_{\widetilde{k}-1}, d_{\widetilde{k}-1}\}$, then we know that $\widetilde{k} \in [2, r]$, and, clearly, $k(l), k_* \in [1, r] \setminus \{\widetilde{k}\}$.) Now it is clear that there exists $k_0 \in [1, r]$ such that

(i) if $k_0 = 1$, one of two pairs $(a_1, d_{n'})$ and $(b_1, c_{n'})$ is a pair of $\mathcal{A}_{r+1}$-similar ultrafilters, and another is a pair of $\mathcal{A}_{r+2}$-similar ultrafilters;
(ii) if $k_0 > 1$, one of two pairs (a_{k_0}, d_{k_0-1}) and (b_{k_0}, c_{k_0-1}) is a pair of $\mathcal{A}_{r+1}$-similar ultrafilters, and another is a pair of $\mathcal{A}_{r+2}$-similar ultrafilters.

Moreover, if $q'_{r+1} \notin L_{n'}$ and $k_0 > 1$, then $q'_{r+1} \notin \{c_{k_0-1}, d_{k_0-1}\}$. Let us assume that $q'_{r+1} \in L_{n'}$, a_{r+2}, q_{r+2} are $\mathcal{A}_r$-similar ultrafilters, $k_0 = 3$, a_3, d_2 are $\mathcal{A}_{r+1}$-similar ultrafilters, and b_3, s_2 are $\mathcal{A}_{r+2}$-similar ultrafilters. Let us construct the corresponding ultrafilters $a_1^*, \dots, a_{n'}^*, b_1^*, \dots, b_{n'}^*$. Let us list them in pairs:

$$(a_1^*, b_1^*) = (a_1, b_1), \quad (a_2^*, b_2^*) = (a_2, b_2), \quad (a_3^*, b_3^*) = (a_4, c_3), \; \dots,$$
$$(a_{r-1}^*, b_{r-1}^*) = (a_r, c_{r-1}), \quad (a_r^*, b_r^*) = (a_{r+2}, q_{r+2}), \quad (a_{r+1}^*, b_{r+1}^*) = (a_3, d_2),$$
$$(a_{r+2}^*, b_{r+2}^*) = (b_3, c_2), \quad (a_{r+3}^*, b_{r+3}^*) = (a_{r+3}, b_{r+3}), \; \dots,$$
$$(a_{n'-1}^*, b_{n'-1}^*) = (a_{n'-1}, b_{n'-1}), \quad (a_{n'}^*, b_{n'}^*) = (a_{r+1}, q'_{r+1}).$$

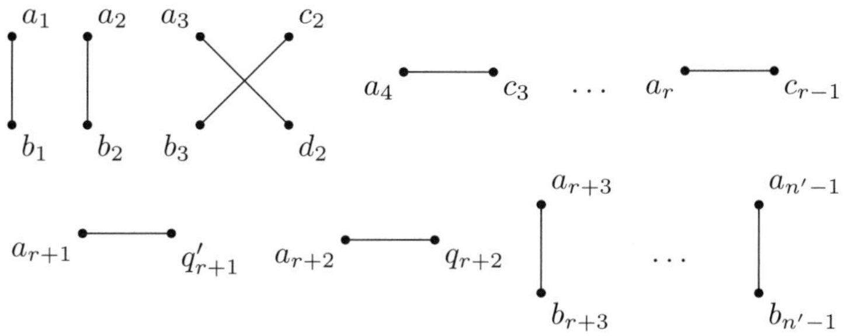

If $q'_{r+1} \notin L_{n'}$, we can construct the corresponding ultrafilters $a_1^*, \dots, a_{n'}^*, b_1^*, \dots, b_{n'}^*$ in a similar way.

4) We have achieved the goal formulated at the beginning of the proof, where we assumed

$$4(n'-1) - \mathfrak{v}(n'-1) \geq m-1.$$

Since this assumption is true for $m = 6$ and $n' = 9$ (see 14.17), the statement is proved for $n = 9, 10$. Now, this assumption is true for $m = 7$ and $n' = 11$, and the statement is proved for $n = 11, 12$, etc. □

14.19. Until now we were estimating the function $\mathfrak{v}(n)$ from above. Let us show that $3n - 2 \leq \mathfrak{v}(n)$ for all n. Let $n > 1$.[5] Let us take $3n-3$ pairwise distinct ultrafilters

$$a_1, \dots, a_{n-1}, b_1, \dots, b_{n-1}, c_1, \dots, c_{n-1}.$$

Let $\mathcal{A}_1 = \cdots = \mathcal{A}_n$, $X \in \mathcal{A}_k$, and $\big\{\{a_i, b_i, c_i\}\big\}_{i \in [1,n-1]}$ be the family of all $\mathcal{A}_k$-similar sets.

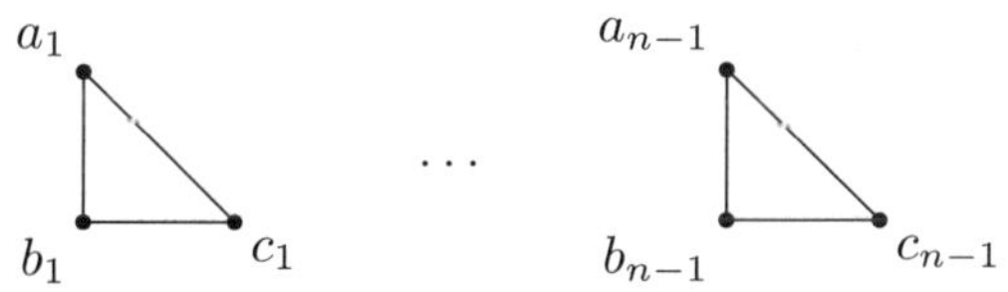

Obviously, for the algebras $\mathcal{A}_1, \dots, \mathcal{A}_n$ the corresponding sets $U_1, \dots, U_n, V_1, \dots, V_n$ do not exist. But since $\#(\ker \mathcal{A}_k) = 3n - 3$, it follows that $3n - 2 \leq \mathfrak{v}(n)$. □

14.20. Let us summarize: in the previous six sections we have proved Theorem 2.38.

[5] If $n = 1$, there is nothing to prove.

CHAPTER 15

Preliminary Notions and Lemmas[1]

15.1. In this chapter we consider matrices which have a finite number of rows, and each row has a finite number of elements.

DEFINITION. A matrix

$$\mathfrak{U} = \begin{pmatrix} U_1^1 & \dots & U_{m_1}^1 \\ \dots & \dots & \dots \\ U_1^k & \dots & U_{m_k}^k \\ \dots & \dots & \dots \\ U_1^n & \dots & U_{m_n}^n \end{pmatrix}$$

is said to be *regular* if the sequence $m_1, \dots, m_k, \dots, m_n$ is nondecreasing.

15.2. Everywhere in this chapter, if a matrix $\mathfrak{U}$ (not necessarily regular) is considered, then it is assumed to contain n rows, with k-th row containing m_k elements.

15.3. DEFINITION. Let $\varphi(k)$ be a function defined on the set $[1, n]$ and taking values in $\mathbb{N}^+$. A matrix $\mathfrak{U}$ is said to be *of type* $\varphi(k)$ if $\varphi(k) \leq \sum_{i=1}^{k} m_i$ for all $k \leq n$.

15.4. DEFINITION. Two matrices $\mathfrak{U}$ and $\mathfrak{D}$ are called *similar* if there exists a one-to-one correspondence between the rows of $\mathfrak{U}$ and the rows of $\mathfrak{D}$ such that any two corresponding rows coincide.

15.5. Obviously, for each matrix there exists a regular matrix similar to it.

15.6. LEMMA. *Let $\mathfrak{U}$ be a regular matrix of type $k(k+1)/2$, and let there exist a natural number $p < n$ such that*

$$\sum_{i=1}^{p} m_i = \frac{p(p+1)}{2}.$$

Consider a matrix $\mathfrak{U}'$ containing $n - p$ rows. For each $k \leq n - p$, let all elements of the k-th row of $\mathfrak{U}'$ be elements of the $(k+p)$-th row of $\mathfrak{U}$, and in this $(k+p)$-th row let there be at most p elements not belonging to $\mathfrak{U}'$. Let $\mathfrak{U}''$ be a regular matrix similar to $\mathfrak{U}'$. Then $\mathfrak{U}''$ is of type $k(k+1)/2$.

[1]These preliminary notions and lemmas are used in the next chapter.

PROOF. Let us construct a regular matrix $\mathfrak{U}^*$ containing $n-p$ rows, as follows: the k-th row of $\mathfrak{U}^*$ is obtained from the $(k+p)$-th row of $\mathfrak{U}$ by removing p elements. (Since $\mathfrak{U}$ is a regular matrix and $\sum_{i=1}^{p} m_i = p(p+1)/2$, the $(p+1)$-th row of $\mathfrak{U}$ has at least $p+1$ elements; hence our removal can be performed.) Let the first k rows of $\mathfrak{U}^*$ contain α elements. Clearly,

$$\alpha \geq \frac{(k+p)(k+p+1)}{2} - \frac{p(p+1)}{2} - kp = \frac{k(k+1)}{2}.$$

Hence $\mathfrak{U}^*$ is of type $k(k+1)/2$. But the first k rows of $\mathfrak{U}''$ contain at least as many elements as the first k rows of $\mathfrak{U}^*$. Hence $\mathfrak{U}''$ is of type $k(k+1)/2$. □

15.7. LEMMA. *Let $\mathfrak{U}$ be a regular matrix of type $k(k+1)/2$, let the minimal natural number p_0 such that $\sum_{i=1}^{p_0} m_i = p_0(p_0+1)/2$ exist, and let $p_0 > 1$.*[2] *Then $p_0 > m_{p_0}$.*

PROOF. Suppose that $m_{p_0} \geq p_0$. Then $\sum_{i=1}^{p_0-1} m_i \leq (p_0-1)p_0/2$. But this is impossible by the minimality of p_0. □

15.8. LEMMA. *Let $\mathfrak{U}$ be a regular matrix of type $k(k+1)/2$, let the minimal natural number p_0 such that $\sum_{i=1}^{p_0} m_i = p_0(p_0+1)/2$ exist, and let $p_0 > 1$. Let $r_* \in [1, p_0]$, and let a matrix $\mathfrak{U}'$ contain $n-1$ rows. If $k < r_*$, then let all elements of the k-th row of $\mathfrak{U}'$ be elements of the k-th row of $\mathfrak{U}$. If $r_* \leq k < n$, then let all elements of the k-th row of $\mathfrak{U}'$ be elements of the $(k+1)$-th row of $\mathfrak{U}$. If $k < r_*$, then let one of the following two cases hold:*

(1) *the k-th row of $\mathfrak{U}'$ coincides with the k-th row of $\mathfrak{U}$;*
(2) *in the k-th row of $\mathfrak{U}$ there is only one element not belonging to $\mathfrak{U}'$.*

If $r_* \leq k < p_0$, then let one of the following two cases hold:

(3) *the k-th row of $\mathfrak{U}'$ coincides with the $(k+1)$-th row of $\mathfrak{U}$;*
(4) *in the $(k+1)$-th row of $\mathfrak{U}$ there is only one element not belonging to $\mathfrak{U}'$.*

If $p_0 \leq k < n$, then let one of the following two cases hold:

(5) *in the $(k+1)$-th row of $\mathfrak{U}$ there are either one or no elements not belonging to $\mathfrak{U}'$;*
(6) *in the $(k+1)$-th row of $\mathfrak{U}$ there are only two elements not belonging to $\mathfrak{U}'$.*[3]

The following is assumed: case (1) can hold for any $k < r_*$; case (3) can hold for any k such that $r_* \leq k < p_0$; case (5) can hold for any k such that $p_0 \leq k < n$; each of the cases (2), (4), (6) can hold only for one k and if one of these cases holds, then the other two do not. Let $\mathfrak{U}''$ be a regular matrix similar to $\mathfrak{U}'$. Then $\mathfrak{U}''$ is of type $k(k+1)/2$.

PROOF. Let us construct a regular matrix $\mathfrak{U}^*$ containing $n-1$ rows.

Let $r_* > 1$. The first row of $\mathfrak{U}^*$ is obtained from the first row of $\mathfrak{U}$ by removing one element (we always have $m_1 > 1$, since $p_0 > 1$); if $1 < k < r_*$, then the k-th row of $\mathfrak{U}^*$ coincides with the k-th row of $\mathfrak{U}$; if $r_* \leq k < p_0$, then the k-th row of $\mathfrak{U}^*$ coincides with the $(k+1)$-th row of $\mathfrak{U}$.

---

[2] This means that if $k < p_0$, then $k(k+1)/2 < \sum_{i=1}^{k} m_i$.

[3] Since $p_0 > 1$, we have $m_{k+1} \geq 3$. (Actually one necessarily has $p_0 \geq 3$ and $m_{k+1} \geq 4$.) Similar simple arguments justify that the conditions in Lemmas 15.9, 15.11, 15.12, and 15.15 are reasonable.

Let $r_* = 1$. The first row of $\mathfrak{U}^*$ is obtained from the second row of $\mathfrak{U}$ by removing one element; if $1 < k < p_0$, then the k-th row of $\mathfrak{U}^*$ is the $(k+1)$-th row of $\mathfrak{U}$.

In either case ($r_* > 1$ or $r_* = 1$), if $p_0 \le k < n$, then the k-th row of $\mathfrak{U}^*$ is obtained from the $(k+1)$-th row of $\mathfrak{U}$ by removing one element. Let us prove that $\mathfrak{U}^*$ is of type $k(k+1)/2$. Let α be the number of elements in the first k rows of $\mathfrak{U}^*$, and let $k < r_*$. We know that $k < r_* \le p_0$. Hence

$$\alpha = \sum_{i=1}^{k} m_i - 1 \ge \frac{k(k+1)}{2}.$$

Let β be the number of elements in the first k rows of $\mathfrak{U}^*$, and let $r_* \le k < p_0$. Taking into account the fact that $\mathfrak{U}$ is a regular matrix, we obtain:

$$\beta = \sum_{i=1}^{k+1} m_i - 1 - m_{r_*} \ge \sum_{i=1}^{k} m_i - 1 \ge \frac{k(k+1)}{2}.$$

Let γ be the number of elements in the first k rows of $\mathfrak{U}^*$, and let $k \ge p_0$. Clearly,

$$\gamma = \sum_{i=1}^{k+1} m_i - 1 - m_{r_*} - (k - p_0 + 1).$$

By Lemma 15.7, $p_0 - 1 \ge m_{p_0}$. Since $m_{p_0} \ge m_{r_*}$, we can write:

$$\gamma \ge \sum_{i=1}^{k+1} m_i - 1 - m_{r_*} - k + m_{p_0} \ge \frac{(k+1)(k+2)}{2} - (k+1) = \frac{k(k+1)}{2}.$$

So $\mathfrak{U}^*$ is of type $k(k+1)/2$. But the first k rows of the matrix $\mathfrak{U}''$ contain at least as many elements as the first k rows of the matrix $\mathfrak{U}^*$. Hence $\mathfrak{U}''$ is of type $k(k+1)/2$. □

15.9. LEMMA. *Let $\mathfrak{U}$ be a regular matrix of type $k(k+1)/2$, and let $m_1 = 1$ and $n > 1$. Let some elements of $\mathfrak{U}$ be called marked; the element U_1^1 is not marked. If $k \in [2, n]$, then let one of the following two cases hold:*

(1) *in the k-th row there are either one or no marked elements;*
(2) *in the k-th row there are only two marked elements.*

It is assumed that: case (1) is possible for any $k \in [2, n]$; case (2) is possible only for one k. Let $\mathfrak{U}'$ be the matrix obtained from $\mathfrak{U}$ by removing the first row and all marked elements. Then there are only four possible cases:

I. $\mathfrak{U}' = \varnothing$;
II. *$n > 2$, and $\mathfrak{U}'$ has $n - 2$ rows;*
III. *$\mathfrak{U}'$ contains $n - 1$ rows, and a regular matrix $\mathfrak{U}''$ similar to it is not of type $k(k+1)/2$;*
IV. *$\mathfrak{U}'$ contains $n - 1$ rows, and a regular matrix $\mathfrak{U}''$ similar to it is of type $k(k+1)/2$.*

Moreover, one of the first three cases holds if and only if there exist $\nu, p \in \mathbb{N}^+$ such that

(a) $1 < \nu \le p \le n$;
(b) *if $k \in [2, p]$ and $k \ne \nu$, then the k-th row of $\mathfrak{U}$ has one marked element;*
(c) *in the ν-th row of $\mathfrak{U}$ there are two marked elements;*
(d) $\sum_{i=1}^{p} m_i = p(p+1)/2$.

PROOF. In Cases I and II, we have $m_2 = 2$, and U_1^2 and U_2^2 are the marked elements, $\nu = 2$. In Case I, $n = p = 2$, and in Case II one can put $p = 2$ and $n > 2$. If Cases I and II do not hold, then either Case III or Case IV holds.

Case III. Let us take $\mu_0 \in \mathbb{N}^+$ such that in the first μ_0 rows of $\mathfrak{U}''$ there are less than $\mu_0(\mu_0+1)/2$ elements. Let $\mu \in [1, \mu_0]$, and let the μ-th row of $\mathfrak{U}''$ either coincide with the k_μ-th row of $\mathfrak{U}$, or be obtained from the k_μ-th row of $\mathfrak{U}$ by removing all marked elements. Put

$$\mathbb{N}^* = \{1\} \cup \bigcup_{\mu=1}^{\mu_0} \{k_\mu\}.$$

Since $\mathfrak{U}$ is a regular matrix of type $k(k+1)/2$,

$$\sum_{i \in \mathbb{N}^*} m_i \geq \frac{(\mu_0+1)(\mu_0+2)}{2}.$$

Let τ be the number of elements in the first μ_0 rows of $\mathfrak{U}''$, and let

$$\xi = \sum_{k \in \mathbb{N}^*} \psi(k),$$

where $\psi(k)$ is the number of marked elements in the k-th row of $\mathfrak{U}$ (if there are no such elements, then $\psi(k) = 0$). Clearly,

$$\xi = \sum_{i \in \mathbb{N}^*} m_i - 1 - \tau \geq \frac{(\mu_0+1)(\mu_0+2)}{2} - 1 - \left(\frac{\mu_0(\mu_0+1)}{2} - 1\right) = \mu_0 + 1.$$

On the other hand, $\xi \leq \mu_0 + 1$ (U_1^1 is not marked, $1 \in \mathbb{N}^*$, $\#(\mathbb{N}^*) = \mu_0 + 1$, each row of $\mathfrak{U}$ contains at most two marked elements, and only one row can contain two marked elements). Now it is clear that

(i) $\xi = \mu_0 + 1$;
(ii) $\sum_{i \in \mathbb{N}^*} m_i = \frac{(\mu_0+1)(\mu_0+2)}{2}$;
(iii) there exists $\nu \in \mathbb{N}^* \setminus \{1\}$ such that in the ν-th row of $\mathfrak{U}$ there are two marked elements;
(iv) if $k \in \mathbb{N}^* \setminus \{1, \nu\}$, then in the k-th row of $\mathfrak{U}$ there is one marked element.

Let us prove that $\mathbb{N}^* = [1, \mu_0 + 1]$. Assume the contrary: there exists a $k_0 \in [1, \mu_0+1] \setminus \mathbb{N}^*$. Let k^* be the maximal number in $\mathbb{N}^*$. Clearly, $m_{k^*} \leq \mu_0 + 1$ (because, assuming the contrary, we come to a contradiction:

$$\frac{(\mu_0+1)(\mu_0+2)}{2} - m_{k^*} = \sum_{i \in \mathbb{N}^* \setminus \{k^*\}} m_i < \frac{\mu_0(\mu_0+1)}{2}$$

and $\#(\mathbb{N}^* \setminus \{k^*\}) = \mu_0$). Clearly, $m_{k_0} \leq m_{k^*}$. Hence

$$\sum_{i \in \mathbb{N}^* \cup \{k_0\}} m_i \leq \frac{(\mu_0+1)(\mu_0+2)}{2} + \mu_0 + 1 < \frac{(\mu_0+2)(\mu_0+3)}{2}.$$

But $\#(\mathbb{N}^* \cup \{k_0\}) = \mu_0 + 2$. Hence

$$\sum_{i \in \mathbb{N}^* \cup \{k_0\}} m_i \geq \frac{(\mu_0+2)(\mu_0+3)}{2}.$$

The contradiction thus obtained implies that $\mathbb{N}^* = [1, \mu_0 + 1]$. Put $p = \mu_0 + 1$; let ν be as in (iii). Then for the numbers ν and p conditions (a), (b), (c), (d) are satisfied.

So we have shown that there are only four possible cases: I, II, III, IV, and in the first three cases there exist numbers ν and p for which conditions (a), (b), (c), (d) are satisfied. Assume that either $m_2 = 2$ and the second row of $\mathfrak{U}$ contains only one marked element, or $m_2 > 2$; assume also that there exist numbers ν and p, for which conditions (a), (b), (c), (d) are satisfied. It remains to prove that then Case III holds. Clearly, $\mathfrak{U}''$ contains $n-1$ rows. Consider the natural one-to-one correspondence between the rows of the matrix $\mathfrak{U}$, from the second row until the n-th row, and all the rows of the matrix $\mathfrak{U}''$: the k-th row of $\mathfrak{U}$ corresponds to the η_k-th row of $\mathfrak{U}''$. This means that the η_k-th row of $\mathfrak{U}''$ either coincides with the k-th row of $\mathfrak{U}$, or is obtained from the k-th row of $\mathfrak{U}$ by removing all marked elements. Let m_j^* be the number of elements in the j-th row of $\mathfrak{U}''$. Clearly,

$$\sum_{1<k\le p} m_{\eta_k}^* = \sum_{i=1}^{p} m_i - 1 - p = \frac{p(p+1)}{2} - 1 - p < \frac{(p-1)p}{2}.$$

Since $\mathfrak{U}''$ is a regular matrix, it is not of type $k(k+1)/2$, and Case III holds. □

15.10. LEMMA. *Let $\mathfrak{U}$ be a regular matrix of type $(k^2+3k-2)/2$, and let there exist the minimal natural number p_0 such that $\sum_{i=1}^{p_0} m_i = (p_0^2+3p_0-2)/2$. Then $p_0 \ge m_{p_0}$.*

PROOF. If $p_0 = 1$, then the statement is obvious. Let $p_0 > 1$. Assume that $m_{p_0} > p_0$. Then

$$\sum_{i=1}^{p_0-1} m_i \le \frac{p_0^2+3p_0-2}{2} - p_0 - 1 = \frac{(p_0-1)^2+3(p_0-1)-2}{2}.$$

But this cannot be true by the minimality of p_0. □

15.11. LEMMA. *Let $\mathfrak{U}$ be a regular matrix of type $(k^2+3k-2)/2$, $n>1$, and let the minimal natural number p_0 such that $\sum_{i=1}^{p_0} m_i = (p_0^2+3p_0-2)/2$ exist. Let $r_* \in [1, p_0]$, and consider a matrix $\mathfrak{U}'$ containing $n-1$ rows. If $k < r_*$, then let all elements of the k-th row of $\mathfrak{U}'$ be elements of the k-th row of $\mathfrak{U}$. If $r_* \le k < n$, then let all elements of the k-th row of $\mathfrak{U}'$ be elements of the $(k+1)$-th row of $\mathfrak{U}$. If $k < r_*$, then let one of the following two cases hold:*

(1) *the k-th row of $\mathfrak{U}'$ coincides with the k-th row of $\mathfrak{U}$;*
(2) *in the k-th row of $\mathfrak{U}$ there is only one element not belonging to $\mathfrak{U}'$.*

If $r_* \le k < p_0$, then let one of the following two cases hold:

(3) *the k-th row of $\mathfrak{U}'$ coincides with the $(k+1)$-th row of $\mathfrak{U}$;*
(4) *in the $(k+1)$-th row of $\mathfrak{U}$ there is only one element not belonging to $\mathfrak{U}'$.*

If $p_0 \le k < n$, then let one of the following two cases hold:

(5) *in the $(k+1)$-th row of $\mathfrak{U}$ there are either one or no elements not belonging to $\mathfrak{U}'$;*
(6) *in the $(k+1)$-th row of $\mathfrak{U}$ there are only two elements not belonging to $\mathfrak{U}'$.*

The following is assumed: case (1) can hold for any $k < r_*$; case (3) can hold for any k such that $r_* \le k < p_0$; case (5) can hold for any k such that $p_0 \le k < n$; each of cases (2), (4), (6) can hold only for one k and if one of these cases holds, then the two other do not hold. Let $\mathfrak{U}''$ be a regular matrix similar to $\mathfrak{U}'$. Then $\mathfrak{U}''$ is of type $(k^2+3k-2)/2$.

PROOF. Let us construct a regular matrix $\mathfrak{U}^*$ containing $n-1$ rows.

Let $r_* > 1$. The first row of $\mathfrak{U}^*$ is obtained from the first row of $\mathfrak{U}$ by removing one element (since $1 < r_* \leq p_0$, we obviously have $m_1 > 1$); if $1 < k < r_*$, then the k-th row of $\mathfrak{U}^*$ coincides with the k-th row of $\mathfrak{U}$; if $r_* \leq k < p_0$, then the k-th row of $\mathfrak{U}^*$ coincides with the $(k+1)$-th row of $\mathfrak{U}$.

Let $r_* = 1$. The first row of $\mathfrak{U}^*$ is obtained from the second row of $\mathfrak{U}$ by removing one element; if $1 < k < p_0$, then the k-th row of $\mathfrak{U}^*$ is the $(k+1)$-th row of $\mathfrak{U}$.

In either case ($r_* > 1$ or $r_* = 1$), if $p_0 \leq k < n$, then the k-th row of $\mathfrak{U}^*$ is obtained from the $(k+1)$-th row of $\mathfrak{U}$ by removing one element. Let us prove that $\mathfrak{U}^*$ is of type $(k^2+3k-2)/2$. Let α be the number of elements in the first k rows of $\mathfrak{U}^*$, and let $k < r_*$. We know that $k < r_* \leq p_0$. Hence

$$\alpha = \sum_{i=1}^{k} m_i - 1 \geq \frac{k^2+3k-2}{2}.$$

Let β be the number of elements in the first k rows of $\mathfrak{U}^*$, and let $r_* \leq k < p_0$. Taking into account the fact that $\mathfrak{U}$ is a regular matrix, we obtain:

$$\beta = \sum_{i=1}^{k+1} m_i - 1 - m_{r_*} \geq \sum_{i=1}^{k} m_i - 1 \geq \frac{k^2+3k-2}{2}.$$

Let γ be the number of elements in the first k rows of $\mathfrak{U}^*$, and let $k \geq p_0$. Then

$$\gamma = \sum_{i=1}^{k+1} m_i - 1 - m_{r_*} - (k - p_0 + 1).$$

By Lemma 15.10, $p_0 \geq m_{p_0} \geq m_{r_*}$. Hence

$$\gamma \geq \sum_{i=1}^{k+1} m_i - 1 - m_{r_*} - k + m_{r_*} - 1 \geq \frac{(k+1)^2+3(k+1)-2}{2} - (k+2) = \frac{k^2+3k-2}{2}.$$

So $\mathfrak{U}^*$ is of type $(k^2+3k-2)/2$. But the first k rows of the matrix $\mathfrak{U}''$ contain at least as many elements as the first k rows of the matrix $\mathfrak{U}^*$. Hence $\mathfrak{U}''$ is of type $(k^2+3k-2)/2$. □

15.12. LEMMA. *Let $\mathfrak{U}$ be a regular matrix of type $(k^2+5k-4)/2$, and let there exist a natural number $p < n$ such that $(p^2+5p)/2 \geq \sum_{i=1}^{p} m_i$. Consider a matrix $\mathfrak{U}'$ containing $n-p$ rows. For each $k \leq n-p$, let all elements of the k-th row of $\mathfrak{U}'$ be elements of the $(k+p)$-th row of $\mathfrak{U}$, and in this $(k+p)$-th row let there be at most p elements not belonging to $\mathfrak{U}'$. Let $\mathfrak{U}''$ be a regular matrix similar to $\mathfrak{U}'$. Then $\mathfrak{U}''$ is of type $(k^2+5k-4)/2$.*

PROOF. Let us construct a regular matrix $\mathfrak{U}^*$ containing $n-p$ rows as follows: the k-th row of $\mathfrak{U}^*$ is obtained from the $(k+p)$-th row of $\mathfrak{U}$ by removing p elements. (Since $\mathfrak{U}$ is a regular matrix and $(p^2+5p)/2 \geq \sum_{i=1}^{p} m_i$, the $(p+1)$-th row of $\mathfrak{U}$ has at least

$$\frac{(p+1)^2+5(p+1)-4}{2} - \frac{p^2+5p}{2} = p+1$$

elements; hence our removal can be performed.) Let the first k rows of $\mathfrak{U}^*$ contain α elements. Obviously,

$$\alpha \geq \frac{(k+p)^2+5(k+p)-4}{2} - \frac{p^2+5p}{2} - kp = \frac{k^2+5k-4}{2}.$$

Hence $\mathfrak{U}^*$ is of type $(k^2+5k-4)/2$. But the first k rows of $\mathfrak{U}''$ contain at least as many elements as the first k rows of $\mathfrak{U}^*$. Hence $\mathfrak{U}''$ is of type $(k^2+5k-4)/2$. □

15.13. LEMMA. *Let $\mathfrak{U}$ be a regular matrix of type $(k^2+5k-4)/2$, let the minimal natural number p_0 such that $\sum_{i=1}^{p_0} m_i = (p_0^2+5p_0-4)/2$ exist, and let there exist a natural number $l < p_0$ such that $m_l = m_{l+1}$ and $\sum_{i=1}^{l} m_i = (l^2+5l-2)/2$. Then $p_0 - 1 = l > 1$ and $m_l = l+2$.*

PROOF. If $l = 1$, then $m_1 = m_2 = 2$. But this contradicts the fact that $\mathfrak{U}$ is of type $(k^2+5k-4)/2$. Hence $l > 1$. Assume that $p_0 - l > 1$. By the minimality of p_0,

$$\frac{l^2+5l-2}{2} + m_{l+1} \geq \frac{(l+1)^2+5(l+1)-2}{2}.$$

Hence $m_{l+1} \geq l+3$, and therefore $m_l \geq l+3$. Further,

$$\sum_{i=1}^{l-1} m_i = \frac{l^2+5l-2}{2} - m_l \leq \frac{l^2+5l-2}{2} - l - 3 = \frac{(l-1)^2+5(l-1)-4}{2}.$$

But this contradicts the fact that p_0 is the corresponding minimal number. Hence $l = p_0 - 1$. But then

$$m_l = m_{l+1} = \frac{(l+1)^2+5(l+1)-4}{2} - \frac{l^2+5l-2}{2} = l+2.$$ □

15.14. LEMMA. *Consider a matrix $\mathfrak{U}$, and numbers $\nu \in \mathbb{N}$ and $r \in \mathbb{N}^+$, $\nu < r \leq n$. If $\nu > 0$, let $\sum_{i=1}^{\nu} m_i \geq (\nu^2+5\nu)/2$; if $\nu < i \leq r$, let $m_i > r$. Then*

$$\sum_{i=1}^{r} m_i > \frac{r^2+5r-4}{2}.$$

PROOF. By assumption,

$$\sum_{i=1}^{r} m_i \geq \frac{\nu^2+5\nu}{2} + (r-\nu)(r+1) = \frac{2r^2-2r\nu+\nu^2+2r+3\nu}{2}.$$

It suffices to show that

$$\frac{2r^2-2r\nu+\nu^2+2r+3\nu}{2} - \frac{r^2+5r-4}{2} = \frac{(r-\nu)^2-3(r-\nu)+4}{2} > 0,$$

which is an easy exercise. □

15.15. LEMMA. *Let $\mathfrak{U}$ be a regular matrix of type $(k^2+5k-4)/2$, $n > 1$, and let the minimal natural number p_0 such that $\sum_{i=1}^{p_0} m_i = (p_0^2+5p_0-4)/2$ exist. Let $r_* \in [1, p_0]$, and consider a matrix $\mathfrak{U}'$ containing $n-1$ rows. If $k < r_*$, then let $\sum_{i=1}^{k} m_i \geq (k^2+5k)/2$, and let all elements of the k-th row of $\mathfrak{U}'$ be elements of the k-th row of $\mathfrak{U}$. If $r_* \leq k < n$, then let all elements of the k-th row of $\mathfrak{U}'$ be elements of the $(k+1)$-th row of $\mathfrak{U}$. If $k < r_*$, then let one of the following two cases hold:*

(1) *the k-th row of $\mathfrak{U}'$ coincides with the k-th row of $\mathfrak{U}$;*
(2) *in the k-th row of $\mathfrak{U}$ there are at least one but at most two elements not belonging to $\mathfrak{U}'$.*

If $r_* \leq k < p_0$, then let one of the following two cases hold:

(3) *the k-th row of $\mathfrak{U}'$ coincides with the $(k+1)$-th row of $\mathfrak{U}$;*

(4) *in the* $(k+1)$*-th row of* $\mathfrak{U}$ *there are at least one but at most two elements not belonging to* $\mathfrak{U}'$.

If $p_0 \leq k < n$, *then let one of the following two cases hold*:

(5) *in the* $(k+1)$*-th row of* $\mathfrak{U}$ *there are either one or no elements not belonging to* $\mathfrak{U}'$;

(6) *in the* $(k+1)$*-th row of* $\mathfrak{U}$ *there are at least two and at most three elements not belonging to* $\mathfrak{U}'$.

The following is assumed: *case* (1) *can hold for any* $k < r_*$; *case* (3) *can hold for any* k *such that* $r_* \leq k < p_0$; *case* (5) *can hold for any* k *such that* $p_0 \leq k < n$; *each of cases* (2), (4), (6) *can hold only for one* k *and if one of these cases holds, then the two other do not hold. Let* $\mathfrak{U}''$ *be a regular matrix similar to* $\mathfrak{U}'$. *Then* $\mathfrak{U}''$ *is of type* $(k^2+5k-4)/2$.

PROOF. Let us construct a regular matrix $\mathfrak{U}^*$.

Let $r_* > 1$. The first row of $\mathfrak{U}^*$ is obtained from the first row of $\mathfrak{U}$ by removing two elements; if $1 < k < r_*$, then the k-th row of $\mathfrak{U}^*$ coincides with the k-th row of $\mathfrak{U}$; if $r_* \leq k < p_0$, then the k-th row of $\mathfrak{U}^*$ is the $(k+1)$-th row of $\mathfrak{U}$. (We have $m_1 \geq 3$, since

$$\sum_{i=1}^{k} m_i \geq \frac{k^2+5k}{2} \quad \text{for} \quad k < r_*,$$

and hence our construction makes sense.)

Let $r_* = 1$. The first row of $\mathfrak{U}^*$ is obtained from the second row of $\mathfrak{U}$ by removing two elements; if $1 < k < p_0$, then the k-th row of $\mathfrak{U}^*$ is the $(k+1)$-th row of $\mathfrak{U}$. (We have $m_2 \geq 3$, since $\mathfrak{U}$ is a regular matrix of type $(k^2+5k-4)/2$, and hence our construction makes sense.)

In either case ($r_* > 1$ or $r_* = 1$), if $p_0 \leq k < n$, then the k-th row of $\mathfrak{U}^*$ is obtained from the $(k+1)$-th row of $\mathfrak{U}$ by removing one element.

Let us prove that $\mathfrak{U}^*$ is of type $(k^2+5k-4)/2$. Let us first show that if $k < p_0$, then in the first k rows of $\mathfrak{U}^*$ there are not less than $(k^2+5k-4)/2$ elements. Suppose the contrary, i.e., let there exist a natural number $l < p_0$ such that in the first l rows of $\mathfrak{U}^*$ there are α elements and

$$\alpha < \frac{l^2+5l-4}{2}.$$

By the assumption of the lemma, if $k < r_*$, then $\sum_{i=1}^{k} m_i \geq \frac{k^2+5k}{2}$. Hence $l \geq r_*$. Obviously,

$$\alpha = \sum_{i=1}^{l+1} m_i - 2 - m_{r_*}.$$

If $m_{l+1} > m_{r_*}$, then $\alpha \geq \sum_{i=1}^{l} m_i - 1$. This implies that $\alpha \geq (l^2+5l-4)/2$ (we have $l < p_0$, and p_0 is the minimal number). But this contradicts our assumption. Thus, under our assumption, we have $m_{r_*} = m_l = m_{l+1}$. By the minimality of p_0,

$$\sum_{i=1}^{l} m_i \geq \frac{l^2+5l-2}{2}.$$

But

$$\sum_{i=1}^{l} m_i \geq \frac{l^2+5l}{2} \quad \text{implies} \quad \alpha = \sum_{i=1}^{l+1} m_i - 2 - m_{r_*} = \sum_{i=1}^{l} m_i - 2 \geq \frac{l^2+5l-4}{2}.$$

Hence

$$\sum_{i=1}^{l} m_i = \frac{l^2+5l-2}{2}.$$

By Lemma 15.13, $l = p_0 - 1 > 1$ and $m_l = l+2$. Therefore

$$\sum_{i=1}^{l-1} m_i = \frac{l^2+5l-2}{2} - l - 2 = \frac{(l-1)^2+5(l-1)-2}{2}.$$

Let us recall again that if $k < r_*$, then $\sum_{i=1}^{k} m_i \geq (k^2+5k)/2$. Hence $r_* \leq l-1$ and $m_{l-1} = m_l$ (since $m_{r_*} = m_l$). Putting $l_* = l-1$, we obtain:

$$l_* < p_0, \quad m_{l_*} = m_{l_*+1}, \quad \sum_{i=1}^{l_*} m_i = \frac{l_*^2+5l_*-2}{2}.$$

By Lemma 15.13, $l_* = p_0 - 1$. But $l_* = p_0 - 2$. The contradiction thus obtained implies that if $k < p_0$, then the first k rows of $\mathfrak{U}^*$ contain not less than $(k^2+5k-4)/2$ elements.

Let β be the number of elements in the first k rows of $\mathfrak{U}^*$, and let $k \geq p_0$. Obviously,

$$\begin{aligned}\beta &\geq \frac{(k+1)^2+5(k+1)-4}{2} - 2 - m_{r_*} - (k+1-p_0)\\ &= \frac{k^2+5k-4+2p_0-2m_{r_*}}{2}.\end{aligned}$$

Let us prove that $p_0 \geq m_{r_*}$. Suppose the contrary and put $\nu = r_* - 1$, $r = p_0$. By Lemma 15.14,

$$\sum_{i=1}^{p_0} m_i > \frac{p_0^2+5p_0-4}{2}.$$

The contradiction thus obtained implies that $p_0 \geq m_{r_*}$. It follows that

$$\beta \geq \frac{k^2+5k-4}{2}.$$

So we have proved that the matrix $\mathfrak{U}^*$ is of type $(k^2+5k-4)/2$. But the first k rows of the matrix $\mathfrak{U}''$ contain at least as many elements as the first k rows of the matrix $\mathfrak{U}^*$. Hence $\mathfrak{U}''$ is of type $(k^2+5k-4)/2$. □

15.16. Let us proceed to the study of other objects.

DEFINITION. By a *system*, or, more exactly, a *finite system*, $\{\mathfrak{A}, \mathfrak{U}\}$ we mean a finite sequence of algebras $\mathfrak{A} = \{\mathcal{A}_1, \dots, \mathcal{A}_k, \dots, \mathcal{A}_n\}$ and a matrix of pairwise disjoint sets

$$\mathfrak{U} = \begin{pmatrix} U_1^1 & \dots & U_{m_1}^1 \\ \dots & \dots & \dots \\ U_1^k & \dots & U_{m_k}^k \\ \dots & \dots & \dots \\ U_1^n & \dots & U_{m_n}^n \end{pmatrix},$$

where $U_i^k \notin \mathcal{A}_k$.[4]

[4]In Chapter 16 we will meet countable systems which have a countable number of algebras and, accordingly, infinite matrices. This is why in Definition 15.16 we say: "more exactly, by a finite system". Unless explicitly specified otherwise, below the term "system" means a finite system.

15.17. The following three definitions correspond to Definitions 15.1, 15.3, and 15.4 given above.

DEFINITION. A system $\{\mathfrak{A}, \mathfrak{U}\}$ is said to be *regular* if $\mathfrak{U}$ is a regular matrix.

15.18. DEFINITION. A system $\{\mathfrak{A}, \mathfrak{U}\}$ is said to be *of type* $\varphi(k)$ if $\mathfrak{U}$ is of type $\varphi(k)$.

15.19. DEFINITION. Two systems $\{\mathfrak{A}, \mathfrak{U}\}$ and $\{\mathfrak{B}, \mathfrak{D}\}$ are said to be *similar* if there exists a one-to-one correspondence between the algebras from $\mathfrak{A}$ and $\mathfrak{B}$ such that if an algebra $\mathcal{A}_i \in \mathfrak{A}$ corresponds to an algebra $\mathcal{B}_j \in \mathfrak{B}$, then $\mathcal{A}_i = \mathcal{B}_j$ and the i-th row of $\mathfrak{U}$ coincides with the j-th row of $\mathfrak{D}$.

15.20. Below, to each set $U_i^k \in \mathfrak{U}$ we assign either an $\mathcal{A}_k$-special ultrafilter $s_i^k \ni U_i^k$, or $\mathcal{A}_k$-similar ultrafilters s_i^k, t_i^k; $s_i^k \ni U_i^k$, $t_i^k \not\ni U_i^k$. For the proof of the general theorems of the next chapter for finite systems we will consider the matrix

$$\mathfrak{W} = \begin{pmatrix} W_1^1 & \dots & W_{m_1}^1 \\ \dots & \dots & \dots \\ W_1^k & \dots & W_{m_k}^k \\ \dots & \dots & \dots \\ W_1^n & \dots & W_{m_n}^n \end{pmatrix},$$

where $s_i^k \ni W_i^k \subset U_i^k$.[5] Also, if the ultrafilter t_i^k is considered and $t_i^k \neq s_j^l$, then $t_i^k \not\ni \bigcup_{l,j} W_j^l$. Below we will move sets from one row to another. Assume that $t_i^k = s_j^l$, $k \neq l$, $m_l > 1$. Then the set W_j^l can be moved to the k-th row. This means the following: denote the set W_j^l by $W_{m_k+1}^k$, the ultrafilter s_j^l by $s_{m_k+1}^k$, and put $t_{m_k+1}^k = s_i^k$. If before the move we had $t_r^q = s_j^l$, then after the move we have $t_r^q = s_{m_k+1}^k$. In particular, $t_i^k = s_{m_k+1}^k$. If $j < p \leq m_l$, then denote U_p^l by U_{p-1}^l, s_p^l by s_{p-1}^l, and if t_p^l is considered, then denote t_p^l by t_{p-1}^l; if before the move we had $t_r^q = s_p^l$, then after the move we have $t_r^q = s_{p-1}^l$. As a result of the moves, one obtains a new matrix of pairwise disjoint sets, which constitutes a system together with the sequence of algebras $\mathfrak{A}$.

15.21. REMARK. Assume that s_i^k, s_j^k are $\mathcal{A}_k$-similar ultrafilters and

$$t_i^k \notin \{s_1^k, \dots, s_{m_k}^k\}.$$

Then we can take another ultrafilter $t_i^k = s_j^k$. Hence we will always assume that if s_i^k, s_j^k are $\mathcal{A}_k$-similar ultrafilters, then $t_i^k \in \{s_1^k, \dots, s_{m_k}^k\}$. We will call this assumption the *ultrafilter requirement*.

[5]In 16.6 and 16.8 we consider the infinite matrix $\mathfrak{W}$.

CHAPTER 16

Finite Sequences of Algebras (4). Proof of Theorems 2.39(1,2), 2.45(1,2)

16.1. THEOREM. *Let $\{\mathfrak{A}, \mathfrak{U}\}$ be a regular system of type $k(k+1)/2$. Then there exists a sequence of pairwise disjoint sets $U_1, \ldots, U_n$ such that*

$$U_k \subset Q \subset \bigcup_{i=1}^{n} U_i \quad \textit{implies} \quad Q \notin \mathcal{A}_k.$$

PROOF. If $n = 1$, then the statement of the theorem is obvious: put $U_1 = W_1^1$. (The sets W_i^k and the matrix $\mathfrak{W}$ are defined in 15.20.) *Let $n > 1$.*

Case 1. *The system $\{\mathfrak{A}, \mathfrak{W}\}$ is of type $k(k+1)/2 + 1$.* There exists a natural number m_1^* with the following property: any matrix with n rows such that the first row contains m_1^* elements and if $1 < k \leq n$, then the k-th row contains m_k elements, is of type $k(k+1)/2$ and is not of type $k(k+1)/2 + 1$. Let us remove all sets W_i^1 for which $i > m_1^*$ from $\mathfrak{W}$. Denote the obtained matrix by $\mathfrak{W}^*$. Put $\mathfrak{A}^* = \mathfrak{A}$. In the next step one considers the regular system $\{\mathfrak{A}^*, \mathfrak{W}^*\}$ of type $k(k+1)/2$.

Case 2. *There exists the minimal natural number p_0 such that $\sum_{i=1}^{p_0} m_i = \frac{p_0(p_0+1)}{2}$, and there exists an ultrafilter $t_{i_0}^{k_0} = s_1^1$, where $1 < k_0 \leq p_0$.*[1] By assumption, $p_0 > 1$. Hence $m_1 > 1$. Let us move W_1^1 to the k_0-th row. Denote the obtained matrix by $\mathfrak{W}'$. Let $\mathfrak{W}^*$ be a regular matrix similar to $\mathfrak{W}'$. It is rather easy to show that $\mathfrak{W}^*$ is of type $k(k+1)/2$. Let us transpose the algebras in $\mathfrak{A}$ if necessary; denote the new sequence of n algebras by $\mathfrak{A}^*$; if the algebras have not been transposed, then $\mathfrak{A}^* = \mathfrak{A}$; the k-th algebra in $\mathfrak{A}^*$ is the algebra corresponding to the k-th row of $\mathfrak{W}^*$. In the next step one considers the regular system $\{\mathfrak{A}^*, \mathfrak{W}^*\}$ of type $k(k+1)/2$.

Case 3. *It is assumed that $m_1 = 1$ and if the ultrafilter t_1^1 is considered, then $t_1^1 \not\ni \bigcup_{k,i} W_i^k$.* Let us call the set W_i^k *marked* if the ultrafilter t_i^k is considered and $t_i^k = s_1^1$. By the ultrafilter requirement, in one *row* there cannot be more than one marked set. Consider the matrix $\mathfrak{W}'$ obtained from $\mathfrak{W}$ by removing the first row and all the marked sets. Let $\mathfrak{W}^*$ be a regular matrix similar to $\mathfrak{W}'$. By Lemma 15.6 (this lemma is used here for $p = 1$), $\mathfrak{W}^*$ is of type $k(k+1)/2$. If $\mathfrak{W}^* = \mathfrak{W}'$, then we put $\mathfrak{A}^* = \{\mathcal{A}_2, \ldots, \mathcal{A}_n\}$. If $\mathfrak{W}^*$ is obtained from $\mathfrak{W}'$ by a permutation of rows, then let us transpose the algebras in the sequence $\{\mathcal{A}_2, \ldots, \mathcal{A}_n\}$ in the same way; denote the new sequence of $n-1$ algebras by $\mathfrak{A}^*$. In the next step one considers the regular system $\{\mathfrak{A}^*, \mathfrak{W}^*\}$ of type $k(k+1)/2$. Put $U_1 = W_1^1$. The following are important:

(i) if the ultrafilter t_1^1 is considered, then $t_1^1 \not\ni \bigcup\{W_i^k \mid W_i^k \in \mathfrak{W}^*\}$;

---

[1] In general, there can be several ultrafilters of the type of $t_{i_0}^{k_0}$. Among them we choose only one.

(ii) if $W_i^k \in \mathfrak{W}^*$ and the ultrafilter t_i^k is considered, then $t_i^k \not\ni U_1$.

Case 4. *The minimal natural number* p_0 *such that* $\sum_{i=1}^{p_0} m_i = p_0(p_0+1)/2$ *exists,* $m_1 > 1$, *and there does not exist an ultrafilter* $t_{i_0}^{k_0} = s_1^1$, *where* $1 < k_0 \le p_0$. Let us call the set W_i^k ($k > p_0$) *marked* if the ultrafilter t_i^k is considered and $t_i^k = s_1^1$. Besides, if $t_1^1 = s_{\lambda_0}^{\nu_0}$ and $\nu_0 > 1$, then the set $W_{\lambda_0}^{\nu_0}$ is also called *marked.* By the ultrafilter requirement, if $p_0 < k \le n$ and $k \ne \nu_0$, then in the k-th row of $\mathfrak{W}$ there cannot be more than one marked set. If $k > p_0$ and $k = \nu_0$, then in the k-th row there are at most two marked sets. Consider the matrix $\mathfrak{W}'$ obtained from $\mathfrak{W}$ by removing the first row and all the marked sets. Let $\mathfrak{W}^*$ be a regular matrix similar to $\mathfrak{W}'$. Then $\mathfrak{W}^*$ has $n-1$ rows, and, by Lemma 15.8 (this lemma is used here for $r_* = 1$), $\mathfrak{W}^*$ is of type $k(k+1)/2$. Now, as in Case 3, we construct a regular system $\{\mathfrak{A}^*, \mathfrak{W}^*\}$ of type $k(k+1)/2$, which is considered in the next step, put $U_1 = W_1^1$, and emphasize that conditions (i), (ii) are satisfied.

Case 5. $m_1 = 1$, $t_1^1 = s_{\lambda_0}^{\nu_0}$. Let us call the set W_i^k *marked* if the ultrafilter t_i^k is considered and $t_i^k = s_1^1$. The set $W_{\lambda_0}^{\nu_0}$ is also called *marked.* Let us divide this case into two subcases:

Case 5-a. *Removing the first row and all the marked sets from* $\mathfrak{W}$, *we obtain a matrix* $\mathfrak{W}'$ *with* $n-1$ *rows; a matrix* $\mathfrak{W}^*$ *similar to* $\mathfrak{W}'$ *is of type* $k(k+1)/2$.

Case 5-b. *Case* 5-a *does not hold.*

In Case 5-a we arrive at the same results as in Cases 3 and 4: we construct a regular system $\{\mathfrak{A}^*, \mathfrak{W}^*\}$ of type $k(k+1)/2$, which is considered in the next step, put $U_1 = W_1^1$, and emphasize that conditions (i), (ii) are satisfied.

Consider Case 5-b. By Lemma 15.9, our case holds if and only if there exists a natural number p such that

(a) $p \in [\nu_0, n]$;
(b) if $k \in [2,p] \setminus \{\nu_0\}$, then the k-th row of $\mathfrak{W}$ contains one marked set $W_{\lambda_k}^k$;
(c) the ν_0-th row of $\mathfrak{W}$ contains two marked sets: $W_{\lambda_0}^{\nu_0}, W_{\lambda_{\nu_0}}^{\nu_0}$;
(d) $\sum_{i=1}^{p} m_i = p(p+1)/2$.[2]

Put $U_1 = W_{\lambda_0}^{\nu_0}$; if $k \in [2,p]$, then put $U_k = W_{\lambda_k}^k$. If $p = n$, then the theorem is proved. Let $p < n$. Let us call the set W_i^k, $k > p$, *distinguished* if the ultrafilter t_i^k is considered and $t_i^k \ni \bigcup_{j \le p} U_j$. By the ultrafilter requirement, in one row there can be at most p distinguished sets. Consider the matrix $\mathfrak{W}'$ obtained from $\mathfrak{W}$ by removing the first p rows and all the distinguished sets. Let $\mathfrak{W}^*$ be a regular matrix similar to $\mathfrak{W}'$. By Lemma 15.6, $\mathfrak{W}^*$ is of type $k(k+1)/2$. In the next step one considers the regular system $\{\mathfrak{A}^*, \mathfrak{W}^*\}$ of type $k(k+1)/2$ ($\mathfrak{A}^*$ contains the algebras $\mathcal{A}_{p+1}, \dots, \mathcal{A}_n$, but the order of these algebras is changed if necessary, so that the k-th algebra in $\mathfrak{A}^*$ corresponds to the k-th row of $\mathfrak{W}^*$). The following are important:

(i′) if $Q \supset U_k$ and $Q \cap W_1^1 = \varnothing$, then $Q \notin \mathcal{A}_k$;
(ii′) if $W_i^k \in \mathfrak{W}^*$ and the ultrafilter t_i^k is considered, then $t_i^k \not\ni \bigcup_{j \le p} U_j$.

We have examined all possible cases. If all the required sets $U_1, \dots, U_n$ have not been found yet, then consider the regular system $\{\mathfrak{A}^*, \mathfrak{W}^*\}$ of type $k(k+1)/2$, etc. Finally we will find all required sets $U_1, \dots, U_n$. □

[2] In general, there can be several numbers p satisfying conditions (a)–(d). Among them we consider only one chosen arbitrarily. But it is natural to consider the maximal number among them, since in Case 5-b one constructs the required sets $U_1, \dots, U_p$.

16.2. REMARK. The proof of Theorem 16.1 implies that one can take the required set U_k to be equal to $W_{j_k}^{i_k}$. This implies a simplest refinement of Theorem 16.1: each required set $U_k \subset U_{j_k}^{i_k}$ belongs to $\mathfrak{U}$, and if $1 \le k < l \le n$, then $U_{j_k}^{i_k} \ne U_{j_l}^{i_l}$.

16.3. A matrix $\mathfrak{U}$ with $m_k = k$ for each $k \le n$ is of type $k(k+1)/2$. By Theorem 16.1, this implies that Theorem 2.39(1) is true.

16.4. PROOF OF THEOREM 2.39(2). Let us show that the theorem is true for $n = 5$. Note that $X \in \mathcal{A}_k$ for all k, and for each k let us list all the $\mathcal{A}_k$-similar sets:

for $k = 1$: $\{a_1^1, a_1^2\}$;
for $k = 2$: $\{a_1^2, a_2^2\}$;
for $k = 3$: $\{a_1^3, a_1^1\}$, $\{a_2^3, a_2^2\}$, $\{a_3^3, a_3^4\}$;
for $k = 4$: $\{a_1^4, a_1^1\}$, $\{a_2^4, a_2^2\}$, $\{a_3^4, a_4^4\}$;
for $k = 5$: $\{a_1^5, a_1^1\}$, $\{a_2^5, a_2^2\}$, $\{a_3^5, a_3^3\}$, $\{a_4^5, a_4^4\}$.

Let us assume that $a_i^k \ni U_i^k$.

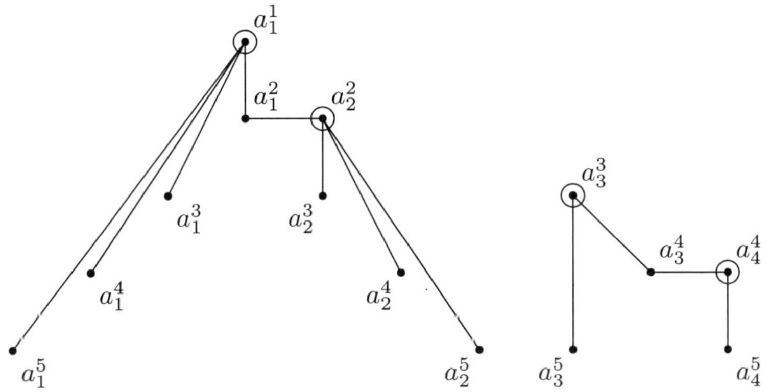

The statement of Theorem 2.39(1) is true for our algebras $\mathcal{A}_1, \mathcal{A}_2, \mathcal{A}_3, \mathcal{A}_4$; we necessarily have

$$U_1 \in a_1^1, \quad U_2 \in a_2^2, \quad U_3 \in a_3^3, \quad U_4 \in a_4^4.$$

But this implies that the statement of Theorem 2.39(1) cannot be true for our algebras $\mathcal{A}_1, \mathcal{A}_2, \mathcal{A}_3, \mathcal{A}_4, \mathcal{A}_5$. Similarly one proves our theorem for any $n > 5$. □

16.5. Let us return to Theorem 16.1. We know that the sets $U_1, \dots, U_n$ can be chosen in such a way that $U_k \subset U_{j_k}^{i_k}$, and if $1 \le k < l \le n$, then $U_{j_k}^{i_k} \ne U_{j_l}^{i_l}$. As will be shown below, if a regular system $\{\mathfrak{A}, \mathfrak{U}\}$ is of type $(k^2 + 3k - 2)/2$, then one can assert more: the sets $U_1, \dots, U_n$ can be chosen so that $i_k \le k$ for each $k \le n$. It will be also shown below that if a regular system $\{\mathfrak{A}, \mathfrak{U}\}$ is of type $(k^2 + 5k - 4)/2$, then one can assert even more: the sets $U_1, \dots, U_n$ can be chosen so that $i_k = k$ for each $k \le n$. If, as in Theorem 16.1, a regular system $\{\mathfrak{A}, \mathfrak{U}\}$ is of type $k(k+1)/2$, then the sets $U_1, \dots, U_n$ can be chosen so that there is, in a sense, some substantial information on the set $\{k \le n \mid i_k > k\}$. This is specified in the following theorem, which is a refinement of Theorem 16.1.

THEOREM. *Let $\{\mathfrak{A}, \mathfrak{U}\}$ be a regular system of type $k(k+1)/2$. Then there exists a sequence of sets $U_1, \dots, U_n$ such that*

$$U_k \subset Q \subset \bigcup_{i=1}^{n} U_i \quad \textit{implies} \quad Q \notin \mathcal{A}_k.$$

Moreover,

1) $U_k \subset U_{j_k}^{i_k}$; *if* $1 \leq k < l \leq n$, *then* $U_{j_k}^{i_k} \neq U_{j_l}^{i_l}$;
2) *if* $i_k > k$, *then there exists an* $r_k \geq i_k$ *such that* $U_{r_k} \subset U_{j_{r_k}}^{r_k}$;
3) *the map* $k \to r_k$ *is one-to-one.*

PROOF. The difference between the proof of Theorem 16.1 and the proof of our theorem is that instead of Case 2 one considers Case 2*, and instead of Case 4 one considers Case 4*. In Cases 1, 3, 5-a, 5-b there are no changes. So assume that at some step we come to considering a regular system $\{\widetilde{\mathfrak{A}}, \widetilde{\mathfrak{W}}\}$ of type $k(k+1)/2$, and $\#(\widetilde{\mathfrak{A}}) > 1$. Let

$$\widetilde{\mathfrak{A}} = \{\widetilde{\mathcal{A}}_1, \dots, \widetilde{\mathcal{A}}_\vartheta\},$$
$$\widetilde{\mathfrak{W}} = \begin{pmatrix} \widetilde{W}_1^1 & \dots & \widetilde{W}_{\widetilde{m}_1}^1 \\ \dots & \dots & \dots \\ \widetilde{W}_1^k & \dots & \widetilde{W}_{\widetilde{m}_k}^k \\ \dots & \dots & \dots \\ \widetilde{W}_1^\vartheta & \dots & \widetilde{W}_{\widetilde{m}_\vartheta}^\vartheta \end{pmatrix};$$

let there correspond to the set $\widetilde{W}_i^k$ either an $\widetilde{\mathcal{A}}_k$-special ultrafilter $\tilde{s}_i^k$, or $\widetilde{\mathcal{A}}_k$-similar ultrafilters $\tilde{s}_i^k, \tilde{t}_i^k$. Consider the natural one-to-one map from the rows of $\widetilde{\mathfrak{W}}$ to the rows of $\mathfrak{W}$. Let the k-th row of $\widetilde{\mathfrak{W}}$ correspond to the ω_k-th row of $\mathfrak{W}$ (clearly, $\widetilde{\mathcal{A}}_k = \mathcal{A}_{\omega_k}$). Naturally, we have $\widetilde{W}_i^k = W_\eta^\xi \in \mathfrak{W}$ and $\tilde{s}_i^k = s_\eta^\xi$. If W_η^ξ has never been moved to another row, then $\xi = \omega_k$; if also in this case the ultrafilter $t_\eta^{\omega_k}$ is considered, then $\tilde{t}_i^k = t_\eta^{\omega_k}$. Denote by Υ the totality of all sets U_k found up to this step. Let us assume that $U_k = W_{j_k}^{i_k}$. Put

$$K = \{k \mid U_k \in \Upsilon\}, \quad \Omega = \{\omega_1, \dots, \omega_\vartheta\}.$$

Let us assume that $K \cap \Omega = \varnothing$ and $K \cup \Omega = [1, n]$. Put

$$V = X \setminus \bigcup \{W_i^k \mid W_i^k \in \Upsilon \cup \widetilde{\mathfrak{W}}\}.$$

Let us assume that

(i″) if $Q \supset U_k$ and $Q \cap V = \varnothing$, then $Q \notin \mathcal{A}_k$;
(ii″) if the ultrafilter $\tilde{t}_i^k$ is considered, then $\tilde{t}_i^k \not\ni \bigcup \Upsilon$.

Let us also assume that if $U_k = W_{j_k}^{i_k}$ and $i_k > k$, then the set $U_{r_k} = W_{j_{r_k}}^{r_k}$ has been found, and for all the sets U_k already found such that $i_k > k$, the map $k \to r_k$ is one-to-one. Clearly, all our assumptions make sense if they remain valid after we proceed to the new system. Our constructions will imply that this is the case.

Case 2*. *The minimal natural number* ρ_0 *such that* $\sum_{i=1}^{\rho_0} \widetilde{m}_i = \rho_0(\rho_0+1)/2$ *exists. Let* $\omega_{r_*} = \min(\omega_1, \dots, \omega_{\rho_0})$, *and let there exist an ultrafilter* $\tilde{t}_{i_0}^{k_0} = \tilde{s}_1^{r_*}$, *where* $k_0 \in [1, \rho_0] \setminus \{r_*\}$.[3] Let us move $\widetilde{W}_1^{r_*}$ from the r_*-th row of $\widetilde{\mathfrak{W}}$ to the k_0-th row

[3] In general, there can exist several ultrafilters of the type of $\tilde{t}_{i_0}^{k_0}$. Among them we consider only one.

of $\widetilde{\mathfrak{W}}$. Denote the obtained matrix by $\widetilde{\mathfrak{W}}'$. Let $\widetilde{\mathfrak{W}}^*$ be a regular matrix similar to $\widetilde{\mathfrak{W}}'$. Clearly, $\widetilde{\mathfrak{W}}^*$ is of type $k(k+1)/2$. Similarly to Case 2 in the proof of Theorem 16.1, in the next step we will consider the regular system $\{\widetilde{\mathfrak{A}}^*, \widetilde{\mathfrak{W}}^*\}$ of type $k(k+1)/2$.

Case 4. The minimal natural number ρ_0 such that $\sum_{i=1}^{\rho_0} \widetilde{m}_i = \rho_0(\rho_0+1)/2$ exists and $\widetilde{m}_1 > 1$. Let $\omega_{r_*} = \min(\omega_1, \dots, \omega_{\rho_0})$, and assume that there does not exist an ultrafilter $\tilde{t}_{i_0}^{k_0} = \tilde{s}_1^{r_*}$, where $k_0 \in [1, \rho_0] \setminus \{r_*\}$.* Let us call the set $\widetilde{W}_i^k$, $k > \rho_0$, *marked* if the ultrafilter $\tilde{t}_i^k$ is considered and $\tilde{t}_i^k = \tilde{s}_1^{r_*}$. If $\tilde{t}_1^{r_*} = \tilde{s}_{\lambda_0}^{\nu_0}$ and $\nu_0 \neq r_*$, then let us also call the set $\widetilde{W}_{\lambda_0}^{\nu_0}$ *marked*. Consider the matrix $\widetilde{\mathfrak{W}}'$ which is obtained from $\widetilde{\mathfrak{W}}$ by removing the r_*-th row and all the marked sets. Let $\widetilde{\mathfrak{W}}^*$ be a regular matrix similar to $\widetilde{\mathfrak{W}}'$. Then $\widetilde{\mathfrak{W}}^*$ has $n-1$ rows, and, by Lemma 15.8, $\widetilde{\mathfrak{W}}^*$ is of type $k(k+1)/2$. One naturally defines the regular system $\{\widetilde{\mathfrak{A}}^*, \widetilde{\mathfrak{W}}^*\}$ of type $k(k+1)/2$, which is considered in the next step. Put $U_{\omega_{r_*}} = \widetilde{W}_1^{r_*}$.

Returning to Case 2*, let m_k^* be the number of elements in the k-th row of $\widetilde{\mathfrak{W}}^*$. The minimal natural number ρ_0^* such that $\sum_{i=1}^{\rho_0^*} m_i^* = \rho_0^*(\rho_0^*+1)/2$ exists. Also we have $\rho_0^* \le \rho_0$, and

$$\left\{\widetilde{W}_i^k \in \widetilde{\mathfrak{W}} \mid \widetilde{W}_i^k \text{ is contained in the first } \rho_0^* \text{ rows of } \widetilde{\mathfrak{W}}^*\right\} \subset \left\{\widetilde{W}_i^k \in \widetilde{\mathfrak{W}} \mid k \le \rho_0\right\}.$$

For the system $\{\widetilde{\mathfrak{A}}^*, \widetilde{\mathfrak{W}}^*\}$, one of the Cases 2*, 3, 4*, 5-a, 5-b holds. If Case 2* is repeated, then let us consider the new regular system of type $k(k+1)/2$, etc. Finally we will come to a regular system of type $k(k+1)/2$ for which one of the Cases 3, 4*, 5-a, 5-b holds. In Cases 3, 4*, 5-a we find U_{r_*} and in Case 5-b we find several of the sets $U_1, \dots, U_n$ and U_{r_*} among them.

Since in Case 2* we move a set from the r_*-th row of $\widetilde{\mathfrak{W}}$ to the k_0-th row of $\widetilde{\mathfrak{W}}$ and $\omega_{r_*} < \omega_{k_0}$, the situation from the statement of our theorem, when $i_k > k$, cannot arise after considering this case. Such a situation cannot arise also after considering Cases 1, 3, 4*, 5-a. It cannot arise also after considering a system $\{\widetilde{\mathfrak{A}}, \widetilde{\mathfrak{W}}\}$, where $\#(\widetilde{\mathfrak{A}}) = 1$. Such a situation can arise only after considering Case 5-b. So let Case 5-b hold for a system $\{\widetilde{\mathfrak{A}}, \widetilde{\mathfrak{W}}\}$. Hence,

$$\widetilde{m}_1 = 1, \quad \tilde{t}_1^1 = \tilde{s}_{\lambda_0}^{\nu_0}, \quad U_{\omega_1} = \widetilde{W}_{\lambda_0}^{\nu_0}, \quad U_{\omega_{\nu_0}} = \widetilde{W}_{\lambda_{\nu_0}}^{\nu_0}.$$ [4]

The situation in question can arise if and only if $k = \omega_1$, $\widetilde{W}_{\lambda_0}^{\nu_0} = W_{j_k}^{i_k}$, and $i_k > k$. Put $r_k = \omega_{\nu_0}$. If $W_{j_k}^{i_k}$ has not been moved, then $i_k = r_k$. If $W_{j_k}^{i_k}$ has been moved, then $i_k < r_k$.[5] We know that $\tilde{t}_{\lambda_{\nu_0}}^{\nu_0} = \tilde{s}_1^1$ and $k < r_k$. Hence $\widetilde{W}_{\lambda_{\nu_0}}^{\nu_0}$ could not come to the r_k-th row of $\mathfrak{W}$ by means of a move. Therefore

$$U_{r_k} = \widetilde{W}_{\lambda_{\nu_0}}^{\nu_0} = W_{j_{r_k}}^{r_k} \in \mathfrak{W}.$$

The map $k \to r_k$ is one-to-one since after considering our Case 5-b we remove the ν_0-th row from the matrix $\widetilde{\mathfrak{W}}$ corresponding to the r_k-th row of $\mathfrak{W}$. $\square$

---

[4] Here we use the notation corresponding to the notation from Case 5-b in the proof of Theorem 16.1.

[5] If $i_k < r_k$, then the set U_{i_k} was found earlier.

16.6. Although the title of the chapter is "Finite sequences of algebras", this and the following two sections are concerned with countable systems.

DEFINITION. By a *countable system* $\{\mathfrak{A}, \mathfrak{W}\}$ we mean a countable sequence of algebras $\mathfrak{A} = \{\mathcal{A}_1, \dots, \mathcal{A}_k, \dots\}$ and an infinite matrix of pairwise disjoint sets

$$\mathfrak{W} = \begin{pmatrix} W_1^1 & \dots & W_{m_1}^1 \\ \dots & \dots & \dots \\ W_1^k & \dots & W_{m_k}^k \\ \dots & \dots & \dots \end{pmatrix},$$

where $W_i^k \notin \mathcal{A}_k$.

16.7. For countable systems one naturally introduces the notions of regular countable system, countable system of type $\varphi(k)$, etc.

16.8. The next statement is used in Chapter 18 for the proof of Theorem 2.44, which is the infinite analog of Theorem 16.5.

STATEMENT. *Let $\{\mathfrak{A}, \mathfrak{W}\}$ be a regular countable system of type $k(k+1)/2$; for each set W_i^k let there exist either an $\mathcal{A}_k$-special ultrafilter s_i^k, or $\mathcal{A}_k$-similar ultrafilters s_i^k, t_i^k with $s_i^k \ni W_i^k$; either $t_i^k = s_j^l$, or $t_i^k \not\ni \bigcup_{l,j} W_j^l$. Then there exists a countable sequence of pairwise distinct sets $U_1, \dots, U_k, \dots$ such that*

$$U_k \subset Q \subset \bigcup_{i=1}^{\infty} U_i \quad \textit{implies} \quad Q \notin \mathcal{A}_k.$$

Moreover:

1) $U_k = W_{j_k}^{i_k}$;
2) *if $i_k > k$, then there exists $r_k \geq i_k$ such that $U_{r_k} = W_{j_{r_k}}^{r_k}$;*
3) *the map $k \to r_k$ is one-to-one.*

PROOF.[6] We look for sets U_k, and, if necessary, sets U_{r_k}, by proceeding, if necessary, from one regular countable system of type $k(k+1)/2$ to another regular countable system of type $k(k+1)/2$. (Below we will see that we will not consider any finite system at any step.) So assume that at some step we come to considering a regular countable system $\{\widetilde{\mathfrak{A}}, \widetilde{\mathfrak{W}}\}$ of type $k(k+1)/2$. Let

$$\widetilde{\mathfrak{A}} = \{\widetilde{\mathcal{A}}_1, \dots, \widetilde{\mathcal{A}}_k, \dots\},$$

$$\widetilde{\mathfrak{W}} = \begin{pmatrix} \widetilde{W}_1^1 & \dots & \widetilde{W}_{\widetilde{m}_1}^1 \\ \dots & \dots & \dots \\ \widetilde{W}_1^k & \dots & \widetilde{W}_{\widetilde{m}_k}^k \\ \dots & \dots & \dots \end{pmatrix}.$$

Case 1°. The system $\{\widetilde{\mathfrak{A}}, \widetilde{\mathfrak{W}}\}$ is of type $k(k+1)/2+1$. We argue exactly as in Case 1 in the proof of Theorem 16.1.

Case 2°. The minimal natural number ρ_0 such that $\sum_{i=1}^{\rho_0} \widetilde{m}_i = \rho_0(\rho_0+1)/2$ exists. Let $\omega_{r_*} = \min(\omega_1, \dots, \omega_{\rho_0})$, and let there exist an ultrafilter $\tilde{t}_{i_0}^{k_0} = \tilde{s}_1^{r_*}$, where $k_0 \in [1, \rho_0] \setminus \{r_*\}$. We argue exactly as in Case 2* from the proof of Theorem 16.5.

[6] The "background" of the proof of our theorem is formed by the proofs of Theorems 16.1 and 16.5; hence we use, without special reference, the methods and notions introduced there.

Case 3°. $\widetilde{m}_1 = 1$, *and if the ultrafilter* $\tilde{t}_1^1$ *is considered, then* $\tilde{t}_1^1 \not\ni \bigcup_{k,i} \widetilde{W}_i^k$. Let us call the set $\widetilde{W}_i^k$ *marked* if the ultrafilter $\tilde{t}_i^k$ is considered and $\tilde{t}_i^k = \tilde{s}_1^1$. In contrast to Case 3 from the proof of Theorem 16.1, if in the k-th row of $\widetilde{\mathfrak{W}}$, where $k > 1$, there are no marked sets, then we will call one arbitrary set $\widetilde{W}_i^k$ *marked.* Put $U_{\omega_1} = \widetilde{W}_1^1$, and in the next step we will proceed to the regular countable system of type $k(k+1)/2$, which is constructed in the natural way.

Case 4°. *The minimal natural number* ρ_0 *such that* $\sum_{i=1}^{\rho_0} \widetilde{m}_i = \frac{\rho_0(\rho_0+1)}{2}$ *exists and* $\widetilde{m}_1 > 1$. *Let* $\omega_{r_*} = \min(\omega_1, \dots, \omega_{\rho_0})$, *and assume that there does not exist an ultrafilter* $\tilde{t}_{i_0}^{k_0} = \tilde{s}_1^{r_*}$, *where* $k_0 \in [1, \rho_0] \setminus \{r_*\}$. Let us call the set $\widetilde{W}_i^k$, where $k > \rho_0$, *marked* if the ultrafilter $\tilde{t}_i^k$ is considered and $\tilde{t}_i^k = \tilde{s}_1^{r_*}$. If $\tilde{t}_1^{r_*} = \tilde{s}_{\lambda_0}^{\nu_0}$ and $\nu_0 \neq r_*$, then the set $\widetilde{W}_{\lambda_0}^{\nu_0}$ is also called *marked.* In contrast to Case 4* from the proof of Theorem 16.5, if in the k-th row of $\widetilde{\mathfrak{W}}$, where $k > \rho_0$, there are no marked sets, then we will call one arbitrary set $\widetilde{W}_i^k$ marked. Put $U_{\omega_{r_*}} = \widetilde{W}_1^{r_*}$, and in the next step let us proceed to the regular countable system of type $k(k+1)/2$, which is constructed in the natural way.

Case 5°. $\widetilde{m}_1 = 1$ *and* $\tilde{t}_1^1 = \tilde{s}_{\lambda_0}^{\nu_0}$. Let us call the set $\widetilde{W}_i^k$ *marked* if the ultrafilter $\tilde{t}_i^k$ is considered and $\tilde{t}_i^k = \tilde{s}_1^1$. The set $\widetilde{W}_{\lambda_0}^{\nu_0}$ is also called *marked.* Put

$$\Phi^k = \big\{\widetilde{W}_i^k \mid k > 1, \text{ the } k\text{-th row of } \widetilde{\mathfrak{W}} \text{ does not contain marked sets;}$$
$$\text{and either } \tilde{s}_i^k \text{ is an } \mathcal{A}_k\text{-special ultrafilter, or } \tilde{t}_i^k \not\ni \bigcup_{l,j} \widetilde{W}_j^l, \text{ or } \tilde{t}_i^k = \tilde{s}_j^k\big\}.$$

If $\Phi^k \neq \varnothing$, then, in contrast to Case 5 in the proof of Theorem 16.1, let us call one arbitrary set from Φ^k *marked.* Let us divide Case 5° into two subcases:

Case 5°-a. There does not exist a natural number ρ such that

(a) $\nu_0 \leq \rho$;
(b) if $k \in [2, \rho] \setminus \{\nu_0\}$, then the k-th row of $\widetilde{\mathfrak{W}}$ contains a marked set;
(c) the ν_0-th row of $\widetilde{\mathfrak{W}}$ contains two marked sets;
(d) $\sum_{i=1}^{\rho} \widetilde{m}_i = \rho(\rho+1)/2$.

Case 5°-b. There exists a natural number ρ such that conditions (a)–(d) hold.

The following arguments, *which apply in Case* 5°-a *only*, are absent in the proof of Theorem 16.1. Put

$$\Gamma = \big\{\gamma \in \mathbb{N}^+ \mid \gamma > 1, \text{ and the } \gamma\text{-th row of } \widetilde{\mathfrak{W}} \text{ does not contain marked sets}\big\}.$$

If γ_* is the minimal number from Γ and $\gamma_* < \gamma \in \Gamma$, then let us call one arbitrary set $\widetilde{W}_i^\gamma$ *marked.* Clearly, after introducing the new marked sets, we still remain in Case 5°-a. Now, as was done in the proof of Theorem 16.1, let us put $U_{\omega_1} = \widetilde{W}_1^1$, and in the next step let us proceed to a regular countable system of type $k(k+1)/2$, which is constructed in the natural way.

The following arguments *apply in Case* 5°-b *only*, and some of them do not make sense or are unnecessary in the proof of Theorem 16.1. Let us take an arbitrary number ρ satisfying conditions (a)–(d). If $k \in [2, \rho] \setminus \{\nu_0\}$, then denote by U_{ω_k} a marked set from the k-th row of $\widetilde{\mathfrak{W}}$. Put $U_{\omega_1} = \widetilde{W}_{\lambda_0}^{\nu_0}$, and denote another marked set in the ν_0-th row of $\widetilde{\mathfrak{W}}$ by $U_{\omega_{\nu_0}}$. If there exist arbitrarily large numbers ρ, then for each $k \in \mathbb{N}^+$ let us take the set U_{ω_k}; this completes the construction of all the required sets $U_1, \dots, U_k, \dots$, and our statement is proved. Otherwise let us take any one of the numbers ρ, and consider the sets $U_{\omega_1}, \dots, U_{\omega_\rho}$. The set $\widetilde{W}_i^k$, where $k > \rho$, is called *distinguished* if the ultrafilter $\tilde{t}_i^k$ is considered and $\tilde{t}_i^k \ni \bigcup_{k \leq \rho} U_{\omega_k}$. If the k-th row of $\widetilde{\mathfrak{W}}$, where $k > \rho$, does not contain distinguished sets, then let us call one arbitrary set $\widetilde{W}_i^k$ distinguished. In the next step, as was done in the proof of Theorem 16.1, let us proceed to considering the new regular countable system of type $k(k+1)/2$.

We have finished the description of one step of our inductive process. If this process has a finite number of steps, then the statement is proved. Assume that this process has a countable number of steps. Let us prove that for each $k \in \mathbb{N}^+$ there exists a natural number $\xi(k) \in \mathbb{N}^+$ such that after $\xi(k)$ steps one finds a set U_k. Suppose the contrary. Put

$$K = \{k \in \mathbb{N}^+ \mid \xi(k) \text{ does not exist}\}.$$

By assumption, $K \neq \varnothing$. Put $k_* = \min\{k \in K\}$. Clearly, $k_* > 1$. If $k \in K$, then there exists a step $\eta(k)$ and a number $\varepsilon_k \in \mathbb{N}^+$ such that at any step of our process whose number is not less than $\eta(k)$, the row of the infinite matrix of pairwise disjoint sets that corresponds to the k-th row of $\mathfrak{W}$ contains ε_k elements. Obviously, $\varepsilon_k > 1$. There exist infinitely many steps in the search for the required sets U_k; i.e., there exist infinitely many steps, in each of which one considers either Case 3°, or Case 4°, or Case 5°-a, or Case 5°-b. Let $\eta'(k_*) > \eta(k_*)$, and assume that before the $\eta'(k_*)$-th step we have found all sets $U_1, \dots, U_{k_*-1}$. If $n \geq \eta'(k_*)$ and at the n-th step one considers either Case 3°, or Case 4°, or Case 5°-b, then either at this step one finds a set U_{k_*} (which is impossible), or at the $(n+1)$-th step the row of the infinite matrix of pairwise disjoint sets, which corresponds to the k_*-th row of $\mathfrak{W}$, contains less than ε_{k_*} elements (which is also impossible).[7] The conclusion is the following: there exist steps with arbitrarily large numbers, in which one considers Case 5°-a. Let $K \ni \widehat{k} > k_*$, and assume that at the l-th step one considers Case 5°-a and $l > \max(\eta(k_*), \eta(\widehat{k}))$. Then at the $(l+1)$-th step at least one of the following two statements holds:

I) in the infinite matrix of pairwise disjoint sets, the row that corresponds to the k_*-th row of $\mathfrak{W}$ contains less than ε_{k_*} elements;
II) in the infinite matrix of pairwise disjoint sets, the row that corresponds to the $\widehat{k}$-th row of $\mathfrak{W}$ contains less than $\varepsilon_{\widehat{k}}$ elements

(see the definition of the set Γ above and the next sentence after the definition). The obtained contradiction shows that $K = \{k_*\}$. Put

$$\Delta = \{k \in \mathbb{N}^+ \setminus \{k_*\} \mid \text{there exist ultrafilters } t_i^{k_*} \text{ and } s_j^k \text{ such that } t_i^{k_*} = s_j^k\}.$$

Obviously, $\#(\Delta) < \aleph_0$. Consider a step $m > \eta(k_*)$ at which Case 5°-a is considered, and such that before this step we have found all sets U_k, where $k \in \Delta$. Then at the $(m+1)$-th step, in the infinite matrix of pairwise disjoint sets, the row that corresponds to the k_*-th row of $\mathfrak{W}$ contains less than ε_{k_*} elements (see the definition of the set Φ^k above and the next sentence after the definition). The obtained contradiction implies that $\xi(k)$ exists for each k. This conclusion and the arguments from the proof of Theorem 16.5 imply that our statement is true. □

[7] In Cases 3° and 5°-b we obtain a contradiction taking $n > \eta(k_*)$.

16.9. The first part of the following theorem (which we mentioned in 2.40) is a trivial corollary of Theorem 16.5.

THEOREM. (1) *Let $\mathcal{A}_1, \dots, \mathcal{A}_n$ be a finite sequence of algebras, and let there exist a matrix of pairwise disjoint sets*

$$\begin{pmatrix} U_1^1 & & \\ \dots & \dots & \dots \\ U_1^k & \dots & U_k^k \\ \dots & \dots & \dots \\ U_1^n & \dots & U_n^n \end{pmatrix}$$

(the k-th row contains k sets) such that $U_i^k \notin \mathcal{A}_k$. Then there exist sets $U_1, \dots, U_n$ such that if $U_k \subset Q \subset \bigcup_{i=1}^n U_i$, then $Q \notin \mathcal{A}_k$. Moreover:

1. $U_k \subset U_{j_k}^{i_k}$; *if* $1 \le k < l \le n$, *then* $U_{j_k}^{i_k} \ne U_{j_l}^{i_l}$;
2. $\#(\{k \mid i_k > k\}) \le n/2$.

(2) *The estimate $\#(\{k \mid i_k > k\}) \le n/2$ is the best possible in the following sense. For any natural number n one can construct algebras $\mathcal{A}_1, \dots, \mathcal{A}_n$ and a matrix of pairwise disjoint sets*

$$\begin{pmatrix} U_1^1 & & \\ \dots & \dots & \dots \\ U_1^k & \dots & U_k^k \\ \dots & \dots & \dots \\ U_1^n & \dots & U_n^n \end{pmatrix}$$

(the k-th row contains k sets and $U_i^k \notin \mathcal{A}_k$), which possess the following required property. Let $U_1, \dots, U_n$ be a sequence of pairwise disjoint sets such that if $U_k \subset Q \subset \bigcup_{i=1}^n U_i$, then $Q \notin \mathcal{A}_k$; and $U_k \subset U_{j_k}^{i_k}$. Then

$$\#(\{k \mid i_k > k\}) = \left[\frac{n}{2}\right].$$

PROOF OF THE SECOND PART OF THE THEOREM. Let us present the construction for $n = 6$:

$\ker \mathcal{A}_1 = \{a_1^1, a_1^2\}$, and $\{a_1^1, a_1^2\}$ is the $\mathcal{A}_1$-similar set;

$\ker \mathcal{A}_2 = \{a_1^2, a_2^2, a_1^1\}$, a_1^2 is the $\mathcal{A}_2$-special ultrafilter, and $\{a_2^2, a_1^1\}$ is the $\mathcal{A}_2$-similar set;

$\ker \mathcal{A}_3 = \{a_1^3, a_2^3, a_3^3, a_1^2, a_2^2, a_3^4\}$, and $\{a_1^3, a_1^2\}$, $\{a_2^3, a_2^2\}$, $\{a_3^3, a_3^4\}$ are the $\mathcal{A}_3$-similar sets;

$\ker \mathcal{A}_4 = \{a_1^4, a_2^4, a_3^4, a_4^4, a_1^2, a_2^2, a_3^3\}$, a_3^4 is the $\mathcal{A}_4$-special ultrafilter, and $\{a_1^4, a_1^2\}$, $\{a_2^4, a_2^2\}$, $\{a_4^4, a_3^3\}$ are the $\mathcal{A}_4$-similar sets;

$\ker \mathcal{A}_5 = \{a_1^5, a_2^5, a_3^5, a_4^5, a_5^5, a_1^2, a_2^2, a_3^4, a_4^4, a_5^6\}$, and $\{a_1^5, a_1^2\}$, $\{a_2^5, a_2^2\}$, $\{a_3^5, a_3^4\}$, $\{a_4^5, a_4^4\}$, $\{a_5^5, a_5^6\}$ are the $\mathcal{A}_5$-similar sets;

$\ker \mathcal{A}_6 = \{a_1^6, a_2^6, a_3^6, a_4^6, a_5^6, a_6^6, a_1^2, a_2^2, a_3^4, a_4^4, a_5^5\}$, a_5^6 is the $\mathcal{A}_6$-special ultrafilter, and $\{a_1^6, a_1^2\}$, $\{a_2^6, a_2^2\}$, $\{a_3^6, a_3^4\}$, $\{a_4^6, a_4^4\}$, $\{a_6^6, a_5^5\}$ are the $\mathcal{A}_6$-similar sets;

$U_i^k \in a_i^k$.

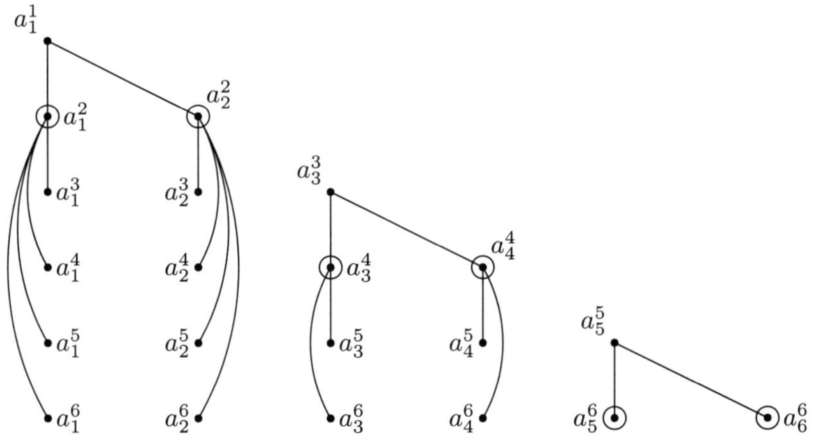

Clearly, we have

$$U_1 \subset U_1^2, \quad U_2 \subset U_2^2, \quad U_3 \subset U_3^4, \quad U_4 \subset U_4^4, \quad U_5 \subset U_5^6, \quad U_6 \subset U_6^6.$$

Hence $\{k \mid i_k > k\} = \{1, 3, 5\}$. It is now clear how the construction works for any natural number n. □

16.10. THEOREM. *Let $\{\mathfrak{A}, \mathfrak{U}\}$ be a regular system of type $(k^2+3k-2)/2$. Then there exists a sequence of sets $U_1, \dots, U_n$ such that*

$$U_k \subset Q \subset \bigcup_{i=1}^{n} U_i \quad \textit{implies} \quad Q \notin \mathcal{A}_k.$$

Moreover, $U_k \subset U_{j_k}^{i_k}$; if $1 \le k < l \le n$, then $U_{j_k}^{i_k} \ne U_{j_l}^{i_l}$; $i_k \le k$.

PROOF.[8] If $n = 1$, then the statement of the theorem is obvious. *Let $n > 1$.*

Case 1. *The system $\{\mathfrak{A}, \mathfrak{W}\}$ is of type $(k^2+3k-2)/2+1$.* As in Case 1 from the proof of Theorem 16.1, let us take a natural number m_1^*; let us remove from $\mathfrak{W}$ all the sets W_i^1, where $i > m_1^*$; and in the next step let us pass to the regular system $\{\mathfrak{A}^*, \mathfrak{W}^*\}$, which is of type $(k^2 + 3k - 2)/2$ and is not of type $(k^2 + 3k - 2)/2 + 1$.

Case 2. *The minimal natural number p_0 such that $\sum_{i=1}^{p_0} m_i = (k^2 + 3k - 2)/2$ exists, and there exists an ultrafilter $t_{i_0}^{k_0} = s_1^1$, where $1 < k_0 \le p_0$.* As in Case 2 from the proof of Theorem 16.1, let us move the set W_1^1 to the k_0-th row, etc. In the next step we consider the regular system $\{\mathfrak{A}^*, \mathfrak{W}^*\}$ of type $(k^2 + 3k - 2)/2$.

Case 3. *The minimal natural number p_0 such that $\sum_{i=1}^{p_0} m_i = (p_0^2 + 3p_0 - 2)/2$ exists, and there does not exist an ultrafilter $t_{i_0}^{k_0} = s_1^1$, where $1 < k_0 \le p_0$.* The set W_i^k, where $k > p_0$, is called *marked* if one considers the ultrafilter t_i^k and $t_i^k = s_1^1$. If $t_1^1 = s_{\lambda_0}^{\nu_0}$, then the set $W_{\lambda_0}^{\nu_0}$ is also called marked. Using Lemma 15.11 (for $r_* = 1$), we will proceed by our usual method to the regular system $\{\mathfrak{A}^*, \mathfrak{W}^*\}$ of type $(k^2 + 3k - 2)/2$, where $\#(\mathfrak{A}^*) = n - 1$. Put $U_1 = W_1^1$. It is important that conditions (i), (ii) from the proof of Theorem 16.1 are satisfied.

We have considered all possible cases. Our further arguments are similar to the arguments from the proof of Theorem 16.5. Assume that at some step we consider

[8]We use, without special reference, the methods and notions introduced in the proofs of Theorems 16.1 and 16.5.

a regular system $\{\widetilde{\mathfrak{A}}, \widetilde{\mathfrak{W}}\}$ of type $(k^2+3k-2)/2$. Let us suppose that we do not deal with the trivial situation, when $\#(\widetilde{\mathfrak{A}}) = 1$. First let us check whether Case 1 holds. If it does not hold, then let us proceed to one of two cases, namely either Case 2*, which occurs instead of Case 2, or Case 3*, which occurs instead of Case 3.

Case 2*. The minimal natural number ρ_0 such that $\sum_{i=1}^{\rho_0} \widetilde{m}_i = (\rho_0^2+3\rho_0-2)/2$ exists. Let $\omega_{r_*} = \min(\omega_1, \dots, \omega_{\rho_0})$, and let there exist an ultrafilter $\tilde{t}_{i_0}^{k_0} = \tilde{s}_1^{r_*}$, where $k_0 \in [1, \rho_0] \setminus \{r_*\}$.

Case 3*. The minimal natural number ρ_0 such that $\sum_{i=1}^{\rho_0} \widetilde{m}_i = (\rho_0^2+3\rho_0-2)/2$ exists. Let $\omega_{r_*} = \min(\omega_1, \dots, \omega_{\rho_0})$, and assume that there does not exist an ultrafilter $\tilde{t}_{i_0}^{k_0} = \tilde{s}_1^{r_*}$ for which $k_0 \in [1, \rho_0] \setminus \{r_*\}$.

Cases 2* and 3* are considered in the usual way. In Case 3*, we use Lemma 15.11. Continuing our constructions by induction, we will find all the required sets $U_1, \dots, U_n$. Clearly, $U_k = W_{j_k}^{i_k} \in \mathfrak{W}$. If $W_{j_k}^{i_k}$ has not been moved, then $i_k = k$; and if it has been moved, then $i_k < k$. □

16.11. Let us proceed to the theorem discussed in 2.41. Its first part is a trivial corollary of Theorem 16.10.

THEOREM. (1) *Assume that for a finite sequence of algebras $\mathcal{A}_1, \dots, \mathcal{A}_n$ there exists a matrix of pairwise disjoint sets*

$$\begin{pmatrix} U_1^1 & & \\ \dots & \dots & \dots \\ U_1^k & \dots & U_{m_k}^k \\ \dots & \dots & \dots \\ U_1^n & \dots & U_{m_n}^n \end{pmatrix},$$

where $m_1 = 1$, and if $1 < k \le n$, then $m_k = k+1$; $U_i^k \notin \mathcal{A}_k$. Then there exist sets $U_1, \dots, U_n$ such that

$$U_k \subset Q \subset \bigcup_{i=1}^{n} U_i \quad \textit{implies} \quad Q \notin \mathcal{A}_k.$$

Also, $U_k \subset U_{j_k}^{i_k}$; if $1 \le k < l \le n$, then $U_{j_k}^{i_k} \neq U_{j_l}^{i_l}$; $i_k \le k$.

(2) *The estimate for m_k in the first part of the theorem is the best possible in the following sense. For any natural number $n > 1$ one can construct algebras $\mathcal{A}_1, \dots, \mathcal{A}_n$, for which the statement of the first part of the theorem is not true, but there exists a matrix of pairwise disjoint sets*

$$\begin{pmatrix} U_1^1 & & \\ \dots & \dots & \dots \\ U_1^k & \dots & U_{m_k}^k \\ \dots & \dots & \dots \\ U_1^n & \dots & U_{m_n}^n \end{pmatrix},$$

where $m_1 = 1$, $m_n = n$, and if $1 < k < n$, then $m_k = k+1$; $U_i^k \notin \mathcal{A}_k$.

PROOF OF THE SECOND PART OF THE THEOREM. Let us present the construction for $n = 4$:

$\ker \mathcal{A}_1 = \{a_1^1, a_1^2\}$, and $\{a_1^1, a_1^2\}$ is the $\mathcal{A}_1$-similar set;

$\ker \mathcal{A}_2 = \{a_1^2, a_2^2, a_3^2, a_1^1, a_1^3\}$, a_1^2 is the $\mathcal{A}_2$-special ultrafilter, and $\{a_2^2, a_1^1\}$, $\{a_3^2, a_1^3\}$ are the $\mathcal{A}_2$-similar sets;

$\ker \mathcal{A}_3 = \{a_1^3, a_2^3, a_3^3, a_4^3, a_1^1, a_3^2, a_1^4\}$, a_1^3 is the $\mathcal{A}_3$-special ultrafilter, and $\{a_2^3, a_1^1\}$, $\{a_3^3, a_3^2\}$, $\{a_4^3, a_1^4\}$ are the $\mathcal{A}_3$-similar sets;

$\ker \mathcal{A}_4 = \{a_1^4, a_2^4, a_3^4, a_4^4, a_1^1, a_3^2, a_4^3\}$, a_1^4 is the $\mathcal{A}_4$-special ultrafilter, and $\{a_2^4, a_1^1\}$, $\{a_3^4, a_3^2\}$, $\{a_4^4, a_4^3\}$ are the $\mathcal{A}_4$-similar sets;

$U_i^k \in a_i^k$.

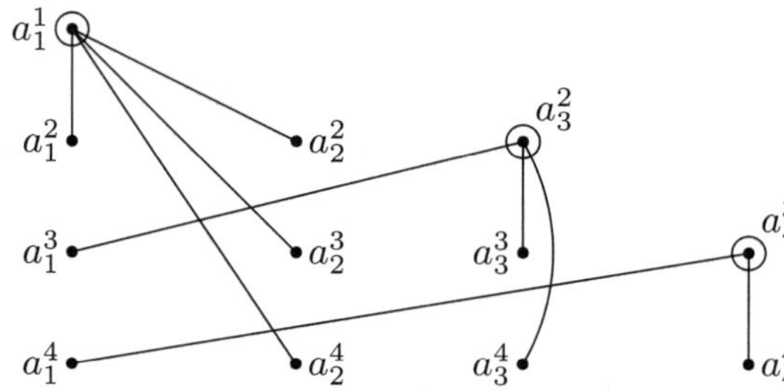

If we try to construct sets U_1, U_2, U_3, U_4 satisfying the conditions of the first part of our theorem, then necessarily $U_1 \in a_1^1$, $U_2 \in a_3^2$, $U_3 \in a_4^3$. But we cannot construct the set U_4. Completely similar arguments give the proof of the second part of our theorem for any natural number $n > 1$. □

16.12. THEOREM. *Let $\{\mathfrak{A}, \mathfrak{U}\}$ be a regular system of type $(k^2+5k-4)/2$. Then there exists a sequence of sets $U_1, \dots, U_n$ such that*

$$U_k \subset Q \subset \bigcup_{i=1}^{n} U_i \quad \textit{implies} \quad Q \notin \mathcal{A}_k;$$

and we have $U_k \subset U_{j_k}^k$.

PROOF.[9] As in the proof of the previous theorems, we will find the required sets $U_1, \dots, U_n$ by proceeding from the system $\{\mathfrak{A}, \mathfrak{W}\}$ to another regular system of type $(k^2+5k-4)/2$, etc. Assume that during this process we came to a system $\{\widetilde{\mathfrak{A}}, \widetilde{\mathfrak{W}}\}$. Assume that the set $\widetilde{W}_i^k \in \widetilde{\mathfrak{W}}$ *has been moved* from one row to another. Our constructions imply that if $\widetilde{W}_i^k = W_\xi^\eta \in \mathfrak{W}$ and the k-th row of $\widetilde{\mathfrak{W}}$ corresponds to the ω_k-th row of $\mathfrak{W}$, then $\eta \neq \omega_k$. In the proof of our theorem, it is assumed that for each moved set $\widetilde{W}_i^k$ there exists a *paired* set $\widetilde{W}_j^k$ for $\widetilde{W}_i^k$, i.e., $\tilde{s}_i^k = \tilde{t}_j^k$, $\tilde{s}_j^k = \tilde{t}_i^k$. (The notion of the paired set only makes sense with the reference to the moved set.) A set $\widetilde{W}_i^k$ which is neither moved nor paired is called *ordinary*. In the proof of our theorem we will follow two rules:

1°. One can move only ordinary sets from one row to another.[10]

2°. It can happen that a moved set will be removed, and the paired set will be left; then we will call this paired set ordinary.

It is clear that the paired set for a moved set is determined uniquely and cannot be a moved set. It cannot happen that the paired set will be removed, and the corresponding moved set will not be removed. Denote by $\mathfrak{D}$ the totality of all sets U_l already found. Let us assume that the following are true:

(a) $U_l = W_{j_l}^l \in \mathfrak{W}$;

(b) if the ultrafilter $t_{j_l}^l$ is considered, then $t_{j_l}^l \not\ni \bigcup\{W_i^k \mid W_i^k \in \mathfrak{D} \cup \widetilde{\mathfrak{W}}\}$;

[9]Again, we use methods and notions introduced in the proofs of Theorems 16.1 and 16.5.

[10]In fact, we always move a set from a row where all the sets are ordinary.

(c) $\widetilde{t}_i^k \not\ni U_l$.

Let $\#(\widetilde{\mathfrak{A}}) = \widetilde{n} > 1$ (if $\widetilde{n} = 1$, then the situation is trivial).

Case 1. $\{\widetilde{\mathfrak{A}}, \widetilde{\mathfrak{W}}\}$ *is of type* $(k^2+5k-4)/2+1$. Let us take a natural number $\widetilde{m}_1^* < \widetilde{m}_1$ such that a matrix with $\widetilde{m}_1^*$ elements in the first row and with $\widetilde{m}_k$ elements in the k-th row $(1 < k \leq \widetilde{n})$ is of type $(k^2+5k-4)/2$ and is not of type $(k^2+5k-4)/2+1$. If in the first row of $\widetilde{\mathfrak{W}}$ there are not less than $\widetilde{m}_1 - \widetilde{m}_1^*$ moved sets, then let us remove from this row $\widetilde{m}_1 - \widetilde{m}_1^*$ moved sets. And if it contains less than $\widetilde{m}_1 - \widetilde{m}_1^*$ moved sets, then let us remove from it all the moved sets, and also other sets, so that the total number of removed sets is $\widetilde{m}_1 - \widetilde{m}_1^*$. Denote the obtained matrix by $\widetilde{\mathfrak{W}}^*$. Put $\widetilde{\mathfrak{A}}^* = \widetilde{\mathfrak{A}}$, and in the next step let us consider the regular system $\{\widetilde{\mathfrak{A}}^*, \widetilde{\mathfrak{W}}^*\}$ of type $(k^2+5k-4)/2$.

Case 2. *There exists a natural number* p *such that* $(p^2+5p)/2 \geq \sum_{i=1}^{p} \widetilde{m}_i$ *and if* $k \leq p$, *then the* k*-th row of* $\widetilde{\mathfrak{W}}$ *contains a moved set.*[11] For each $k \leq p$ consider a moved set $\widetilde{W}^k_{i_k^1}$ and the paired set $\widetilde{W}^k_{i_k^2}$. Put

$$U_{\omega_k} = \widetilde{W}^k_{i_k^2} = W^{\omega_k}_{j_k} \in \mathfrak{W}.$$

Clearly, the ultrafilter $t^{\omega_k}_{j_k}$ is considered and $t^{\omega_k}_{j_k} \ni \widetilde{W}^k_{i_k^1}$. If $p = \widetilde{n}$, then the proof is over. *Let* $p < \widetilde{n}$. The set $\widetilde{W}_i^k$, where $k > p$, is said to be *marked* if the ultrafilter $\widetilde{t}_i^k$ is considered and $\widetilde{t}_i^k \ni \bigcup_{l \leq p} U_{\omega_l}$. Let us remove the first p rows and all the marked sets from $\widetilde{\mathfrak{W}}$. For the obtained matrix, consider a regular matrix $\widetilde{\mathfrak{W}}^*$ similar to it. One naturally defines the regular system $\{\widetilde{\mathfrak{A}}^*, \widetilde{\mathfrak{W}}^*\}$. By Lemma 15.12, this system is of type $(k^2+5k-4)/2$, and it is considered in the next step.

Case 3. *The minimal natural number* p_0 *such that* $\sum_{i=1}^{p_0} \widetilde{m}_i = (p_0^2+5p_0-4)/2$ *exists. There exists a natural number* $r_* \leq p_0$ *such that in the* r_**-th row of* $\widetilde{\mathfrak{W}}$ *all sets are ordinary*; *if* $k < r_*$, *then the* k*-th row of* $\widetilde{\mathfrak{W}}$ *contains a moved set and* $(k^2+5k)/2 < \sum_{i=1}^{k} \widetilde{m}_i$. *There exists an ultrafilter* $\widetilde{t}^{k_0}_{i_0} = \widetilde{s}_1^{r_*}$, *where* $k_0 \in [1, p_0] \setminus \{r_*\}$. Let us move the set $\widetilde{W}_1^{r_*}$ to the k_0-th row of $\widetilde{\mathfrak{W}}$ and naturally define the regular system $\{\widetilde{\mathfrak{A}}^*, \widetilde{\mathfrak{W}}^*\}$ of type $(k^2+5k-4)/2$, where $\#(\widetilde{\mathfrak{A}}^*) = \#(\widetilde{\mathfrak{A}})$. This system is considered in the next step, and for this system either Case 3 or Case 4 holds.

Case 4. *The minimal number* p_0 *such that* $\sum_{i=1}^{p_0} \widetilde{m}_i = (p_0^2+5p_0-4)/2$ *exists. There exists a natural number* $r_* \leq p_0$ *such that in the* r_**-th row of* $\widetilde{\mathfrak{W}}$ *all sets are ordinary*; *if* $k < r_*$, *then the* k*-th row of* $\widetilde{\mathfrak{W}}$ *contains a moved set and* $(k^2+5k)/2 < \sum_{i=1}^{k} \widetilde{m}_i$. *There does not exist an ultrafilter* $\widetilde{t}^{k_0}_{i_0} = \widetilde{s}_1^{r_*}$, *where* $k_0 \in [1, p_0] \setminus \{r_*\}$. Since $\widetilde{W}_1^{r_*}$ is an ordinary set, $\widetilde{W}_1^{r_*} = W^{\omega_{r_*}}_{j_{r_*}} \in \mathfrak{W}$. Put $U_{\omega_{r_*}} = W^{\omega_{r_*}}_{j_{r_*}}$. The set $\widetilde{W}_i^k$ is called *marked* if the ultrafilter $\widetilde{t}_i^k$ is considered and $\widetilde{t}_i^k = \widetilde{s}_1^{r_*}$. If the ultrafilter $\widetilde{t}_1^{r_*}$ is considered and $\widetilde{t}_1^{r_*} = \widetilde{s}^{\nu_0}_{\lambda_0}$, then the set $\widetilde{W}^{\nu_0}_{\lambda_0}$ is also called *marked*. Further, if $\widetilde{W}^{\nu_0}_{\lambda_0}$ is a paired set, then the corresponding moved set is also called *marked*. Denote by $\widetilde{\mathfrak{W}}'$ the matrix obtained from $\widetilde{\mathfrak{W}}$ by removing the r_*-th row and all the marked sets. Clearly, $\widetilde{\mathfrak{W}}'$ contains $\widetilde{n} - 1$ rows, and if $\widetilde{\mathfrak{W}}'$ contains a moved set, then it contains the paired set also. Clearly, conditions (a), (b), (c) hold if appropriately

[11] In general, there can be several such numbers p. Among them we choose an arbitrary one. But it is natural to choose the largest among them, since in our case we construct p sets from the family of required sets $U_1, \ldots, U_n$.

corrected. Let $\widetilde{\mathfrak{W}}^*$ be a regular matrix similar to $\widetilde{\mathfrak{W}}'$. One naturally defines the regular system $\{\widetilde{\mathfrak{A}}^*, \widetilde{\mathfrak{W}}^*\}$, which, by Lemma 15.15, is of type $(k^2+5k-4)/2$. This system is considered in the next step.

We have examined all possible cases. Continuing our constructions by induction, we will find all the required sets $U_1, \ldots, U_n$. □

16.13. Let us proceed to the detailed statement and the proof of the theorem mentioned in 2.42.

THEOREM. (1) *Assume that for a finite sequence of algebras* $\mathcal{A}_1, \ldots, \mathcal{A}_n$ *there exists a matrix of pairwise disjoint sets*

$$\begin{pmatrix} U_1^1 & & \\ \cdots & \cdots & \cdots \\ U_1^k & \ldots & U_{m_k}^k \\ \cdots & \cdots & \cdots \\ U_1^n & \ldots & U_{m_n}^n \end{pmatrix},$$

where $m_1 = 1$; *if* $k = 2$, *then* $m_k = 3$, *and if* $k > 2$, *then* $m_k = k+2$; $U_i^k \notin \mathcal{A}_k$. *Then there exist sets* $U_1, \ldots, U_n$ *such that*

$$U_k \subset Q \subset \bigcup_{i=1}^n U_i \quad \textit{implies} \quad Q \notin \mathcal{A}_k.$$

Also, we have $U_k \subset U_{j_k}^k$.

(2) *The estimate* $k+2$ *is the best possible in the following sense. For each natural number* $n > 1$ *one can construct a sequence of algebras* $\mathcal{A}_1, \ldots, \mathcal{A}_n, \mathcal{A}_{n+1}$ *and a matrix of pairwise disjoint sets*

$$\begin{pmatrix} U_1^1 & & \\ \cdots & \cdots & \cdots \\ U_1^k & \ldots & U_{m_k}^k \\ \cdots & \cdots & \cdots \\ U_1^n & \ldots & U_{m_n}^n \end{pmatrix}$$

such that $m_1 = 1$, $m_k = k+1$ *for each* $k \in [2, n+1]$, *and* $U_i^k \notin \mathcal{A}_k$. *Further, there exist* $U_1, \ldots, U_n$ *such that*

$$U_k \subset Q \subset \bigcup_{i=1}^n U_i \quad \textit{implies} \quad Q \notin \mathcal{A}_k,$$

and we have $U_k \subset U_{j_k}^k$. *But for the algebras* $\mathcal{A}_1, \ldots, \mathcal{A}_n, \mathcal{A}_{n+1}$, *the corresponding sets* $U_1', \ldots, U_n', U_{n+1}'$ *do not exist.*

PROOF. **(1)** If $n = 1$, then the statement is obvious.

Let $n > 1$. Instead of our matrix, let us consider the matrix $\mathfrak{W}$ as usual. Put $U_1 = W_1^1$. The set W_i^k is said to be *marked* if the ultrafilter t_i^k is considered and $t_i^k = s_1^1$. If the ultrafilter t_1^1 is considered and $t_1^1 = s_{\lambda_0}^{\nu_0}$, then the set $W_{\lambda_0}^{\nu_0}$ is also called *marked.* Denote by $\mathfrak{W}'$ the matrix obtained from $\mathfrak{W}$ by removing the first row and all the marked sets. A regular matrix $\mathfrak{W}^*$ similar to $\mathfrak{W}'$ contains $n-1$ rows and is of type $(k^2+5k-4)/2$. By Theorem 16.12, one can find the other required sets $U_2, \ldots, U_n$.

(2) Let us show that the statement is true for $n = 5$. Let us describe the kernels of the algebras $\mathcal{A}_1, \mathcal{A}_2, \mathcal{A}_3, \mathcal{A}_4, \mathcal{A}_5, \mathcal{A}_6$:

$\ker \mathcal{A}_1 = \{a_1^1, a_1^2\}$, and $\{a_1^1, a_1^2\}$ is the $\mathcal{A}_1$-similar set;

$\ker \mathcal{A}_2 = \{a_1^2, a_2^2, a_3^2, a_1^1, a_1^3\}$, a_1^2 is the $\mathcal{A}_2$-special ultrafilter, and $\{a_2^2, a_1^1\}$, $\{a_3^2, a_1^3\}$ are the $\mathcal{A}_2$-similar sets;

$\ker \mathcal{A}_3 = \{a_1^3, a_2^3, a_3^3, a_4^3, a_1^1, a_3^2, a_1^4\}$, a_1^3 is the $\mathcal{A}_3$-special ultrafilter, and $\{a_2^3, a_1^1\}$, $\{a_3^3, a_3^2\}$, $\{a_4^3, a_1^4\}$ are the $\mathcal{A}_3$-similar sets;

$\ker \mathcal{A}_4 = \{a_1^4, a_2^4, a_3^4, a_4^4, a_5^4, a_1^1, a_3^2, a_4^3, a_1^6\}$, a_1^4 is the $\mathcal{A}_4$-special ultrafilter, and $\{a_2^4, a_1^1\}$, $\{a_3^4, a_3^2\}$, $\{a_4^4, a_4^3\}$, $\{a_5^4, a_1^6\}$ are the $\mathcal{A}_4$-similar sets;

$\ker \mathcal{A}_5 = \{a_1^5, a_2^5, a_3^5, a_4^5, a_5^5, a_6^5, a_1^1, a_3^2, a_4^3, a_5^4, a_6^6, a_7^6\}$, and $\{a_1^5, a_1^1\}$, $\{a_2^5, a_3^2\}$, $\{a_3^5, a_4^3\}$, $\{a_4^5, a_5^4\}$, $\{a_5^5, a_6^6\}$, $\{a_6^5, a_7^6\}$ are the $\mathcal{A}_5$-similar sets;

$\ker \mathcal{A}_6 = \{a_1^6, a_2^6, a_3^6, a_4^6, a_5^6, a_6^6, a_7^6, a_1^1, a_3^2, a_4^3, a_5^4, a_5^5, a_6^5\}$, a_1^6 is the $\mathcal{A}_6$-special ultrafilter, and $\{a_2^6, a_1^1\}$, $\{a_3^6, a_3^2\}$, $\{a_4^6, a_4^3\}$, $\{a_5^6, a_5^4\}$, $\{a_6^6, a_6^5\}$, $\{a_7^6, a_5^5\}$ are the $\mathcal{A}_6$-similar sets.

Let us assume that $a_i^k \ni U_i^k$.

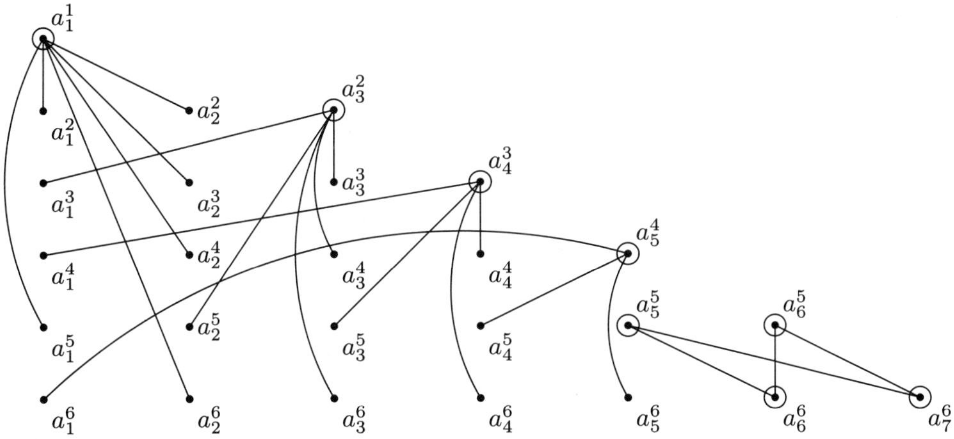

Let us find sets U_1, U_2, U_3, U_4, U_5. One necessarily has $U_1 \subset U_1^1$, $U_2 \subset U_3^2$, $U_3 \subset U_4^3$, $U_4 \subset U_5^4$. Clearly, either $U_5 \subset U_5^5$, or $U_5 \subset U_6^5$. It is impossible to construct sets $U_1', U_2', U_3', U_4', U_5', U_6'$. Here one should draw attention to the "bunch" of ultrafilters $a_5^5, a_6^5, a_6^6, a_7^6$.

It is clear how our statement is proved for $n = 2$ and for each $n > 2$. □

16.14. Comparing Theorems 2.39 and 16.11 with Theorem 16.13, one can see that the nonimprovability of the estimate in the latter theorem is in a sense "weaker" than in the former two theorems. The following statement is a refinement of Theorem 16.13(1) for $n = 4$.

STATEMENT. *Assume that for algebras $\mathcal{A}_1, \mathcal{A}_2, \mathcal{A}_3, \mathcal{A}_4$ there exists a matrix of pairwise disjoint sets*

$$\begin{pmatrix} U_1^1 & & & & \\ U_1^2 & U_2^2 & U_3^2 & & \\ U_1^3 & U_2^3 & U_3^3 & U_4^3 & U_5^3 \\ U_1^4 & U_2^4 & U_3^4 & U_4^4 & U_5^4 \end{pmatrix},$$

where $U_i^k \notin \mathcal{A}_k$. Then there exist sets U_1, U_2, U_3, U_4 such that

$$U_k \subset Q \subset \bigcup_{i=1}^{4} U_i \quad \textit{implies} \quad Q \notin \mathcal{A}_k.$$

Also, $U_k \subset U_{j_k}^k$.

PROOF. As usual, let us consider the matrix $\mathfrak{W}$. Put $U_1 = W_1^1$, and proceed by our already habitual way to considering the regular system $\{\widetilde{\mathfrak{A}}, \widetilde{\mathfrak{W}}\}$, where $\widetilde{\mathfrak{A}} = \{\mathcal{A}_2, \mathcal{A}_3, \mathcal{A}_4\}$. We must examine only two cases.

Case 1.

$$\widetilde{\mathfrak{W}} = \begin{pmatrix} \widetilde{W}_1^1 & & & \\ \widetilde{W}_1^2 & \widetilde{W}_2^2 & \widetilde{W}_3^2 & \widetilde{W}_4^2 \\ \widetilde{W}_1^3 & \widetilde{W}_2^3 & \widetilde{W}_3^3 & \widetilde{W}_4^3 \end{pmatrix}.$$

Case 2.

$$\widetilde{\mathfrak{W}} = \begin{pmatrix} \widetilde{W}_1^1 & \widetilde{W}_2^1 & & \\ \widetilde{W}_1^2 & \widetilde{W}_2^2 & \widetilde{W}_3^2 & \\ \widetilde{W}_1^3 & \widetilde{W}_2^3 & \widetilde{W}_3^3 & \widetilde{W}_4^3 \end{pmatrix}.$$

In Case 1, we put $U_2 = \widetilde{W}_1^1$ and proceed to the system $\{\mathring{\mathfrak{A}}, \mathring{\mathfrak{W}}\}$, where

$$\mathring{\mathfrak{W}} = \begin{pmatrix} \mathring{W}_1^1 & \mathring{W}_2^1 & \\ \mathring{W}_1^2 & \mathring{W}_2^2 & \mathring{W}_3^2 \end{pmatrix}.$$

Now it is easy to find the required sets U_3, U_4. In Case 2 let us assume that $\tilde{t}_1^1 = \tilde{s}_i^3$ (for example, $i = 1$), and $\tilde{t}_j^2 = \tilde{s}_2^1$ for some j. Otherwise it is rather easy to find the required sets U_2, U_3, U_4. Put $U_2 = \widetilde{W}_1^1$. Another required set is $\widetilde{W}_j^2$. Now it is easy to find the last required set among the three sets $\widetilde{W}_2^3, \widetilde{W}_3^3, \widetilde{W}_4^3$. □

16.15. The following interesting example has to do with Theorem 16.13 and Statement 16.14.

EXAMPLE. One can construct algebras $\mathcal{A}_1, \mathcal{A}_2, \mathcal{A}_3, \mathcal{A}_4, \mathcal{A}_5$ and a matrix of pairwise disjoint sets

$$\begin{pmatrix} U_1^1 & & & & \\ U_1^2 & U_2^2 & U_3^2 & & \\ U_1^3 & U_2^3 & U_3^3 & U_4^3 & U_5^3 \\ U_1^4 & U_2^4 & U_3^4 & U_4^4 & U_5^4 \\ U_1^5 & U_2^5 & U_3^5 & U_4^5 & U_5^5 \end{pmatrix}$$

such that

1) $U_i^k \notin \mathcal{A}_k$;
2) for these algebras and for this matrix, the statement of Theorem 16.13(1) does not hold.

CONSTRUCTION. $\ker \mathcal{A}_1 = \{a_1^1, a_1^2\}$, and $\{a_1^1, a_1^2\}$ is the $\mathcal{A}_1$-similar set;

$\ker \mathcal{A}_2 = \{a_1^2, a_2^2, a_3^2, a_1^1, a_1^3\}$, a_1^2 is the $\mathcal{A}_2$-special ultrafilter, and $\{a_2^2, a_1^1\}$, $\{a_3^2, a_1^3\}$ are the $\mathcal{A}_2$-similar sets;

$\ker \mathcal{A}_3 = \{a_1^3, a_2^3, a_3^3, a_4^3, a_5^3, a_1^1, a_3^2, a_3^4, a_3^5\}$, a_1^3 is the $\mathcal{A}_3$-special ultrafilter, and $\{a_2^3, a_1^1\}$, $\{a_3^3, a_3^2\}$, $\{a_4^3, a_3^4\}$, $\{a_5^3, a_3^5\}$ are the $\mathcal{A}_3$-similar sets;

$\ker \mathcal{A}_4 = \{a_1^4, a_2^4, a_3^4, a_4^4, a_5^4, a_1^1, a_3^2, a_5^3, a_4^5, a_5^5\}$, and $\{a_1^4, a_1^1\}$, $\{a_2^4, a_3^2\}$, $\{a_3^4, a_5^3\}$, $\{a_4^4, a_4^5\}$, $\{a_5^4, a_5^5\}$ are the $\mathcal{A}_4$-similar sets;

$\ker \mathcal{A}_5 = \{a_1^5, a_2^5, a_3^5, a_4^5, a_5^5, a_1^1, a_3^2, a_4^3, a_4^4, a_5^4\}$, and $\{a_1^5, a_1^1\}$, $\{a_2^5, a_3^2\}$, $\{a_3^5, a_4^3\}$, $\{a_4^5, a_5^4\}$, $\{a_5^5, a_4^4\}$ are the $\mathcal{A}_5$-similar sets;

$U_i^k \in a_i^k$.

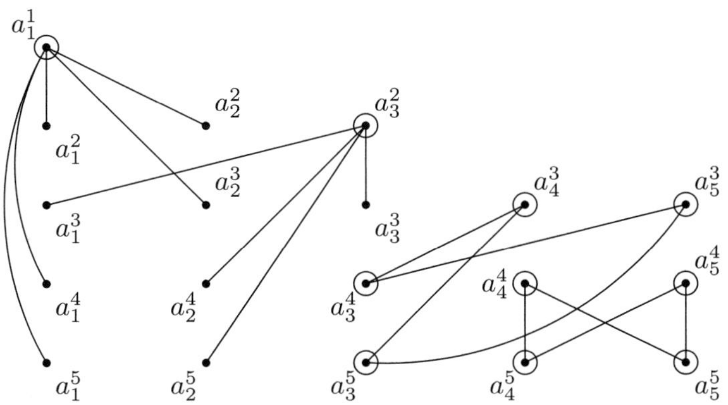

It is easy to check that condition 2) holds. □

16.16. REMARK. Whereas in the proof of Theorem 16.13(2) we constructed one "bunch" of ultrafilters, in Example 16.15 we had to construct two "bunches" of ultrafilters: $a_4^3, a_5^3, a_3^4, a_3^5$ and $a_4^4, a_5^4, a_4^5, a_5^5$.

16.17. This section completes the series of results that started in 16.12. Let us present a statement which it is natural to try to generalize.

STATEMENT. *Assume that for some algebras $\mathcal{A}_1, \mathcal{A}_2, \mathcal{A}_3, \mathcal{A}_4$ there exists a matrix of pairwise disjoint sets*

$$\begin{pmatrix} U_1^1 & U_2^1 & & & \\ U_1^2 & U_2^2 & U_3^2 & & \\ U_1^3 & U_2^3 & U_3^3 & U_4^3 & \\ U_1^4 & U_2^4 & U_3^4 & U_4^4 & U_5^4 \end{pmatrix}$$

such that $U_i^k \notin \mathcal{A}_k$. Then there exist sets U_1, U_2, U_3, U_4 such that $U_k \subset U_{j_k}^k$ and

$$U_k \subset Q \subset \bigcup_{i=1}^{4} U_i \quad \textit{implies} \quad Q \notin \mathcal{A}_k.\text{[12]}$$

PROOF. Consider the system $\{\widetilde{\mathfrak{A}}, \widetilde{\mathfrak{W}}\}$, where $\widetilde{\mathfrak{A}} = \{\widetilde{\mathcal{A}}_1, \widetilde{\mathcal{A}}_2, \widetilde{\mathcal{A}}_3\}$ and

$$\widetilde{\mathfrak{W}} = \begin{pmatrix} \widetilde{W}_1^1 & \widetilde{W}_2^1 & \\ \widetilde{W}_1^2 & \widetilde{W}_2^2 & \widetilde{W}_3^2 \\ \widetilde{W}_1^3 & \widetilde{W}_2^3 & \widetilde{W}_3^3 \end{pmatrix}.$$

Also, as usual, let us consider the ultrafilters $\tilde{s}_i^k$ and $\tilde{t}_i^k$, and either $\tilde{t}_i^k = \tilde{s}_j^l$, or $\tilde{t}_i^k \not\ni \bigcup_{l,j} \widetilde{W}_j^l$. It is not true that there always exist sets $\widetilde{U}_1, \widetilde{U}_2, \widetilde{U}_3$ such that

(a) $\widetilde{U}_k = \widetilde{W}_{j_k}^k$;
(b) $\widetilde{U}_k \subset Q \subset \bigcup_{i=1}^3 \widetilde{U}_i$ implies $Q \notin \widetilde{\mathcal{A}}_k$.

---

[12]In the proof of Statement 16.14 we have shown that a similar statement for three algebras holds.

For example, let $\tilde{t}_1^1 = \tilde{s}_1^2$, $\tilde{t}_2^1 = \tilde{s}_1^3$, $\tilde{t}_1^2 = \tilde{s}_2^1$, $\tilde{t}_2^2 = \tilde{s}_2^3$, $\tilde{t}_3^2 = \tilde{s}_3^3$, $\tilde{t}_1^3 = \tilde{s}_1^1$, $\tilde{t}_2^3 = \tilde{s}_3^2$, $\tilde{t}_3^3 = \tilde{s}_3^2$.

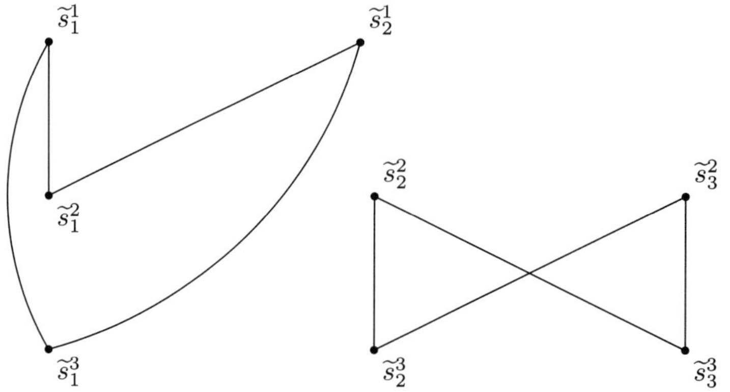

It is clear that in such a situation the sets $\widetilde{U}_1, \widetilde{U}_2, \widetilde{U}_3$ may fail to exist. It is worth noticing that here we also have two "bunches" of ultrafilters: $\tilde{s}_1^1, \tilde{s}_2^1, \tilde{s}_1^2, \tilde{s}_1^3$ and $\tilde{s}_2^2, \tilde{s}_3^2, \tilde{s}_2^3, \tilde{s}_3^3$. The following is important here: in the situation where one cannot construct the sets $\widetilde{U}_1, \widetilde{U}_2, \widetilde{U}_3$, there does not exist an ultrafilter $\tilde{t}_j^i$ different from all ultrafilters $\tilde{s}_j^i$.

Let us now turn to the sets U_i^k. As usual, let us consider the ultrafilters s_i^k, t_i^k and the sets W_i^k. If $t_i^1 = s_j^4$, then one can find the required sets U_1, U_2, U_3, U_4 using the corresponding result about three algebras in the proof of Statement 16.14 and applying the previous arguments. *Assume that always* $t_i^1 \neq s_j^4$. Consider the system $\{\mathfrak{A}', \mathfrak{W}'\}$, where $\mathfrak{A}' = \{\mathcal{A}_1, \mathcal{A}_2, \mathcal{A}_3\}$ and

$$\mathfrak{W}' = \begin{pmatrix} W_1^1 & W_2^1 & & \\ W_1^2 & W_2^2 & W_3^2 & \\ W_1^3 & W_2^3 & W_3^3 & W_4^3 \end{pmatrix}.$$

We know that for this system there exist the corresponding sets U_1, U_2, U_3. It is convenient to denote them by U_1', U_2', U_3', respectively, because these sets (or some of them) may be absent among the required sets U_1, U_2, U_3, U_4. Let

$$U_1' = W_2^1, \quad U_2' = W_3^2, \quad U_3' = W_4^3.$$

The situation is quite simple unless $t_3^2, t_4^3 \in \{s_1^4, s_2^4, s_3^4, s_4^4, s_5^4\}$. For example, let $t_3^2 = s_4^4$, $t_4^3 = s_5^4$. Moreover, the situation is not quite simple only if

$$\{t_1^4, t_2^4, t_3^4\} = \{s_2^1, s_3^2, s_4^3\}.$$

For example, let $t_1^4 = s_2^1$, $t_2^4 = s_3^2$, $t_3^4 = s_4^3$.

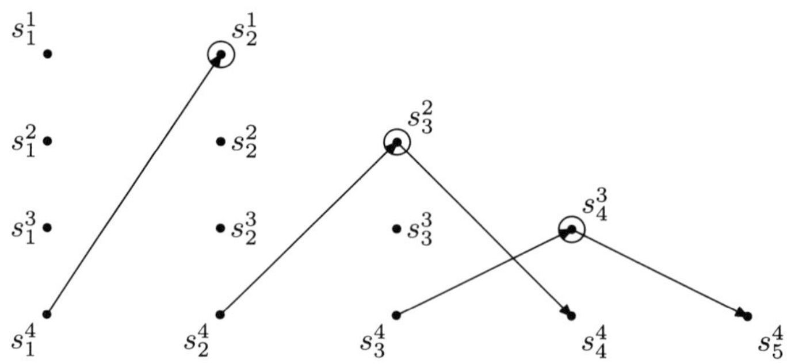

Put $U_1 = W_1^1$. Further, taking into account that $t_1^1 \neq s_j^4$, we will find sets U_2, U_3, U_4, and $U_k = W_{j_k}^k$. □

16.18. The following theorem, whose obvious corollary is Theorem 2.45(1), differs from the other general theorems of this chapter.

THEOREM. *Let $\{\mathfrak{A}, \mathfrak{U}\}$ be a system of type $k(k+1)-1$ such that $m_k \leq 2k$ for all k.*[13] *Then there exists a sequence of pairwise disjoint sets $U_1, \dots, U_n, V_1, \dots, V_n$ such that if $U_k \subset Q$ and $Q \cap V_k = \varnothing$, then $Q \notin \mathcal{A}_k$.*

PROOF. *Let* $n > 1$ (if $n = 1$, then the situation is trivial). Let us list all the possible cases.

Case 1. $m_1 = 1$. Then, if $1 < k \leq n$, we have $m_k = 2k$.

Case 2. $m_k = 2k$ for all k. This case amounts to Case 1 after removing the set W_2^1.

Case 3. $m_1 = 2$, and there exists a k_0 such that $m_{k_0} = 2k_0 - 1$. Then, if $k \in [1, n] \setminus \{k_0\}$, we have $m_k = 2k$.

Case 3 is divided into two subcases.

Case 3-a. There exists an ultrafilter $t_{i_0}^{k_0} = s_1^1$.

Case 3-b. There does not exist an ultrafilter $t_{i_0}^{k_0} = s_1^1$.

In Case 3-a, let us move the set W_1^1 to the k_0-th row. Then Case 3-a amounts to Case 1. Thus in order to prove the theorem we must examine only two cases: Case 1 and Case 3-b.

Case 1. The set W_i^k is called *marked* if the ultrafilter t_i^k is considered and either $t_i^k = s_1^1$, or $t_i^k = t_1^1$. If $t_1^1 = s_{\lambda_0}^{\nu_0}$, then the set $W_{\lambda_0}^{\nu_0}$ is also called *marked.* If in the k-th row of $\mathfrak{W}$ $(k > 1)$ there are no marked sets, then let us call two arbitrary sets from this row *marked.* If the k-th row contains only one marked set, then let us call another arbitrary set from this row *marked.* So the k-th row of $\mathfrak{W}$ $(k > 1)$ contains two marked sets and there can exist only one row with three marked sets. Denote by $\mathfrak{W}^*$ the matrix which is obtained from $\mathfrak{W}$ by removing the first row and all the marked sets. Put $\mathfrak{A}^* = \{\mathcal{A}_2, \dots, \mathcal{A}_n\}$. The system $\{\mathfrak{A}^*, \mathfrak{W}^*\}$ is of type $k(k+1)-1$, and the k-th row of $\mathfrak{W}^*$ contains at most $2k$ sets. This system will be considered in the next step. We know that

$$W_1^1 \notin \mathfrak{W}^*, \quad t_1^1 \not\ni W_i^k \in \mathfrak{W}^*, \quad s_1^1, t_1^1 \notin T = \{t_i^k \mid W_i^k \in \mathfrak{W}^*\}.$$

If t_1^1 is considered, then we take $V_1 \in t_1^1$ such that

$$V_1 \cap \left(W_1^1 \cup \bigcup\{W_i^k \mid W_i^k \in \mathfrak{W}^*\}\right) = \varnothing, \quad \overline{V}_1 \cap T = \varnothing.$$

If t_1^1 is not considered, then we put $V_1 = \varnothing$. Let us take $U_1 = W_1^1$.

Case 3-b. Marked sets are defined exactly as in Case 1. We will make only one change compared with Case 1. Namely, if the k_0-th row contains no marked sets, we mark an arbitrary set in this row, and if the k_0-th row contains a marked set, we will not mark any other set of this row. As in Case 1, we define the sets $U_1 = W_1^1$, V_1, and in the next step consider the system $\{\mathfrak{A}^*, \mathfrak{W}^*\}$ of type $k(k+1)-1$; the k-th row of $\mathfrak{W}^*$ contains at most $2k$ sets and $\#(\mathfrak{A}^*) = n - 1$.

Continuing our constructions by induction, we will find the required sets $U_1, \dots, U_n, V_1, \dots, V_n$. □

[13]It is easy to check that such a system is regular.

16.19. REMARK. It can happen that the set U_k in Theorem 16.18 equals $W_{p_k}^{i_k}$ and $i_k \neq k$. Clearly, $W_{p_k}^{i_k}$ is a moved set. (Since in the proof of Theorem 16.18 we do not transpose rows in the matrix, we have $i_k < k$.) But then we can put $V_k = W_{j_k}^k$. Nothing will happen if we put $U_k = W_{j_k}^k$ and $V_k = W_{p_k}^{i_k}$. So the sets $U_1, \ldots, U_n$ in Theorem 16.18 can be chosen so that $U_k \subset U_{j_k}^k$.

16.20. Using the idea of the proof of Theorem 16.18, one can easily prove the following result.

THEOREM. *Let $\{\mathfrak{A}, \mathfrak{U}\}$ be a system of type $(k^2+3k-2)/2$ such that $m_k \leq k+1$ for all k.*[14] *Then there exists a sequence of sets $U_1, \ldots, U_n$ such that*

$$U_k \subset Q \subset \bigcup_{i=1}^{n} U_i \quad \textit{implies} \quad Q \notin \mathcal{A}_k.$$

Also $U_k \subset U_{j_k}^{i_k}$; if $1 \leq k < l \leq n$, then $U_{j_k}^{i_k} \neq U_{j_l}^{i_l}$; $i_k \leq k$.

Of course, this theorem is a particular case of Theorem 16.10. But its proof is simpler than the proof of Theorem 16.10. On the other hand, Theorem 16.11(1) is an obvious corollary of our theorem.

16.21. To conclude this chapter, let us present the *proof of Theorem* 2.45(2). Let us describe the construction for $n = 5$. First we find the kernels of all the algebras, taking into account the fact that $X \in \mathcal{A}_k$ for all k. That is, we list all $\mathcal{A}_k$-similar sets for each $k \leq 5$:

for $k=1$: $\{a_1^1, a_1^2\}$;
for $k=2$: $\{a_1^2, a_2^2\}$, $\{a_3^2, a_1^1\}$, $\{a_4^2, a_3^3\}$;
for $k=3$: $\{a_1^3, a_1^1\}$, $\{a_2^3, a_1^2\}$, $\{a_3^3, a_4^3\}$, $\{a_5^3, a_4^2\}$, $\{a_6^3, a_5^4\}$;
for $k=4$: $\{a_1^4, a_1^1\}$, $\{a_2^4, a_1^2\}$, $\{a_3^4, a_4^2\}$, $\{a_4^4, a_3^3\}$, $\{a_5^4, a_6^4\}$, $\{a_7^4, a_6^3\}$, $\{a_8^4, a_7^5\}$;
for $k=5$: $\{a_1^5, a_1^1\}$, $\{a_2^5, a_1^2\}$, $\{a_3^5, a_4^2\}$, $\{a_4^5, a_3^3\}$, $\{a_5^5, a_6^3\}$, $\{a_6^5, a_5^4\}$, $\{a_7^5, a_8^5\}$, $\{a_9^5, a_8^4\}$.

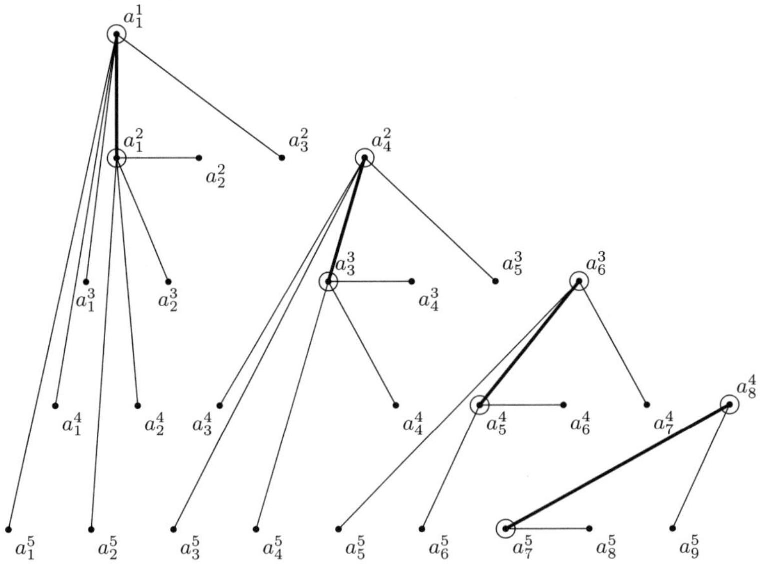

[14]It is easy to check that such a system is regular.

Let us assume that $a_i^k \ni U_i^k$.

For the algebras $\mathcal{A}_1, \mathcal{A}_2, \mathcal{A}_3, \mathcal{A}_4$, the statement of Theorem 2.45(1) is valid, and we necessarily have

$$U_1 \cup V_1 \in a_1^1, a_1^2; \quad U_2 \cup V_2 \in a_4^2, a_3^3;$$
$$U_3 \cup V_3 \in a_6^3, a_5^4; \quad U_4 \cup V_4 \in a_8^4, a_7^5.$$

Clearly, the statement of Theorem 2.45(1) does not hold for the algebras $\mathcal{A}_1, \mathcal{A}_2, \mathcal{A}_3, \mathcal{A}_4, \mathcal{A}_5$.

The construction in the general case does not differ, in principle, from the construction for $n = 5$. □

CHAPTER 17

Countable Sequences of Algebras (2). Proof of Theorems 2.29, 2.32, 2.46

17.1. In this chapter the following theorem is proved:

THEOREM. *Let* $\mathcal{A}_1, \dots, \mathcal{A}_k, \dots$ *and* $\mathcal{B}_1, \dots, \mathcal{B}_k, \dots$ *be two countable sequences of* σ*-algebras. Let all the algebras* $\mathcal{A}_k$ *be simple, and among the algebras* $\mathcal{B}_k$ *let there be no simple algebras. Then there exist pairwise disjoint sets* $W, U_1, \dots, U_k, \dots, V_1, \dots, V_k, \dots$ *such that*

(1) $\ker \mathcal{A}_k \subset \overline{W}$ *for each* k;
(2) *if* $U_k \subset Q$ *and* $Q \cap V_k = \varnothing$, *then* $Q \notin \mathcal{B}_k$.

17.2. The proof of Theorem 17.1 is preceded by two definitions and five lemmas.

DEFINITION. An algebra $\mathcal{A}$ is said to be *absolutely nonseparable* if there exists an externally nonseparable set $M \subset \ker \mathcal{A}$ such that any two distinct ultrafilters from M are $\mathcal{A}$-similar.

17.3. LEMMA. *Suppose* $\mathcal{A}$ *is not an absolutely nonseparable* σ*-algebra; let* M *be an externally separable subset of* βX, *and let*

$$K = \{x \in \ker \mathcal{A} \mid \textit{there exists an ultrafilter } y \in \overline{M}\ \mathcal{A}\textit{-similar to } x\}.$$

Then K *is an externally separable set.*

PROOF. Let

$$M^* = \{y \in \overline{M} \mid \text{there exist } \mathcal{A}\text{-similar ultrafilters } x, y\}.$$

If $M^* = \varnothing$, then $K = \varnothing$ and the lemma is proved. Let $M^* \neq \varnothing$, $Z \subset \beta X$, $\#(Z) \leq \aleph_0$, $\overline{Z} \supset M^*$, and

$$Z^* = \{z \in Z \mid z \text{ is not an } \mathcal{A}\text{-special ultrafilter}\}.$$

Clearly, $\overline{Z^*} \supset M^*$. Let $Z^* = \{z_1, \dots, z_n, \dots\}$ and

$$C_n = \{x \in \ker \mathcal{A} \mid x, z_n \text{ are } \mathcal{A}\text{-similar ultrafilters}\} \cup \{z_n\}.$$

Let us show that $K \subset \overline{\bigcup_n C_n}$. Suppose the contrary. Thus there exist sets $G, W_1, \dots, W_n, \dots \subset X$ such that

$$\overline{G} \cap K \neq \varnothing; \qquad G \cap W_n = \varnothing,\ C_n \subset \overline{W}_n,\ W_n \in \mathcal{A} \text{ for each } n.$$

Clearly, we have

$$\mathcal{A} \ni W = \bigcup_n W_n; \quad M^* \subset \overline{W}, \quad W \cap G = \varnothing.$$

There exist $\mathcal{A}$-similar ultrafilters x, y such that $x \ni G$, $y \ni W$, a contradiction. Since the algebra $\mathcal{A}$ is not absolutely nonseparable, C_n is an externally separable set. Hence K is also an externally separable set. □

17.4. DEFINITION. Consider an algebra $\mathcal{A}$. Ultrafilters a, b are called an $\mathcal{A}$-*nonseparable pair* if one of the following two conditions holds:

(1) $a = b$, and for each $G \in a$ the set $\overline{G} \cap \ker \mathcal{A}$ is externally nonseparable, while the set

$$\{y \in \overline{X \setminus G} \mid \text{there exists an ultrafilter } x \ni G \ \mathcal{A}\text{-similar to } y\}$$

is externally separable;

(2) $a \neq b$, and for each $G_1 \in a$, $G_2 \in b$ the sets $\overline{G}_1 \cap \ker \mathcal{A}$ and

$$\{y \in \overline{G}_2 \mid \text{there exists an ultrafilter } x \ni G_1 \ \mathcal{A}\text{-similar to } y\}$$

are externally nonseparable.

17.5. LEMMA. *For any nonseparable σ-algebra $\mathcal{A}$ there exists an $\mathcal{A}$-nonseparable pair.*

PROOF. Suppose c is a point of inseparability of $\ker \mathcal{A}$, and c, c is not an $\mathcal{A}$-nonseparable pair. Then there exist ultrafilters a, b such that $a \neq b$ and if $G_1 \in a$, $G_2 \in b$, then the set

$$\{y \in \overline{G}_2 \mid \text{there exists an ultrafilter } x \ni G_1 \ \mathcal{A}\text{-similar to } y\}$$

is externally nonseparable. If the algebra $\mathcal{A}$ is not absolutely nonseparable, then, by Lemma 17.3, a, b is an $\mathcal{A}$-nonseparable pair. And if the algebra $\mathcal{A}$ is absolutely nonseparable, then the assertion of Lemma 17.5 is obvious. □

17.6. LEMMA. *Let $\mathcal{A}_1, \dots, \mathcal{A}_k, \dots$ and $\mathcal{B}_1, \dots, \mathcal{B}_k, \dots$ be two countable sequences of σ-algebras. Let all algebras $\mathcal{A}_k$ be separable, and let all algebras $\mathcal{B}_k$ be nonseparable. Then there exist pairwise disjoint sets*

$$W, U_1, \dots, U_k, \dots, V_1, \dots, V_k, \dots$$

such that

(1) $\ker \mathcal{A}_k \subset \overline{W}$ *for each* k;
(2) *if* $U_k \subset Q$ *and* $Q \cap V_k = \varnothing$, *then* $Q \notin \mathcal{A}_k$.

PROOF. For each algebra $\mathcal{B}_k$ let us choose a $\mathcal{B}_k$-nonseparable pair b_1^k, b_2^k. Let

$$Z \subset \beta X, \quad \#(Z) \le \aleph_0; \quad b_1^k, b_2^k \in Z, \quad \ker \mathcal{A}_k \subset \overline{Z}$$

for all k. Using Lemma 17.3 if necessary (it is indeed necessary if $b_1^k = b_2^k$ for some numbers k, and for these k the algebras $\mathcal{B}_k$ are not absolutely nonseparable), let us construct two matrices of pairwise disjoint sets

$$\mathfrak{U} = \begin{pmatrix} U_1^1 & & \\ \dots\dots & \dots & \dots\dots \\ U_1^k & \dots & U_k^k \\ \dots\dots & \dots & \dots\dots \end{pmatrix}, \qquad \mathfrak{D} = \begin{pmatrix} V_1^1 & & \\ \dots\dots & \dots & \dots\dots \\ V_1^k & \dots & V_k^k \\ \dots\dots & \dots & \dots\dots \end{pmatrix}$$

such that

$$\bigcup_{k,i} U_i^k \cap \bigcup_{k,i} V_i^k = \varnothing.$$

If $V_i^k = \varnothing$, then U_i^k belongs to a $\mathcal{B}_k$-special ultrafilter. If $V_i^k \neq \varnothing$, then there exist two $\mathcal{B}_k$-similar ultrafilters, one of which contains U_i^k and the other contains V_i^k. Moreover, $\overline{U_i^k \cup V_i^k} \cap Z = \varnothing$. By Lemma 5.3, one can choose sets

$$U_{i_1}^1, V_{i_1}^1; \; \dots \; ; U_{i_k}^k, V_{i_k}^k; \; \dots \quad \text{such that} \quad Z \cap \overline{\bigcup_k (U_{i_k}^k \cup V_{i_k}^k)} = \varnothing.$$

Put $U_k = U_{i_k}^k$, $V_k = V_{i_k}^k$. Let us take a subset $W \subset X$ such that $Z \subset \overline{W}$ and

$$W \cap \bigcup_k (U_k \cup V_k) = \varnothing. \qquad \square$$

17.7. LEMMA. *Suppose $\mathcal{A}$ is a σ-algebra, $Z \subset \beta X$, $\#(Z) \leq \aleph_0$, all points of Z are irregular, q, r are $\mathcal{A}$-similar ultrafilters, $Q \in q$, $r \in \overline{Z}$, and all points of $\overline{Q} \cap \ker \mathcal{A}$ are regular. Then there exists a set $\mathfrak{L} \subset \overline{Q} \cap \ker \mathcal{A}$ such that*

(i) *any two different ultrafilters in $\mathfrak{L}$ are $\mathcal{A}$-similar;*
(ii) *defining $\mu(M)$ to be 0 if and only of $\overline{M} \cap \mathfrak{L} = \varnothing$, we obtain a σ-two-valued measure on X.*

PROOF. Let us first assume that $\overline{Q} \cap Z = \varnothing$. Next, we can assume that $r \in Z$, since there exist $\mathcal{A}$-similar ultrafilters q', r' with $Q \in q'$, $r' \in Z$, because $r \in \overline{Z}$. Set

$$\mathfrak{L} = \{x \in \overline{Q} \cap \ker \mathcal{A} \mid x, r \text{ are } \mathcal{A}\text{-similar ultrafilters}\}. \qquad \square$$

17.8. Let $\mathcal{A}$ be a σ-algebra, $K, L \subset X$, and $K \cap L = \varnothing$. Assume that

$$P = \{x \in \overline{L} \mid \text{there exists an ultrafilter } y \ni K \ \mathcal{A}\text{-similar to } x\} \neq \varnothing, \quad P \subset \beta X \setminus X.$$

If $M \subset X$ and $\overline{M} \cap P = \varnothing$, then put $\mu_{\mathcal{A}}^{K,L}(M) = 0$. The proof of the following lemma is very simple.

LEMMA. *$\mu_{\mathcal{A}}^{K,L}$ defines a σ-two-valued measure on X.*

17.9. Let $\mathcal{A}$ be an almost σ-algebra,

$$U \subset X, \quad \overline{U} \cap \ker \mathcal{A} \neq \varnothing, \quad (\overline{U} \cap \ker \mathcal{A}) \cap X = \varnothing.$$

Then, as in 5.13, one can define a σ-two-valued measure $\mu_{\mathcal{A}}^U$ on X: $\mu_{\mathcal{A}}^U(M) = 0$ if and only if $\mathfrak{P}(M \cap U) \subset \mathcal{A}$ (this is equivalent to the equality $\overline{M \cap U} \cap \ker \mathcal{A} = \varnothing$).

17.10. PROOF OF THEOREM 17.1. Let us assume that all algebras $\mathcal{B}_k$ are strictly separable. This is possible by Lemma 17.6. Put

$$Z = \bigcup_k \operatorname{sp} \mathcal{A}_k \cup \bigcup_k \operatorname{sp} \mathcal{B}_k.$$

Let us assign to each algebra $\mathcal{B}_k$ a set $Q_k \subset X$ satisfying the following conditions:

1) $\overline{Q}_k \cap Z = \varnothing$;
2) $\overline{Q}_k \cap \ker \mathcal{B}_k \neq \varnothing$.

Let us assume, without loss of generality, that there does not exist a $\mathcal{B}_k$-special ultrafilter containing Q_k. Consider σ-two-valued measures $\mu_{\mathcal{B}_k}^{Q_k}$. As in Chapter 5, using the Gitik–Shelah theorem, let us choose pairwise disjoint sets $Q_1', \dots, Q_k', \dots$ such that $Q_k' \subset Q_k$ and $Q_k' \notin \mathcal{B}_k$. Let us assume that

$$Z \cap \overline{\bigcup_k Q_k'} = \varnothing.$$

This is possible by Lemma 5.3. Put

$$\mathbb{N}_1 = \big\{k \in \mathbb{N}^+ \mid k > 1, \text{ there exist } \mathcal{B}_k\text{-similar ultrafilters } q_k, r_k \text{ such that } Q'_k \in q_k,\ Q'_1 \in r_k\big\}.$$

Assume that $\mathbb{N}_1 \neq \varnothing$, and let $\mathbb{N}_1 = \{k_1^1, \dots, k_n^1, \dots\}$. Consider the sequence of σ-two-valued measures

$$\mu_0^1 = \mu_{\mathcal{B}_1}^{Q'_1},\ \mu_1^1 = \mu_{\mathcal{B}_{k_1^1}}^{Q'_{k_1^1}, Q'_1},\ \dots,\ \mu_n^1 = \mu_{\mathcal{B}_{k_n^1}}^{Q'_{k_n^1}, Q'_1},\ \dots.$$

Using the Gitik–Shelah theorem, let us take pairwise disjoint sets $F_0^1, F_1^1, \dots, F_n^1 \dots \subset Q'_1$ such that each set F_n^1 is μ_n^1-nonmeasurable ($n \geq 0$). For each set F_n^1 ($n > 0$) we consider the following σ-two-valued measure:

$$\mu_n^{(1)} = \mu_{\mathcal{B}_{k_n^1}}^{F_n^1, Q'_{k_n^1}}.$$

In further considerations, we can replace $Q'_{k_n^1}$ with its subset, and this subset should contain a $\mu_n^{(1)}$-nonmeasurable set. It is possible to choose the set F_0^1 so that there exist $\mathcal{B}_1$-similar ultrafilters a_1, b_1 for which $F_0^1 \in a_1$, $F_0^1 \notin b_1$, $b_1 \notin \overline{Z}$ (see Lemma 17.7). Replacing, if necessary, the sets $Q'_2, \dots, Q'_k, \dots$ by appropriate subsets, let us assume that

$$b_1 \not\ni \bigcup_{k>1} Q'_k.$$

Put

$$\mathbb{N}_2 = \big\{k \in \mathbb{N}^+ \setminus \mathbb{N}_1 \mid k > 2 \text{ and there exist } \mathcal{B}_k\text{-similar ultrafilters } q_k, r_k \text{ such that } Q'_k \in q_k,\ Q'_2 \in r_k\big\}.$$

Assume that $\mathbb{N}_2 \neq \varnothing$, and let $\mathbb{N}_2 = \{k_1^2, \dots, k_n^2, \dots\}$. Consider the sequence of σ-two-valued measures

$$\mu_0^2 = \mu_{\mathcal{B}_2}^{Q'_2},\ \mu_1^2 = \mu_{\mathcal{B}_{k_1^2}}^{Q'_{k_1^2}, Q'_2},\ \dots,\ \mu_n^2 = \mu_{\mathcal{B}_{k_n^2}}^{Q'_{k_n^2}, Q'_2},\ \dots.$$

In the same way as above, let us take pairwise disjoint sets

$$F_0^2, F_1^2, \dots, F_n^2, \dots \subset Q'_2$$

such that each set F_n^2 is μ_n^2-nonmeasurable ($n \geq 0$). As above, for each set F_n^2 ($n > 0$) we consider the following σ-two-valued measure

$$\mu_n^{(2)} = \mu_{\mathcal{B}_{k_n^2}}^{F_n^2, Q'_{k_n^2}},$$

and so on.

Let $2 \in \mathbb{N}_1$ *and* $k_1^1 = 2$. We can assume that F_0^2 is a $\mu_1^{(1)}$-nonmeasurable set. Therefore there exist $\mathcal{B}_2$-similar ultrafilters a_2, b_2 such that $F_0^2 \in a_2$, $F_1^1 \in b_2$.

Let $2 \notin \mathbb{N}_1$. As above, let us assume that there exist $\mathcal{B}_2$-similar ultrafilters a_2, b_2 with $F_0^2 \in a_2$, $F_0^2 \notin b_2$, $b_2 \notin \overline{Z}$, and

$$b_2 \not\ni \bigcup_{k>2} Q'_k.$$

Clearly, $F_0^1 \notin b_2$, since $2 \notin \mathbb{N}_1$.

Finally we will have constructed pairwise disjoint sets $F_0^1, \dots, F_0^k, \dots$ and ultrafilters $a_1, b_1, \dots, a_k, b_k, \dots$ such that a_k, b_k are $\mathcal{B}_k$-similar and

$$F_0^k \in a_k, \quad b_k \not\ni \bigcup_m F_0^m, \quad b_k \notin \overline{Z}.$$

For each k let us choose a $G_k \in b_k$ such that

$$G_k \cap \bigcup_m F_0^m = \varnothing, \quad \overline{G}_k \cap Z = \varnothing.$$

By the Gitik–Shelah theorem and Lemma 17.8, for each k there exists a $V_k \subset G_k$ and $\mathcal{B}_k$-similar ultrafilters a'_k, b'_k such that $F_k^0 \in a'_k$, $V_k \in b'_k$. In addition, $V_1, \dots, V_k, \dots$ are pairwise disjoint ultrafilters. Let us assume that $Z \cap \overline{\bigcup_k V_k} = \varnothing$. This is possible by Lemma 5.3. So we have constructed the required sets V_k. It remains to put

$$U_k = F_0^k, \quad W = X \setminus \bigcup_k (U_k \cup V_k). \qquad \square$$

17.11. PROOF OF THEOREM 2.29. Let us assume that all the algebras $\mathcal{A}_k$ are simple. This is possible by Theorem 17.1. For each algebra $\mathcal{A}_k$ take an ω-saturated σ-algebra $\mathcal{A}'_k > \mathcal{A}_k$ such that

$$\#(\ker \mathcal{A}'_k) > \frac{5}{2}(k-1)$$

(see Lemmas 7.4, 7.5). Arguing as in the proof of Lemma 7.12, construct a countable sequence of pairwise distinct ultrafilters $S = \{s_1, \dots, s_k, \dots\}$ such that $s_k \in \ker \mathcal{A}'_k$, and either s_k is an $\mathcal{A}_k$-special ultrafilter, or there exists an ultrafilter t_k that is $\mathcal{A}'_k$-similar to s_k, and $t_k \notin S$. But all points of $\ker \mathcal{A}'_k$ are irregular. Hence we can take pairwise disjoint sets $U_1, \dots, U_k, \dots$ ($U_k \in s_k$) such that if $U_k \subset Q \subset \bigcup U_i$, then $Q \notin \mathcal{A}_k$. $\square$

17.12. We can prove Theorems 2.32 and 2.46 in a similar way. Infinite analogs of Theorems 16.1 and 16.18, in which all the algebras are σ-algebras, can readily be stated. These infinite analogs can be very easily justified using Theorems 16.1, 16.18, and 17.1, and the arguments in the proof of Lemma 7.12.

CHAPTER 18

A Refinement of Theorems on σ-Algebras. Proof of Theorems 2.34, 2.44

18.1. THEOREM. *Let $\mathcal{A}_1, \dots, \mathcal{A}_k, \dots$ be a countable sequence of σ-algebras, and let*

$$\begin{pmatrix} U_1^1 & \dots & U_{m_1}^1 \\ \dots & \dots & \dots \\ U_1^k & \dots & U_{m_k}^k \\ \dots & \dots & \dots \end{pmatrix}$$

be an infinite matrix of sets such that the sets from each row are pairwise disjoint and $U_i^k \notin \mathcal{A}_k$. Then:

(1) *Each set U_i^k contains a subset W_i^k, and the following hold:*
  (1–i) *either there exists an $\mathcal{A}_k$-special ultrafilter s_i^k, or there exist $\mathcal{A}_k$-similar ultrafilters s_i^k, t_i^k, and $s_i^k \ni W_i^k$;*
  (1–ii) *either $W_i^k = W_j^p$, or $W_i^k \cap W_j^p = \varnothing$;*
  (1–iii) *if $W_i^k = W_j^p$ and $k \neq p$, then the ultrafilters s_i^k, s_j^p are irregular and $s_i^k = s_j^p$.*

(2) *t_i^k contains either a set $D'_{k,i}$, or a set $D''_{k,i}$, and the following hold:*
  (2–i) *if $D'_{k,i} \in t_i^k$, then the ultrafilters s_i^k, t_i^k are irregular;*
  (2–ii) *if $D'_{k,i} = W_j^p$, then $t_i^k = s_j^p$;*
  (2–iii) *either $D'_{k,i} = D'_{p,j}$, and then $t_i^k = t_j^p$, or $D'_{k,i} \cap D'_{p,j} = \varnothing$.*

(3) *The family consisting of all sets $D''_{k,i}$, the union of all sets W_i^k, and the union of all sets $D'_{k,i}$ that do not coincide with any sets W_i^k is a family of pairwise disjoint sets.*

PROOF. The proof will not be presented in detail because all the auxiliary statements can be established without too much difficulty using the techniques worked out earlier in this monograph. First let us assign to each set U_i^k a subset $U_{k,i} \subset U_i^k$ so that

(a) either $\overline{U}_{k,i} \cap \ker \mathcal{A}_k$ contains an irregular point s_i^k, or
(b) all points of the set $\overline{U}_{k,i} \cap \ker \mathcal{A}_k$ are regular;
(c) either $U_{k,i} = U_{p,j}$, or $U_{k,i} \cap U_{p,j} = \varnothing$;
(d) if $U_{k,i} = U_{p,j}$ and $k \neq p$, then these sets satisfy condition (a) and $s_i^k = s_j^p$.

All required irregular ultrafilters s_i^k are defined in (a). Assume for simplicity that $X \in \mathcal{A}_k$ for all k. Let the irregular ultrafilter s_i^k have an $\mathcal{A}_k$-similar irregular ultrafilter t_i^k. Then let us assume that either $t_i^k = s_j^p$, or $t_i^k \not\ni \bigcup_{p,j} U_{p,j}$. Let the irregular ultrafilter s_i^k have no $\mathcal{A}_k$-similar irregular ultrafilters. Then there exists a set $U^*_{k,i}$ which belongs to a regular ultrafilter $\mathcal{A}_k$-similar to s^k_i. Let us assume that either $U^*_{k,i}$ is contained in a set $U_{p,j}$ which satisfies condition (b), or

$$U^*_{k,i} \cap \bigcup_{p,j} U_{p,j} = \varnothing.$$

Denote by Θ the set of all ultrafilters t^k_i defined up to now (as we know, they all are irregular, and all the other ultrafilters t^k_i, not defined yet, will be regular). Let us assume that $\Theta \cap \overline{U^*_{k,i}} = \varnothing$ for each set $U^*_{k,i}$. Let $U_{k,i}$ satisfy condition (b). There are two possible cases:

I. $U_{k,i}$ belongs to an ultrafilter which has an $\mathcal{A}_k$-similar irregular ultrafilter.
II. $U_{k,i}$ does not belong to an ultrafilter which has an $\mathcal{A}_k$-similar irregular ultrafilter.

In Case II, let us assume that there exists a set $U^\circ_{k,i}$ satisfying the following conditions:

(i) $U_{k,i} \cap U^\circ_{k,i} = \varnothing$;
(ii) either $U^\circ_{k,i} \subset U_{p,j}$ which satisfies condition (b), or $U^\circ_{k,i} \cap \bigcup_{p,j} U_{p,j} = \varnothing$;
(iii) $\Theta \cap \overline{U^\circ_{k,i}} = \varnothing$;
(iv) the set $P_{k,i} = \{x \in \overline{U^\circ_{k,i}} \mid x$ has an $\mathcal{A}_k$-similar ultrafilter containing $U_{k,i}\}$ is nonempty (and contains, as we know, only regular points).

In Case I there exist $\mathcal{A}_k$-similar ultrafilters, each of which contains $U_{k,i}$. Hence in Case I as well, replacing the set $U_{k,i}$ by its subset, we will find a set $U^\circ_{k,i}$ which satisfies all conditions (i)–(iv). Denote by $\mathcal{U}$ the family of all sets of the type of $U^*_{k,i}$ and $U^\circ_{k,i}$. Let us assume that the sets from this family are pairwise disjoint. Let $U_{k,i}$ satisfy condition (b). Put

$$W^k_i = U_{k,i} \setminus \bigcup \{U \in \mathcal{U} \mid U \subset U_{k,i}\}.$$

One can ensure that there exist $\mathcal{A}_k$-similar ultrafilters s^k_i, t^k_i such that $W^k_i \in s^k_i$, $U^\circ_{k,i} \in t^k_i$. If $U_{k,i}$ satisfies condition (a), then put $W^k_i = U_{k,i}$. The sets $D'_{k,i}, D''_{k,i}$ and the (not yet defined) ultrafilters t^k_i, each of which contains the corresponding set $U^*_{k,i}$, can be defined in the obvious way. □

18.2. Theorem 18.1 and Statement 16.8 imply Theorem 2.44. All other corresponding statements are proved using Theorem 18.1 and the arguments from the proof of Lemma 7.12. Among these statements are refinements of Theorems 2.29 and 2.32 (taking into account Remark 14.9), as well as Theorem 2.34, and infinite analogs (for σ-algebras) of Theorems 16.10, 16.12, 16.13(1), and 16.18 (taking into account Remark 16.19). Let us present the statement and the proof of the infinite analog of the result in 16.2 for σ-algebras.

THEOREM. *Let $\{\mathfrak{A}, \mathfrak{U}\}$ be a regular countable system of type $k(k+1)/2$, let all the algebras $\mathcal{A}_k \in \mathfrak{A}$ be σ-algebras, and let*

$$\mathfrak{U} = \begin{pmatrix} U^1_1 & \dots & U^1_{m_1} \\ \dots & \dots & \dots \\ U^k_1 & \dots & U^k_{m_k} \\ \dots & \dots & \dots \end{pmatrix}.$$

Then there exists a countable sequence of sets $U_1, \dots, U_k, \dots$ *such that the following hold*:

(i) $U_k \subset Q \subset \bigcup U_i$ *implies* $Q \notin \mathcal{A}_k$;

(ii) $U_k \subset U_{j_k}^{i_k}$; $U_{j_k}^{i_k} \neq U_{j_l}^{i_l}$, *if* $k \neq l$.

PROOF. By Theorem 18.1, we can consider sets W_i^k, etc. Put

$$W^{(k)} = \{W_j^p \mid \text{either } p = k, \text{ or there exists an ultrafilter } t_i^k = s_j^p\}.$$

Consider the finite system $\{\mathfrak{A}_n, \mathfrak{W}_n\}$, where $\mathfrak{A}_n = \{\mathcal{A}_1, \dots, \mathcal{A}_n\}$ and

$$\mathfrak{W}_n = \begin{pmatrix} W_1^1 & \dots & W_{m_1}^1 \\ \dots & \dots & \dots \\ W_1^n & \dots & W_{m_n}^n \end{pmatrix}.$$

By Theorem 16.1, for this system the corresponding sets, which we denote by $U_1^{(n)}, \dots, U_n^{(n)}$, exist. One can assume that $U_k^{(n)} \in W^{(k)}$. Now since $\#(W^{(k)}) < \aleph_0$, the arguments from the proof of Lemma 7.12 imply that our theorem is true. □

18.3. REMARK. The last two results on σ-algebras stated at the end of 2.34 follow from Theorem 4.29(1), Statement 8.21, and Theorem 18.1.

CHAPTER 19

Semistructures and Structures of Sets. Proof of Theorem 2.48

19.1. DEFINITION. A set $\mathcal{A} \subset \mathfrak{P}(X)$ is called a *semistructure* if $\varnothing \in \mathcal{A}$, and $M_1, M_2 \in \mathcal{A}$ implies $M_1 \cup M_2 \in \mathcal{A}$.

19.2. Let $\mathcal{A}$ be a semistructure and $Q \notin \mathcal{A}$. Then there exists an ultrafilter $s \ni Q$ such that $(s \cap Q) \cap \mathcal{A} = \varnothing$.

19.3. This simple fact implies the following theorem, whose proof raises no difficulties.

THEOREM. *Let $\mathcal{A}_1, \dots, \mathcal{A}_n$ be a finite sequence of semistructures, and for each $k \in [1, n]$ let there exist k pairwise disjoint sets not belonging to $\mathcal{A}_k$. Then there exist pairwise disjoint sets $U_1, \dots, U_n$ such that $U_k \notin \mathcal{A}_k$ for each $k \leq n$.*[1]

19.4. Clearly, one can generalize Theorem 19.3 using Theorem 3.27 to a system of distinct representatives.

19.5. DEFINITION. Let $\mathcal{A}$ be a semistructure. The set

$$\{s \in \beta X \mid \text{there exists } Q \in s \text{ such that } (s \cap Q) \cap \mathcal{A} = \varnothing\}$$

is called the *kernel* of $\mathcal{A}$ and denoted $\operatorname{kr} \mathcal{A}$.

19.6. Recall what was said in 3.19: if $\mathcal{A}$ is an algebra, then $\ker \mathcal{A} = \operatorname{kr} \mathcal{A}$.[2]

19.7. One naturally introduces the notion of almost σ-semistructure.

DEFINITION. A semistructure $\mathcal{A}$ is called an *almost σ-semistructure* if for any countable sequence of sets $M_1, \dots, M_k, \dots$ such that $\mathfrak{P}(M_k) \subset \mathcal{A}$ for all k, we have $\mathcal{A} \ni \bigcup_k M_k$.

19.8. The following theorem is a generalization of Statement 5.15.

THEOREM. *Let $\mathcal{A}_1, \dots, \mathcal{A}_k, \dots$ be a countable sequence of almost σ-semistructures, and for each k let there exist k pairwise disjoint sets not belonging to $\mathcal{A}_k$. Then there exist pairwise disjoint sets $U_1, \dots, U_k, \dots$ such that $U_k \notin \mathcal{A}_k$.*

---

[1] By what was said in 3.3, in the proof of Theorem 19.3 one can assume that $\#(X) < \aleph_0$, and manage without using ultrafilters.

[2] The properties of the kernel of a semistructure differ from the properties of the kernel of an algebra. Let $\#(X) \geq \aleph_0$. Define the semistructure $\mathcal{A} = \{M \in \mathfrak{P}(X) \mid \#(M) \neq 1\}$. Obviously, $\operatorname{kr} \mathcal{A} = X$ is not a countably compact set. If $\#(X) < 2^{\mathfrak{c}}$, then $\aleph_0 \leq \#(\operatorname{kr} \mathcal{A}) < 2^{\mathfrak{c}}$. In 5.14 it is shown that the kernel of an algebra is a countably compact set and its cardinality cannot be simultaneously $\geq \aleph_0$ and $< 2^{\mathfrak{c}}$.

PROOF. For each k let us define a σ-algebra $\mathcal{A}'_k$ as follows: $M \in \mathcal{A}'_k$ if and only if one of the following two conditions holds:

(1) $\mathfrak{P}(M) \subset \mathcal{A}_k$;
(2) $\mathfrak{P}(X \setminus M) \subset \mathcal{A}_k$.

Using the techniques developed in Chapter 5, it is easy to prove the existence of pairwise disjoint sets $L_1, \dots, L_k, \dots$ such that $L_k \notin \mathcal{A}'_k$. However, we have $\overline{L}_k \cap \operatorname{kr} \mathcal{A}_k \neq \varnothing$. Hence there exists a $U_k \subset L_k$ such that $U_k \notin \mathcal{A}_k$. □

19.9. It is clear that one can formulate the corresponding theorems on countable sequences of almost σ-semistructures, using Theorems 7.20 and 7.22 on systems of distinct representatives.

19.10. The proof of the following theorem is similar to the proof of Theorem 5.2.

THEOREM. *Let $\mathcal{A}_1, \dots, \mathcal{A}_k, \dots$ be a countable sequence of structures, and let*

$$\begin{pmatrix} U_1^1 & & & \\ U_1^2 & U_2^2 & U_3^2 & \\ \dots & \dots & \dots & \\ U_1^k & \dots & U_k^k & U_{k+1}^k \\ \dots & \dots & \dots & \dots \end{pmatrix}$$

be an infinite matrix of pairwise disjoint sets; the first row contains one element, and the k-th row, where $k > 1$, contains $k+1$ elements; $U_i^k \notin \mathcal{A}_k$. Then there exists a $U = \bigcup_{l=1}^{\infty} U_{i_l}^l$ such that

(i) *$\mathcal{A}_k \not\ni \bigcup_{l \geq k} U_{i_l}^l$ for all k;*
(ii) *if $U \in \mathcal{A}_k$, then $\mathcal{A}_k \ni \bigcup_{l<k} U_{i_l}^l$.*

PROOF. Of course, we put $U_{i_1}^1 = U_1^1$. Assume that there exists a countable sequence of sets

$$U_{i_2^1}^2, U_{i_3^1}^3, \dots, U_{i_j^1}^j, \dots \quad \text{such that} \quad \mathcal{A}_1 \ni U_{i_1}^1 \cup \bigcup_{j>1} U_{i_j^1}^j.$$

Then we say that all the sets $U_{i_j^1}^j$ are marked. The following property is important. If $U_1, \dots, U_k, \dots$ is a countable sequence of sets and U_k is a nonmarked set from the k-th row of our matrix, then

$$\mathcal{A}_1 \not\ni \bigcup_{k=1}^{\infty} U_k.$$

Indeed, if the sets $U_{i_j^1}^j$ do not exist, then, obviously, $\mathcal{A}_1 \not\ni \bigcup_{k=1}^{\infty} U_k$. If such sets exist, then $\mathcal{A}_1 \not\ni \bigcup_{k=1}^{\infty} U_k$ because

$$\left(U_{i_1}^1 \cup \bigcup_{j>1} U_{i_j^1}^j \right) \cap \bigcup_{k \geq 1} U_k = U_{i_1}^1 \notin \mathcal{A}_1.$$

Let us proceed to the sets from the second row of our matrix. Among them two are nonmarked sets, say, U_2^2 and U_3^2. If $U_{i_1}^1 \in \mathcal{A}_2$, then denote one of the two sets U_2^2 and U_3^2, say U_2^2, by $U_{i_2}^2$. And if $U_{i_1}^1 \notin \mathcal{A}_2$, then at least one of the two sets $U_{i_1}^1 \cup U_2^2$ and $U_{i_1}^1 \cup U_3^2$, say $U_{i_1}^1 \cup U_2^2$, does not belong to $\mathcal{A}_2$, since

$$(U_{i_1}^1 \cup U_2^2) \cap (U_{i_1}^1 \cup U_3^2) = U_{i_1}^1 \notin \mathcal{A}_2;$$

denote $U_{i_2}^2 = U_2^2$. Under the condition $U_{i_1}^1 \cup U_{i_2}^2 \notin \mathcal{A}_2$, let us verify the *first assumption*: there exists a countable sequence of sets

$$U_{i_3^2}^3, U_{i_4^2}^4, \dots, U_{i_j^2}^j, \dots \quad \text{such that} \quad \mathcal{A}_2 \ni U_{i_1}^1 \cup U_{i_2}^2 \cup \bigcup_{j>2} U_{i_j^2}^j.$$

If the first assumption does not hold, or if $U_{i_1}^1 \cup U_{i_2}^2 \in \mathcal{A}_2$ (this can happen only if $U_{i_1}^1 \in \mathcal{A}_2$), then let us verify the *second assumption*: there exists a countable sequence of sets

$$U_{i_3^2}^3, U_{i_4^2}^4, \dots, U_{i_j^2}^j, \dots \quad \text{such that} \quad \mathcal{A}_2 \ni U_{i_2}^2 \cup \bigcup_{j>2} U_{i_j^2}^j.$$

Under any of these assumptions, the sets $U_{i_j^2}^j$ are called *marked*. We also call marked all the sets of the second row except $U_{i_2}^2$. The following observation is important. If $U_1, \dots, U_k, \dots$ is a countable sequence of sets and U_k is a nonmarked set from the k-th row of our matrix, then

$$\mathcal{A}_2 \not\ni \bigcup_{k=2}^{\infty} U_k;$$

and if $\mathcal{A}_2 \ni \bigcup_{k=1}^{\infty} U_k$, then $\mathcal{A}_2 \ni U_{i_1}^1$. Indeed, if the first assumption is true, then

$$\mathcal{A}_2 \not\ni \bigcup_{k=1}^{\infty} U_k, \quad \bigcup_{k=2}^{\infty} U_k,$$

since

$$\bigcup_{k=1}^{\infty} U_k \cap \left(U_{i_1}^1 \cup U_{i_2}^2 \cup \bigcup_{j>2} U_{i_j^2}^j \right) = U_{i_1}^1 \cup U_{i_2}^2 \notin \mathcal{A}_2,$$

$$\bigcup_{k=2}^{\infty} U_k \cap \left(U_{i_1}^1 \cup U_{i_2}^2 \cup \bigcup_{j>2} U_{i_j^2}^j \right) = U_{i_2}^2 \notin \mathcal{A}_2.$$

If the second assumption is true, and therefore always, we have

$$\mathcal{A}_2 \not\ni \bigcup_{k=2}^{\infty} U_k.$$

If $U_{i_1}^1 \cup U_{i_2}^2 \notin \mathcal{A}_2$, then we always have

$$\mathcal{A}_2 \not\ni \bigcup_{k=1}^{\infty} U_k.$$

If $\mathcal{A}_2 \ni \bigcup_{k=1}^{\infty} U_k$, then $U_{i_1}^1 \cup U_{i_2}^2 \in \mathcal{A}_2$; hence $U_{i_1}^1 \in \mathcal{A}_2$.

Let us proceed to the third row of our matrix. This row contains two nonmarked sets, say U_1^3 and U_2^3. If $U_{i_1}^1 \cup U_{i_2}^2 \in \mathcal{A}_3$, then denote one of the two sets U_1^3 and U_2^3, say U_1^3, by $U_{i_3}^3$. If $U_{i_1}^1 \cup U_{i_2}^2 \notin \mathcal{A}_3$, then at least one of the two sets $U_{i_1}^1 \cup U_{i_2}^2 \cup U_1^3$ and $U_{i_1}^1 \cup U_{i_2}^2 \cup U_2^3$, say $U_{i_1}^1 \cup U_{i_2}^2 \cup U_2^3$, does not belong to $\mathcal{A}_3$; denote $U_{i_3}^3 = U_2^3$, etc. Continuing our constructions in the natural way, we will find all the required sets $U^l_{i_l}$. $\square$

19.11. PROOF OF THEOREM 2.48. As in the proof of Theorem 19.8, let us define a σ-two-valued measure μ_k for each k: $\mu_k(M) = 0$ if and only if $\mathfrak{P}(M) \subset \mathcal{A}_k$. By the Gitik–Shelah theorem, there exist pairwise disjoint sets $L_1, \dots, L_k, \dots$ such that L_k is μ_k-nonmeasurable. Let us take $U^1_1 \subset L_1$ such that $U^1_1 \notin \mathcal{A}_1$. For each $k > 1$ let us take pairwise disjoint sets

$$U^k_1, \dots, U^k_k, U^k_{k+1} \subset L_k \quad \text{such that} \quad U^k_i \notin \mathcal{A}_k.$$

It remains to use Theorem 19.10 and put $V_l = U^l_{i_l}$. $\square$

19.12. DEFINITION. An almost σ-structure $\mathcal{A}$ is called a *σ^*-structure* if for any countable sequence of sets $M_1, \dots, M_k, \dots \in \mathcal{A}$ we have $\mathcal{A} \ni \bigcap_k M_k$.

19.13. REMARK. If in Theorem 2.48 one replaces almost σ-structures by σ^*-structures, then in its proof instead of Theorem 19.10 one can use simpler arguments similar to those presented in the proof of Statement 5.22.

19.14. REMARK. In Theorem 2.48 we can assume that $\#(X)$ is a σ-nonmeasurable cardinal.

19.15. Let us present a simple generalization of Theorem 2.48.

THEOREM. *Let $\mathcal{A}_1, \dots, \mathcal{A}_k, \dots$ be a countable sequence of almost σ-structures, each being different from $\mathfrak{P}(X)$ and containing all one-point sets. Assume that $\#(X)$ is a σ-nonmeasurable cardinal. Then there exists a matrix of sets*

$$\begin{pmatrix}
V^1_1 \supset V^1_2 \supset V^1_3 \supset V^1_4 \supset \dots \supset V^1_k \supset V^1_{k+1} \supset V^1_{k+2} \supset \dots \supset V^1_i \supset \dots \\
V^2_1 \supset \dots \supset V^2_3 \supset V^2_4 \supset \dots \supset V^2_k \supset V^2_{k+1} \supset V^2_{k+2} \supset \dots \supset V^2_i \supset \dots \\
V^3_1 \supset \qquad \dots \quad \supset V^3_4 \supset \dots \supset V^3_k \supset V^3_{k+1} \supset V^3_{k+2} \supset \dots \supset V^3_i \supset \dots \\
\cdots\cdots\cdots\cdots\cdots\cdots\cdots\cdots\cdots\cdots\cdots\cdots\cdots\cdots\cdots\cdots \\
V^k_1 \supset \qquad\qquad\qquad\qquad \dots \supset V^k_{k+1} \supset V^k_{k+2} \supset \dots \supset V^k_i \supset \dots \\
\cdots\cdots\cdots\cdots\cdots\cdots\cdots\cdots\cdots\cdots\cdots\cdots\cdots\cdots\cdots\cdots
\end{pmatrix}$$

such that V^k_i exists for each natural k if $i = 1$ and for each natural $i > k$; $V^k_i \notin \mathcal{A}_k$, $V^k_1 \cap V^m_1 = \varnothing$ for $k \neq m$, $V^k_i \supset V^k_j$ for $i < j$, and

(a) *for a natural $k > 1$ and an arbitrary set $M \subset [1, k-1]$ (including $M = \varnothing$) we have*

$$\Big(\bigcup_{l \geq k} V^l_1\Big) \bigcup \Big(\bigcup_{l \in M} V^l_k\Big) \notin \mathcal{A}_k;$$

(b) $\mathcal{A}_1 \not\ni \bigcup_{l \geq 1} V^l_1$;

(c) *if $\mathcal{A}_k \ni \bigcup_{l \geq 1} V^l_1$, then $\mathcal{A}_k \ni \bigcup_{l < k} V^l_1$.*

PROOF. As in 19.11, take the sets $V_1, \dots, V_k, \dots$ and set $V^k_1 = V_k$. Take $m \in N^+$ and assume that all sets V^k_i with $i \leq m$ are already constructed. Our goal is to construct the corresponding sets

$$V^1_{m+1}, \dots, V^m_{m+1}.$$

Take m pairs of disjoint sets. If $i \leq m - 1$, take disjoint sets

$$W^i_1, W^i_2 \subset V^i_m.$$

Also, take disjoint sets

$$W^m_1, W^m_2 \subset V^m_1.$$

Let $W_j^i \notin \mathcal{A}_i$ for all $i \le m$ and $j = 1, 2$. Let us assume that there exists $M_0 \subset [1, m]$ such that

$$\mathcal{A}_{m+1} \ni \bigcup_{k>m} V_1^k \cup \bigcup_{i \in M_0} W_1^i.$$

Let $M \subset [1, m]$. Clearly,

$$\mathcal{A}_{m+1} \not\ni \left(\bigcup_{k>m} V_1^k \cup \bigcup_{i \in M_0} W_1^i \right) \cap \left(\bigcup_{k>m} V_1^k \cup \bigcup_{i \in M} W_2^i \right) = \bigcup_{k>m} V_1^k.$$

Therefore,

$$\mathcal{A}_{m+1} \not\ni \bigcup_{k>m} V_1^k \cup \bigcup_{i \in M} W_2^i.$$

It remains to set $V_{m+1}^i = W_2^i$ for all i. □

CHAPTER 20

Final Comments. Generalization of Theorem 2.1

20.1. DEFINITION. The *characteristic* $\chi(\mathcal{A})$ of an algebra $\mathcal{A}$ is the maximal number (which is either a natural number ≥ 2 or $\aleph_0$) such that if $Q \subset U$, $Q \notin \mathcal{A}$, $U \in \mathcal{A}$, then there exists a family $\{U_\lambda\}_{\lambda \in \Lambda}$ of pairwise disjoint subsets of U, each not belonging to $\mathcal{A}$ and $\#(\Lambda) \geq \chi(\mathcal{A})$. This definition shows that if there do not exist sets $Q \subset U$ such that $Q \notin \mathcal{A}$ and $U \in \mathcal{A}$, then $\chi(\mathcal{A}) = \aleph_0$.

20.2. For an algebra $\mathcal{A}$ and $n \in \mathbb{N}^+$ we set

$$\chi(\mathcal{A}, n) = \begin{cases} n - 1 & \text{if } \chi(\mathcal{A}) = \aleph_0, \\ \frac{2 \cdot \chi(\mathcal{A})}{2 \cdot \chi(\mathcal{A}) - 1} \cdot n & \text{if } \chi(\mathcal{A}) \neq \aleph_0. \end{cases}$$

Since

$$\chi(\mathcal{A}, n) \leq \frac{4}{3} n,$$

the following theorem is a clear generalization of Theorem 2.1.

THEOREM. *Let $\mathcal{A}_1, \dots, \mathcal{A}_n, \mathcal{A}_{n+1}$, $n \in \mathbb{N}^+$, be a sequence of algebras such that $\bigcup_{k=1}^{n} \mathcal{A}_k \neq \mathfrak{P}(X)$ and there exist more than $\chi(\mathcal{A}_{n+1}, n)$ pairwise disjoint sets not belonging to $\mathcal{A}_{n+1}$. Then $\bigcup_{k=1}^{n+1} \mathcal{A}_k \neq \mathfrak{P}(X)$.*

PROOF. For the algebras $\mathcal{A}_1, \dots, \mathcal{A}_n$ take the corresponding sets of ultrafilters S_n, T_n.

(1) *First assume that* $\chi(\mathcal{A}_{n+1}) \neq \aleph_0$. One can easily show that at least one of the following three cases occurs:

1) $\ker \mathcal{A}_{n+1} \setminus (S_n \cup T_n) \neq \varnothing$;
2) $\ker \mathcal{A}_{n+1}$ contains the root of the bush K of (S_n, T_n) and $\#(K) = \|K\| + 1$;
3) there exists a bush K of (S_n, T_n) of the fourth kind, $K \subset \ker \mathcal{A}_{n+1}$, and
   (i) $\#(K) = \|K\| + 1$,
   (ii) $\#(K) < 2 \cdot \chi(\mathcal{A}_{n+1})$.

In Cases 1) and 2) one can construct the corresponding sequences of ultrafilters S_{n+1}, T_{n+1} (see the proof of Theorem 2.1 in 4.15). Let us consider Case 3). There can be trivial situations:

(a) there exists an $\mathcal{A}_{n+1}$-special ultrafilter $a \in K$;
(b) there exist $\mathcal{A}_{n+1}$-similar ultrafilters s and t such that $s \in K \cap S_n$ and $t \in K \cap T_n$.

Otherwise, there exists an ultrafilter $a \in K$ that has an $\mathcal{A}_{n+1}$-similar ultrafilter $r \notin K$. Now it is clear how to construct the corresponding sets of ultrafilters S_{n+1}, T_{n+1}.

(2) *Now assume that* $\chi(\mathcal{A}_{n+1}) = \aleph_0$. If there exist $\aleph_0$ pairwise distinct sets not belonging to $\mathcal{A}_{n+1}$, everything is clear. Otherwise, there exist n pairwise distinct special ultrafilters forming a set S^*. If $S^* \setminus T_n \neq \varnothing$, everything is again simple.

Otherwise, $S^* = T_n$ and S_n does not contain an $\mathcal{A}_i$-special ultrafilter s_i, $i \leq n$. But then we can take

$$S_{n+1} = T_n, \quad T_{n+1} = S_n. \qquad \square$$

APPENDIX

On a Question of Grinblat

Saharon Shelah[1,2]

We prove the consistency of: there is a $\varkappa$-complete ideal on $\varkappa$ for some uncountable $\varkappa < 2^{\aleph_0}$ such that the Boolean algebra $\mathcal{P}(\varkappa)/I$ is σ-centered and there are (uncountable) Q-sets of reals.

In set theoretic language, Grinblat has been asking for some time

A.1. PROBLEM. Is it consistent with ZFC that:

(a) there is an $\aleph_1$-complete ideal I on some $\varkappa < 2^{\aleph_0}$ such that $\mathcal{P}(\varkappa)/I$ is σ-centered (see below, and also Section 6.1 of this monograph)?
(b) there is a Q-set?

We answer positively.

REMARK. Of course, $\alpha < \varkappa \Rightarrow \{\alpha\} \in I$.

A.2. CLAIM. *Assume that $\varkappa < \chi = \chi^{\varkappa}$ and $\varkappa$ is measurable.*
Then for some c.c.c. forcing notion $\mathbb{P}$ of cardinality χ we have in $\mathbf{V}^{\mathbb{P}}$:

(i) $2^{\aleph_0} = \chi$;
(ii) $\mathrm{MA}_{<\varkappa,<\mathrm{cf}(\chi)}(\sigma\text{-centered})$ *holds (i.e.,* MA *for σ-centered forcing notions of cardinality $< \varkappa$ and $< \mathrm{cf}(\chi)$ dense open subsets); hence*
(ii)$^-$ *every uncountable set of reals of cardinality $< \varkappa$ is a Q-set;*
(iii) *assume D is a normal ultrafilter on $\varkappa$ and I is the dual ideal. The Boolean algebra $\mathcal{P}(\varkappa)/I$ is σ-centered, i.e., $\mathcal{P}(\varkappa)\backslash I^{\mathbf{V}^{\mathbb{P}}}$ is the union of countably many filters where*

$$I^{\mathbf{V}^{\mathbb{P}}} = \{A \in \mathbf{V}^P : A \subseteq \varkappa \text{ and } A \text{ is included in some member of } I\};$$

note that $I^{\mathbf{V}^{\mathbb{P}}}$ is a normal ideal on $\varkappa$.

A.3. REMARK. Why "hence (ii)$^-$" in (ii)?
Because of the following result.

---

[1]Institute of Mathematics, The Hebrew University, Jerusalem, Israel, and Rutgers University, Department of Mathematics, New Brunswick, New Jersey, USA.

[2]I would like to thank Alice Leonhardt for the beautiful typing.

A.4. Claim. *For $X \subseteq Y \subseteq {}^{\omega}2$ the natural forcing $\mathbb{Q} = \mathbb{Q}_{X,Y}$ (defined below) satisfies*

(i) *it adds subtrees $\underset{\sim}{T}_n \subseteq {}^{\omega>}2$ for $n < \omega$ such that $\cup_{n<\omega} \lim(\underset{\sim}{T}_n) \cap Y = X$;*

(ii) *$\mathbb{Q}$ is σ-centered of cardinality $\leq |Y| + \aleph_0$;*

(iii) *if we can find a directed $G \subseteq \mathbb{Q}$ intersecting $|Y| + \aleph_0$ dense subsets $\mathcal{J}^{\ell}_{\eta,n}$ ($\eta \in Y \backslash X$), $\mathcal{J}_m$ ($m < \omega$), I_η (for $\eta \in X$) (defined below), then there is $\langle T_n : n < \omega \rangle$ as above.*

A.5. Definition. 1) For $X \subseteq Y \subseteq {}^{\omega}2$ let $\mathbb{Q} = \mathbb{Q}_{X,Y}$ be defined as follows:

(A)

$$\mathbb{Q} = \Big\{ p = (\bar{t}, \bar{u}) : \text{for some } n = n(p) < \omega \text{ we have:}$$

$$\begin{aligned} &\text{(a)} \quad \bar{t} = \langle t_\ell : \ell < n \rangle, \text{ each } t_\ell \text{ has the form } T \cap {}^{m_\ell(p) \geq}2, \\ &\qquad T \text{ a subtree}^3 \text{ of } {}^{\omega>}2 \text{ with } m_\ell(p) < \omega, \\ &\text{(b)} \quad \bar{u} = \langle u_\ell : \ell < n(p) \rangle, u_\ell \subseteq X \text{ is finite} \\ &\text{(c)} \quad \text{if } \ell < n(p),\ n \in u_\ell, \text{ then } \eta \restriction m_\ell(p) \in t_\ell \Big\}. \end{aligned}$$

(B) The order is natural: $p \leq q$ iff $\{n(p) \leq n(q)$ and $\bigwedge_{\ell<n(p)} [t^p_\ell \subseteq t^q_\ell$ and $m_\ell(p) \leq m_\ell(q)$ and $t^p_\ell = t^q_\ell \cap {}^{m_\ell(p)\geq}2]$ and $\bigwedge_{\ell<n(p)} u^p_\ell \subseteq u^q_\ell\}$.

2) For $\eta \in Y \backslash X$ and $m < \omega$ let $\mathcal{J}_m = \{p \in \mathbb{Q} : m < n(p)$ and $\ell < m \Rightarrow m \leq m_\ell(p)\}$ and $\mathcal{I}^\ell_\eta = \{p \in \mathbb{Q} : \ell < n(p)$ and $\eta \restriction m_\ell(p) \notin t^p_\ell\}$. For $\eta \in X$ let $\mathcal{I}_\eta$ be $\{p \in \mathbb{Q} : \eta \in u^p_\ell$ for some $\ell < n(p)\}$. Clearly, these sets are dense and open.

Proof of A.4. Let $\underset{\sim}{T}_\ell[G] = \bigcup \{t^p_\ell : p \in \underset{\sim}{G}\}$. So any directed $G \subseteq \mathbb{Q}$ meeting all the $\mathcal{I}^\ell_\eta$ (for $\eta \in Y \backslash X$), I_η (for $\eta \in X$), and $\mathcal{J}_m (m < \omega)$ gives $\underset{\sim}{T}_\ell[G]$ as required. □

Proof of A.2. Let $\mathbb{P} = \mathbb{P}_\chi$, where $\bar{\mathbb{Q}} = \langle \mathbb{P}_i, \underset{\sim}{\mathbb{Q}}_j : i \leq \chi, j < \chi \rangle$, be an FS iteration such that in $\mathbf{V}^{\mathbb{P}_i}, \underset{\sim}{\mathbb{Q}}_j$ is a σ-centered forcing notion of cardinality $< \varkappa$ and its set of elements is an ordinal $< \varkappa$ (not just a $\mathbb{P}_j$-name of such an ordinal), and any such forcing notion appears unboundedly often even χ times, more exactly, if $i_0 < \chi$, $\underset{\sim}{\mathbb{Q}}$ is a $\mathbb{P}_{i_0}$-name of a forcing notion with a set of elements (forced to be) $\subset \underset{\sim}{\alpha}_{\mathbb{Q}} < \varkappa$, then for χ many (hence unboundedly many) ordinals $j \in (i, \chi)$ we have: $\Vdash_{\mathbb{P}_j}$ "if $\underset{\sim}{\mathbb{Q}}$ is σ-centered, then $\underset{\sim}{\mathbb{Q}} \cong \underset{\sim}{\mathbb{Q}}_j$". Also as usual, $\mathbb{P}_i \subseteq \mathcal{H}_{<\aleph_1}(\varkappa + i)$, i.e., the members are sets which are hereditarily countable over the set of ordinals $< \varkappa + i$ (see [**S4**, Ch. III] or just use $\mathbb{P}_{1,i}$ as below).

As each $\underset{\sim}{\mathbb{Q}}_j$ is σ-center (in $\mathbf{V}^{\mathbb{P}_j}$) there is $\underset{\sim}{\bar{f}} = \langle f_j : j < \chi \rangle$ such that $\Vdash_{\mathbb{P}_j}$ "$\underset{\sim}{f}_j$ is a function from $\underset{\sim}{\mathbb{Q}}_j$ to ω such that each $\{p \in \underset{\sim}{\mathbb{Q}}_j : \underset{\sim}{f}_j(p) = n\}$ is directed". We can explicate further. Letting the universe of $\underset{\sim}{Q}_i$ be α_i for any $i < \chi$ and $\alpha, \beta < \alpha_i$ let $\langle g^i_{\alpha,\beta,n} : n < \omega \rangle$ be a maximal antichain of $\mathbb{P}_{1,i}$ such that $g^i_{\alpha,\beta} \Vdash_{\mathbb{P}_i}$ "$\alpha \leq_{\underset{\sim}{\mathbb{Q}}_i} \beta$

[3] T is nonempty, $\nu \lhd \eta$, and $\eta \in T \Rightarrow \nu \in T$ and no $\eta \in T$ is $\lhd$-maximal.

iff $\mathbf{t}^i_{\alpha,\beta,n}$ = truth". So $\langle\langle(t^i_{\alpha,\beta,n}, g^i_{\alpha,\beta,n}) : n < \omega, \alpha, \beta < \alpha_i\rangle : i < \chi\rangle$ gives all the information on $\bar{\mathbb{Q}}$.

Therefore $\mathbb{P}_{1,i} =: \{p \in \mathbb{P}_i :$ if $j \in \mathrm{Dom}(p)$, then $p \restriction j$ forces a value to $\underset{\sim}{f}_j(p(j))$ and a value to $p(j)\}$ is a dense subset of $\mathbb{P}_i$; recall that the set of members of $\underset{\sim}{\mathbb{Q}}_j$ is an ordinal $< \varkappa$. Now clearly clauses (i) and (ii) in Claim A.2 hold in $\mathbf{V}^{\mathbb{P}}$. As D is a normal ultrafilter on $\varkappa$ there is a transitive class M such that $M^\varkappa \subseteq M$ and there is an elementary embedding $\mathbf{j}$ from $\mathbf{V}$ to M with critical ordinal $\varkappa$ such that $D = \{A : A \in \mathbf{V}, A \subseteq \varkappa$ and $\varkappa \in \mathbf{j}(A)\}$. Let $\mathbf{j}(\bar{\mathbb{Q}})$ be $\bar{\mathbb{Q}}' = \langle \mathbb{P}'_i, \underset{\sim}{\mathbb{Q}}'_j : i \leq \mathbf{j}(\varkappa), j < \mathbf{j}(\chi)\rangle$, and let $\underset{\sim}{\bar{f}}' = \mathbf{j}(\underset{\sim}{\bar{f}}') = \langle \underset{\sim}{f}'_j : j < \mathbf{j}(\varkappa)\rangle$; so M "thinks" that $(\bar{\mathbb{Q}}', \underset{\sim}{\bar{f}}')$ satisfies all the properties listed above, hence in $\mathbf{V}$ it satisfies all of those properties replacing $\varkappa, \chi$ by $\mathbf{j}(\varkappa), \mathbf{j}(\chi)$.

Let $\mathbb{P}^* = \{\mathbf{j}(p) : p \in \mathbb{P}'_\chi\}$; so it is well known that $\mathbb{P}^* \lessdot \mathbb{P}'_{\mathbf{j}(\chi)}$, $\mathbb{P}^*$ is isomorphic to $\mathbb{P}_\chi$, all those forcing notions satisfy the c.c.c., and (some complete subalgebra of) the completion of the Boolean algebra corresponding to $\mathbb{P}'_{\mathbf{j}(\chi)}/\mathbb{P}^*$ is isomorphic to $\mathcal{P}(\varkappa)/I^{\mathbf{V}^{\mathbb{P}}}$. Therefore it is enough to prove that $\Vdash_{P^*}$ "$\mathbb{P}'_{\mathbf{j}(\chi)}/\mathbb{P}^*$ is σ-centered" (in $\mathbf{V}^{\mathbb{P}^*}$, which is the same as $\mathbf{V}^{\mathbb{P}_\chi}$). Note also that $\mathbb{P}'_{\mathbf{j}(\chi)}$, hence $\mathbb{P}'_{\mathbf{j}(\chi)}/\mathbb{P}^*$, has cardinality $\leq |\mathbb{P}_{1,\chi}|^\varkappa = \chi^\varkappa = \chi$.

Now the point is that we shall prove below that

⊠ we can reorder the iteration $\bar{\mathbb{Q}}'$: first do $\langle \mathbb{Q}_{\mathbf{j}(j)} : j < \chi\rangle$ and then the rest, and in this reordering, each $\mathbb{Q}_j$ is still σ-centered and $\bar{\mathbb{Q}}'$ is FS iteration.

Note first that proving ⊠ suffices since

(a) the limit of FS iteration of σ-centered forcing notion each of cardinality $\leq 2^{\aleph_0}$ and of length $< (2^{\aleph_0})^+$ (in $\mathbf{V}^{\mathbb{P}_{1,\chi}}$!) is σ-centered;
(b) $\mathbb{P}'_{j(\varkappa)}/\mathbb{P}^*$ is in the universe $\mathbf{V}^{\mathbb{P}^*}$, the limit of such iteration.

They are easy.

Second, this reordering is possible. [Why? The set of elements of $\underset{\sim}{\mathbb{Q}}_j$ is $\alpha_{\underset{\sim}{\mathbb{Q}}_j}$, $\alpha_{\underset{\sim}{\mathbb{Q}}_j}$ is a $\mathbb{P}'_j$-name of an ordinal $< \varkappa$, and $\mathbb{P}'_j$ satisfies the c.c.c.; hence for some $\alpha^*_j < \varkappa$ we have $\Vdash_{P^*_{1,j}}$ "$\alpha^*_{\underset{\sim}{\mathbb{Q}}_j} \leq \alpha^*_j$" so the quasiorder $\underset{\sim}{\mathbb{Q}}_j$ is a subset of $\alpha^*_j \times \alpha^*_j$. For any $\beta, \gamma < \alpha^*$, there is a maximal antichain $\mathcal{I}_{j,\beta,\gamma}$ of $\mathbb{P}^*_{1,j}$ of conditions forcing "$\underset{\sim}{\mathbb{Q}}_j \models \beta \leq \gamma$, so $\beta \in \underset{\sim}{\mathbb{Q}}_j$, $\gamma \in \underset{\sim}{\mathbb{Q}}_j$" or forcing its negation. We choose $\bar{\mathcal{I}} = \langle \mathcal{I}_{j,\beta,\gamma} : j < \chi, \beta, \gamma < \alpha^*_j\rangle$. Let $A_j = \bigcup_{\beta,\gamma<\alpha^*_j} \bigcup_{p\in\mathcal{I}_{j,\beta,\gamma}} \mathrm{Dom}(p)$, so $|A_j| < \varkappa$; all this for every $j < \chi$. Let $A \subseteq \chi$ be called $\bar{\mathbb{Q}}$-closed if $(\forall j \in A)(A_j \subseteq A)$ and clearly for a permutation π of χ, changing the order of the iteration to $\underset{\sim}{Q}_{\pi(0)}, Q_{\pi(1)}, \ldots$ is OK *provided that* $i \in A_j \Rightarrow \pi(i) \in \pi(j)$. Now in M we can compute $\mathbf{j}(\bar{\mathcal{I}})$, hence $\langle A^M_i : i < \mathbf{j}(\chi)\rangle$; now easily $A^M_{\mathbf{j}(j)} = \{\mathbf{j}(i) : i \in A_j\}$ as $|A_j| < \varkappa$. Now in $\mathbf{V}$ all the relevant properties of $\bar{\mathbb{Q}}'$ are preserved. In $\mathbf{V}$ let $A = \{\mathbf{j}(i) : i < \chi\} \subseteq j(\chi)$, which is $\bar{\mathbb{Q}}'$-closed by the form of $A^M_{\mathbf{j}(j)}$. Hence the reordering suggested above ($\pi(\mathbf{j}(i)) = i$ for $i \in A$ and $\pi(j) = \chi + \mathrm{otp}\{\varepsilon : \varepsilon < i, \varepsilon \notin A\}$) is OK.]

So we are done. □

A.6. REMARKS. 1) If $\varkappa < \lambda = \mathrm{cf}(\lambda) \leq \varkappa$ we can even get $\mathrm{MA}_{<\varkappa,<\lambda}$ in A.2; just use iteration of length $\chi \times \lambda$.

2) If D is a normal ultrafilter on $[\theta]^{<\varkappa}$, to which $\mathcal{A} = \{A_\alpha : \alpha < \theta\}$ belongs, $A_\alpha \in [\alpha]^{<\varkappa}$, then in $\mathbf{V}^{\mathbb{P}}$, D generates a normal filter D' with $\mathcal{P}(\mathcal{A})/D$ σ-centered.

Bibliography

[A] P. S. Alexandroff. *Introduction to the general theory of sets and functions*, GITTL, Moscow, 1948. (Russian)

[B] S. Banach. *Sur le problème de la mesure*, Fund. Math. **4**(1923), 7–33.

[B-Ku] S. Banach and C. Kuratowski. *Sur une généralization du problème de la mesure*, Fund. Math. **14**(1929), 127–131.

[Ba-D] J. Baumgartner and E. Dowen. *Strong real compactness and weakly measure cardinals*, Topology and Appl. **35**(1990), 239–251.

[Bi] G. Birkhoff. *Lattice theory*, Amer. Math. Soc., New York, 1948.

[Bo1] N. Bourbaki. *Éléments de Mathématique* XXII, *Théorie des ensembles*, Hermann, Paris, 1957.

[Bo2] ______, *Éléments de Mathématique* XXIX, *Intégration*, Hermann, Paris, 1963.

[Bu] L. Bukovsky. *Borel subsets of metric separable spaces*, Proc. 2-nd Prague Topological Sympos. Academia, Prague, 1966, pp. 83–86.

[C1] H. Cartan. *Théorie des filters*, C. R. Acad. Paris **205**(1937), 595–598.

[C2] ______. *Filters et ultrafilters*, C. R. Acad. Paris **205**(1937), 777–779.

[Co1] P. J. Cohen. *The independence of the Continuum hypothesis*, I, II. Proc. Nat. Acad. Sci. U.S.A. **50**(1963), 1143–1148; **51**(1964), 105–110.

[Co2] ______. *Set theory and the continuum hypothesis*, Benjamin, New York, 1966.

[E] R. Engelking. *General topology*, PWN, Warszawa, 1977; Second Edition, 1985.

[Er] P. Erdös. *Some remarks on set theory*, Proc. Amer. Math. Soc. **1**(1950), 127–141.

[Er-Ha] P. Erdös and A. Hajnal. *Unsolved problems in set theory*, Proc. Symp. Pure Math., Vol. 13, Part I, Amer. Math. Soc., Providence, RI, 1971, pp. 17–48.

[Ev Wh] C. J. Everett and G. Whaples. *Representations of sequence of sets*, Amer. J. Math. **71**(1949), 287–293.

[F] W. G. Fleissner. *Current research on Q-sets*, Topology, Coll. Math. Soc. János Bolyai, vol. 23, Academiai Kiado, Budapest, 1978, pp. 413–431.

[Fr-Gut] R. Frankiewicz and A. Gutek. *Some remarks on embeddings of Boolean algebras and the topological spaces*. I, Bull. Acad. Polon. Sci. (Math.) **29**(1982), 471–476.

[Fre] M. Fréchet. *Sur quelques points du calcul fonctionnel*, Rend. Circ. Mat Palermo **22**(1906), 1–74.

[Frem1] D. H. Fremlin. *The Gitik–Shelah theorem on real-valued-measurables cardinals*, Circulated notes, November 7, 1988.

[Frem2] ______. *Real-valued-measurable cardinals*, Set Theory of the Reals (Ramat Gan, 1991), Israel Math. Conf. Proc., vol. 6, 1993, pp. 151–304.

[Fu] H. Furstenberg. *Ergodic behavior of diagonal measures and a theorem of Szemerédi on arithmetic progressions*, J. Analyse Math. **31**(1977), 204–256.

[G-Je] L. Gillman and M. Jerison. *Rings of continuous functions*, van Nostrand, New York, 1960.

[Gi-S1] M. Gitik and S. Shelah. *Forcings with ideals and simple forcing notions*, Israel J. Math. **68**(1989), 129–160.

[Gi S2] ______. *More on simple forcing notions and forcings with ideals*, Ann. Pure Appl. Logic **59**(1993), 219–238.

[Gl] A. M. Gleason. *Projective topological spaces*, Ill. J. Math. **2**(1958), 482–489.

[Go] K. Gödel. *Consistency-proof for the generalized continuum hypothesis*, Proc. Nat. Acad. Sci. U.S.A. **25**(1939), 220–224.

[Gol] M. Goldstern. *Letter to the author*, March 3, 2000.

[Gow] W. T. Gowers. *A new proof of Szemerédi's theorem*, Geom. Funct. Anal. **11**(2001), 468–588.

[Gr] L. Š. Grinblat. *On sets not belonging to algebras of subsets*, Mem. Amer. Math. Soc. **480**(1992).

[Grz] E. Grzegorek. *On saturated sets of Boolean rings and Ulam's problem on sets of measures*, Fund. Math. **60**(1980), 153–161.

[H] A. Haar. *Der Massbegriff in der Theorie der Kontinuierlichen Gruppen*, Ann. Math. **34**(1933), 147–169.

[Hal1] M. Hall. *Distinct representatives of subsets*, Bull. Amer. Math. Soc. **54**(1948), 922–926.

[Hal2] ———. *Combinatorial theory*, Blaisdell, Waltham, MA, 1967.

[Hall] Ph. Hall. *On representatives of subsets*, J. London Math. Soc. **10**(1935), 26–30.

[Halm-V] P. R. Halmos and H. E. Vaughan. *The marriage problem*, Amer. J. Math. **72**(1950), 214–215.

[Hau1] F. Hausdorff. *Grundzüege der Mengenlehre*, Veit, Leipzig, 1914.

[Hau2] ———. *Probleme 58*, Fund. Math. **20**(1933), 286.

[He] E. Hewitt. *Rings of real-valued continuous functions.* I, Trans. Amer. Math. Soc. **64**(1948), 45–99.

[Hi] D. Hilbert. *Mathematische Problem*, Nach. Acad. Wiss. Göttingen (1900), 253–297.

[Ho-Hor] J. Hodges and A. Horn. *On Maharam's condition for measure*, Trans. Amer. Math. Soc. **64**(1948), 594–595.

[Hor-T] A. Horn and A. Tarski. *Measures in Boolean algebras*, Trans. Amer. Math. Soc. **64**(1948), 467–497.

[J] T. J. Jech. *Lectures on set theory with particular emphasis on the method of forcing*, Springer-Verlag, Berlin, 1971.

[J-Pr] T. Jech and K. Prikry. *Cofinality of the partial ordering of functions from ω_1 to ω and eventual domination*, Math. Proc. Camb. Phil. Soc. **95**(1984), 25–32.

[Ka] A. Kamburelis. *A new proof of the Gitik–Shelah theorem*, Israel J. Math. **72**(1990), 373–380.

[Kat] M. Katětov. *On nearly discrete spaces*, Časopis Pěst. Mat. Fys. **75**(1950), 69–78.

[Ke-T] H. J. Keisler and A. Tarski. *From accessible to inaccessible cardinals (Results holding for whole accessible cardinal numbers and the problem of their extension for inaccessible ones)*, Fund. Math. **53**(1964), 225–308.

[Kel] J. T. Kelley. *Measures in Boolean algebras*, Pacific J. Math. **9**(1959), 1165–1177.

[Ko] P. Komjáth. *On second-category sets*, Proc. Amer. Math. Soc. **107**(1989), 653–654.

[Kr-P] A. Krawczyk and A. Pelc. *On families on σ-complete ideals*, Fund. Math. **59**(1980), 155–161.

[Ku] K. Kuratowski. *Topology* I, Academic Press, New York, 1966.

[Ku-Mo] K. Kuratowski and A. Mostowski. *Set theory*, North-Holland, Amsterdam, 1967.

[L1] H. Lebesgue. *Intégrale, longueur, aire*, Annali di Mat. (3) **VII**(1902), 231–359.

[L2] ———. *Leçons sur l'intégration et la recherche des fonctions primitives*, Gauthier–Villars, Paris, 1904.

[Lu] N. N. Lusin. *Sur la décomposition des ensembles*, C. R. Acad. Sci. Paris **198**(1934), 1671–1674.

[M] G. W. Mackey. *Equivalence of a problem in measure theory to a problem in the theory of vector lattices*, Bull. Amer. Math. Soc. **50**(1944), 719–722.

[Ma] D. Mahaman. *An algebraic characterization of measure algebras*, Ann. Math. **48**(1947), 154–167.

[Mar-So] D. A. Martin and R. M. Solovay. *Internal Cohen extensions*, Ann. Math. Logic **2**(1970), 143–178.

[Mr1] S. Mrówka. *Some properties of Q-spaces*, Bull. Acad. Pol. Sci. C1. III **5**(1957), 947–950.

[Mr2] ———. *On the potency of βN*, Coll. Math. **7**(1959–60), 23–25.

[N-K] M. Nakamura and S. Kakutani. *Banach limits on the Čech compactification of a countable discrete set*, Proc. Imp. Acad. Tokyo (1943), 224–229.

[Ne1] J. von Neumann. *Zur allgemeinen Theorie des Masses*, Fund. Math. **13**(1929), 73–116.

[Ne2] ———. *Die Einführung analytischer Parameter in topologischen Gruppen*, Ann. Math. **34**(1933), 170–190.

[Po] B. Pospíšil. *Remark on bicompact spaces*, Ann. Math. **38**(1937), 845–846.

[R] F. Riesz. *Stetigkeitsbegriff und abstrakte Mengenlehre*, Atti del IV Congresso Intern. Mat. (Roma, 1908), Vol. II, Roma, pp. 18–24.

[Ro] F. Rothberger. *On some problems of Hausdorff and of Sierpiński*, Fund. Math. **35**(1948), 29–46.

[S1] S. Shelah. *On normal ideals and Boolean algebras*, Around Classification Theory of Models, Lecture Notes in Math., vol. 1182, Springer-Verlag, Berlin, 1986, pp. 247–260.

[S2] ______. *Iterated forcing and normal ideals on* ω_1, Israel J. Math. **60**(1987), 345–380.

[S3] ______. *Primitive recursive bounds for van der Waerden numbers*, J. Amer. Math. Soc. **1**(1988), 683–697.

[S4] ______. *Proper and improper forcing*, Perspectives in Mathematical Logic, Springer-Verlag, Berlin, 1998.

[S5] ______. *The null ideal restricted to a non-null set may be saturated*, Fund. Math. (accepted for publication).

[Sh-Gu] G. E. Shilov and B. L. Gurevich. *Integral, measure and derivative*, "Nauka", Moscow, 1967. (Russian)

[Si1] W. Sierpiński. *Sur une décomposition d'ensembles*, Monatsh. Math. and Phys. **35**(1928), 239–242.

[Si2] ______. *Sur un problème de M. Hausdorff*, Fund. Math. **30**(1938), 1–7.

[Si3] ______. *Hypothèse du continu*, Chelsea, New York, 1956.

[Sik] R. Sikorski. *Boolean algebras*, Springer-Verlag, Heidelberg, 1964.

[So] R. Solovay. *Real-valued measurable cardinals*, Axiomatic Set Theory, Proc. Symp. Pure Math., vol. 13, Part 1, Amer. Math. Soc., Providence, RI, 1971, pp. 397–428.

[St1] M. H. Stone. *Applications of the theory of Boolean rings to general topology*, Trans. Amer. Math. Soc. **41**(1937), 375–481.

[St2] ______. *Algebraic characterizations of special Boolean rings*, Fund. Math. **29**(1937), 223–302.

[Sz] E. Szemerédi. *On acts of integers containing no k elements in arithmetic progression*, Acta Arith. **27**(1975), 199–245.

[T] A. Tarski. *Some problems and results relevant to the foundations on set theory*, Logic, Methodology and Philosophy of Science (E. Nagel, P. Suppes, and A. Tarski, eds.), Stanford, CA, 1962, pp. 125–135.

[Ta] A. D. Taylor. *On saturated sets of ideals and Ulam's problem*, Fund. Math. **59**(1980), 37–53.

[Ty] A. Tychonoff. *Über die topologische Erweiterungen von Räumen*, Math. Ann. **102**(1929), 544–561.

[U1] S. Ulam. *Zur Masstheorie in der allgemeinen Mengenlehre*, Fund. Math. **16**(1930), 140–150.

[U2] ______. *Über gewisse Zerlegungen von Mengen*, Fund. Math. **20**(1933), 221–223.

[Vi] G. Vitali. *Sul problema della misura dei gruppi de punti di una retta*, Bologna, 1905.

[W] B. L. van der Waerden. *Beweis einer Bautschen Vermutung*, Nieuw. Arch. Wisk. **15**(1928), 212–216.

[Wa] H. Wallman. *Lattices and topological spaces*, Ann. Math. **39**(1938), 112–126.

[We] H. Weyl. *Almost periodic invariant vector sets in a metric vector space*, Amer. J. Math. **71**(1949), 178–205.

[Wo] W. H. Woodin. *The axiom of determinacy, forcing axioms, and the nonstationary ideal*, de Gruyter, Berlin, 1999.

Index

Titles in This Series

214 **L. Š. Grinblat,** Algebras of sets and combinatorics, 2002
213 **V. N. Sachkov and V. E. Tarakanov,** Combinatorics of nonnegative matrices, 2002
212 **A. V. Mel′nikov, S. N. volkov, and M. L. Nechaev,** Mathematics of financial obligations, 2002
211 **Takeo Ohsawa,** Analysis of several complex variables, 2002
210 **Toshitake Kohno,** Conformal field theory and topology, 2002
209 **Yasumasa Nishiura,** Far-from-equilibrium dynamics, 2002
208 **Yukio Matsumoto,** An introduction to Morse theory, 2002
207 **Ken'ichi Ohshika,** Discrete groups, 2002
206 **Yuji Shimizu and Kenji Ueno,** Advances in moduli theory, 2002
205 **Seiki Nishikawa,** Variational problems in geometry, 2001
204 **A. M. Vinogradov,** Cohomological analysis of partial differential equations and Secondary Calculus, 2001
203 **Te Sun Han and Kingo Kobayashi,** Mathematics of information and coding, 2002
202 **V. P. Maslov and G. A. Omel′yanov,** Geometric asymptotics for nonlinear PDE. I, 2001
201 **Shigeyuki Morita,** Geometry of differential forms, 2001
200 **V. V. Prasolov and V. M. Tikhomirov,** Geometry, 2001
199 **Shigeyuki Morita,** Geometry of characteristic classes, 2001
198 **V. A. Smirnov,** Simplicial and operad methods in algebraic topology, 2001
197 **Kenji Ueno,** Algebraic geometry 2: Sheaves and cohomology, 2001
196 **Yu. N. Lin′kov,** Asymptotic statistical methods for stochastic processes, 2001
195 **Minoru Wakimoto,** Infinite-dimensional Lie algebras, 2001
194 **Valery B. Nevzorov,** Records: Mathematical theory, 2001
193 **Toshio Nishino,** Function theory in several complex variables, 2001
192 **Yu. P. Solovyov and E. V. Troitsky,** C^*-algebras and elliptic operators in differential topology, 2001
191 **Shun-ichi Amari and Hiroshi Nagaoka,** Methods of information geometry, 2000
190 **Alexander N. Starkov,** Dynamical systems on homogeneous spaces, 2000
189 **Mitsuru Ikawa,** Hyperbolic partial differential equations and wave phenomena, 2000
188 **V. V. Buldygin and Yu. V. Kozachenko,** Metric characterization of random variables and random processes, 2000
187 **A. V. Fursikov,** Optimal control of distributed systems. Theory and applications, 2000
186 **Kazuya Kato, Nobushige Kurokawa, and Takeshi Saito,** Number theory 1: Fermat's dream, 2000
185 **Kenji Ueno,** Algebraic Geometry 1: From algebraic varieties to schemes, 1999
184 **A. V. Mel′nikov,** Financial markets, 1999
183 **Hajime Sato,** Algebraic topology: an intuitive approach, 1999
182 **I. S. Krasil′shchik and A. M. Vinogradov, Editors,** Symmetries and conservation laws for differential equations of mathematical physics, 1999
181 **Ya. G. Berkovich and E. M. Zhmud′,** Characters of finite groups. Part 2, 1999
180 **A. A. Milyutin and N. P. Osmolovskii,** Calculus of variations and optimal control, 1998
179 **V. E. Voskresenskiĭ,** Algebraic groups and their birational invariants, 1998

For a complete list of titles in this series, visit the AMS Bookstore at **www.ams.org/bookstore/**.